THE

QUADRUPEDS

OF

NORTH AMERICA.

BY

JOHN JAMES AUDUBON, F.R.S., &c., &c.

AND

THE REV. JOHN BACHMAN, D.D., LL.D., &c. &c.

VOL. II.

NEW·YORK:

PUBLISHED BY V. G. AUDUBON.

M DCCC LI.

H. LUDWIG. PRINTER, 53, VESEY-ST., N. Y.

QUADRUPEDS OF NORTH AMERICA.

GENUS LUTRA.—Ray., Cuv., Mustela spec., Linn., Aonyx, Lesson.

DENTAL FORMULA.

$$Incisive \; \frac{6}{6}; \; Canine \; \frac{1-1}{1-1}; \; Molar \; \frac{5-5}{5-5} = 36$$

The second inferior incisor on each side, a little receding in most of the species; the canine much dilated, hooked; first superior molar, small, blunt, and sometimes deciduous; the second, cutting; the third, of similar form, but larger; the fourth, with two external points, but furnished with a strong spur on the inner side; the fifth has externally three small points, with a broad spur internally. The inferior molars in this genus vary from five to six, the first being wanting in some of the species.

Head large and flattish, terminating in a blunt muzzle; ears short and round; tongue slightly papillous. Body long and slender; legs short; toes five on each foot. In some of the species the fifth toe on the hind foot is rudimental. Toes webbed, armed with short claws which are not retractile. Tail, not as long as the body, thick, and flattened horizontally.

Body covered externally with long, rigid and glossy hair, with a softer, shorter, downy fur intermixed.

On each side of the anus, there is a small gland secreting fetid matter.

All the species are good swimmers, live along the banks of rivers and ponds, and feed on fish.

The generic appellation is derived from Lutra—an Otter: from the Greek λουω (lous), wash.

There are eleven species enumerated by authors, inhabiting the following countries : Europe 1, Island of Trinidad 1, Guyana 1, Brazil 1, Kamtschatka 1, Java 1, Malay 1, Pondicherry 1, The Cape of Good Hope 1, and North America 2.

LUTRA CANADENSIS.—Sabine.

Canada Otter.

PLATE LI.—Male.

L. vellere nitido, saturate fusco ; mento gulâque fusco albis ; L. vulgare major.

CHARACTERS.

Larger than the European Otter, L. Vulgaris. Dark glossy brown ; chin and throat dusky white ; five feet in length.

SYNONYMES.

Loutre de Canada, Buffon, vol. xiii., p. 326, t. 44.
Common Otter, Pennant, Arctic Zoolog., vol. i., p. 653.
Land Otter, Warden's Hist. U. S., p. 206.
Lutra Canadensis, Sabine, Franklin's Journ., p. 653.
   "   Brasiliensis, Harlan, Fauna, p. 72.
   "   "   Godman, Nat. Hist., vol. i., p. 222.
   "   Canadensis, Dekay, Zool., p. 1., p. 39.

DESCRIPTION.

Head, large and nearly of a globular form ; nose, blunt and naked ; lips, thick ; ears, round, slightly ovate, and closer together than in *L. Vulgaris,* clothed densely with short hair on both surfaces ; body, long, cylindrical ; neck, long ; legs, short and stout ; moustaches, very rigid, like bristles ; soles of the feet, thinly clothed with hair between the toes, tubercles at the roots of the claws, naked ; feet, webbed to the nails ; Tail, stout, gradually tapering toward the extremity, depressed at the base, continuing flattened through half its length ; at the base there are two oval glands. The longer hairs covering the fur, are glossy and rigid ; fur, soft, dense, and nearly as fine as that of the Beaver, continuing through the whole extent of the body, even to the extremity of the tail, but shorter on the forehead and extremities.

Plate LI

Drawn from Nature by J. Audubon FRSFLS

Lith. Printed & Cold by J.T. Bowen, Phila.ᵈᵃ

Canada Otter.

We overlooked the opportunity of instituting a careful comparison between the skulls and teeth of the European and American Otters, and have now no access to specimens of the former. We therefore quote the language of Dr. DEKAY, whose observations in this respect correspond with our recollections of a general comparison made at the Berlin Museum, eleven years ago. "In their dentition the Otters are eminently characterized by the enormous dilation of the two posterior cheek teeth in the upper jaw. Our species, in this particular, offers some variations from the European Otter. The penultimate jaw tooth, in our species, has a broad internal heel directed obliquely forward, with a deep fissure dividing the surface into two rounded and elevated portions; and the pointed tubercle is broad, with a high shoulder posteriorly, and comparatively little elevated. The last tubercular tooth subquadrate, nearly as large as the preceding, and its greater axis directed obliquely backwards with four or rather six distinct elevated points; but the outer raised margin, which is so conspicuous in the European Otter, appears to be indistinct or simply elevated into two pointed tubercles, or wanting entirely, in the American."

In age, the canine as well as the anterior molars become much worn. In a specimen from Carolina, the incisors are worn down to the upper surface of the jaw teeth; in another from Georgia, all the teeth are worn down to the gums. A specimen from Canada and another from Texas have the teeth very pointed, and the canine projecting beyond the lips. These were evidently younger animals. In older specimens we have on several occasions found the two anterior jaw teeth entirely wanting, as well as some of the incisors, the former appearing to have dropped out at about the fourth year.

COLOUR.

A specimen from Lower Canada. Moustaches very light brown, many being white, those on the sides of the face dingy white; upper lip and chin light grayish brown, a shade darker under the throat; the long hairs covering the fur are in one half of their length from their roots dingy white, gradually deepening into brown. The general colour on the upper surface is that of a rich dark chesnut brown, a shade lighter on the whole of the under surface. RICHARDSON states: "The Canada Otter may be distinguished from the European species by the fur of its belly being of the same shining brown colour with that of the back." In this particular our observations do not correspond with those of our distinguished friend. Out of more than a hundred specimens of American Otters which we have examined, many of which came from Canada and the Rocky Mountains, we have but with one or two ex . ceptions found the colour on the under surface lighter than on the back.

A specimen from Carolina, an old male, teeth much worn.

Upper lip from the nostrils, chin and throat to near the chest, grayish white ; the fur on the back, although not quite so long as that of specimens from Canada, is quite dense and silky, and very nearly equal in fineness. It is whitish at the roots, with a bluish tinge towards the extremities. The longer hairs which conceal the fur and present the external colouring are very nearly of the same tint as in those procured in Canada, so that the specimens from these widely separated localities can scarcely be regarded even as varieties.

A specimen from Colorado, Texas.

(The form is precisely similar to the Otters of Canada and those existing in various intermediate States. The palms are naked, with a little less hair between the toes on the upper and under surfaces.) The colour is throughout two shades lighter than that of specimens from Canada, but the markings are similarly distributed. Fur on the back from the roots soiled white, inclining to brown at the tips. The long and rigid hairs on the upper surface lightish brown at the roots, then dark brown, tipped with lightish brown.

DIMENSIONS.

Specimen from Canada.—Adult male.

	Feet.	Inches.
From point of nose to root of tail, - - -	2	5
Tail, - - - - - - - - -	1	7
From point of nose to eye, - - - - -	0	1¾
From point of nose to ear, - - - - -	0	4
Height of ear, - - - - - - -	0	0¾
Breadth of ear at base, - - - - -	0	0¾

Specimen from Carolina.

	Feet.	Inches.
From point of nose to root of tail, - - -	2	7
Tail, - - - - - - - - -	1	5
Point of nose to eye, - - - - - -	0	1¾
" " to ear, - - - - - -	0	3¾
Height of ear, - - - - - - -	0	0¾
Breadth at base, - - - - - - -	0	0¾

Weight, 23 lbs.

Specimen from the Colorado, in Texas.

	Feet	Inches.
From nose to root of tail, - - - - -	2	7
Length of tail, - - - - - - -	1	6
From point of nose to eye, - - - - -	0	$1\frac{3}{4}$
" " to ear, - - - - -	0	$3\frac{3}{4}$
Between the ears, - - - - - -	0	$3\frac{3}{4}$
Height, - - - - - - - -	0	10
Around the body behind the shoulder, - -	1	$5\frac{5}{8}$
Around the body, (middle,) - - - - -	1	$7\frac{1}{8}$

Weight 20 lbs.

HABITS.

We concluded our first volume with a brief account of *Spermophilus Richardsonii*, the last animal figured in plates 1 to 50 inclusive, of our illustrations of the Quadrupeds of North America. Having, since that volume was written, published about 60 more plates, we now take up our pen to portray the habits and describe the forms and colours of the species figured in plates 51 to 100 inclusive, and shall, we hope, be able to give our readers tolerably good accounts of them; although, alas! the days of our youth are gone, when, full of enthusiasm, and anxious to examine every object in nature within our reach, the rising sun never found us slumbering away the fresh hours of the morning, but beamed upon our path through the deep forest, or lighted up to joy and gladness the hill side or mountain top, which we had already gained in quest of the birds or the beasts that were to be met with; and where we often prolonged our rambles until the shades of evening found us yet at a distance from our camp, loaded with wild turkeys, ducks, geese, and perchance an Otter.

Fresh and pleasant in our mind is the recollection of our early expeditions among the wild woods, and along the unvisited shores of our new country; and although more than forty years of varied and busy life have passed since the Otter was shot and drawn, whose figure we have given, we will try to take you with us to a spot on the eastern banks of the fair Ohio. It is a cold wintry morning: the earth concealed by a slight covering of snow, and the landscape in all its original wildness. Here let us proceed cautiously, followed by that constant companion, our faithful dog. Whilst we are surveying the quiet waters as they roll onward toward the great Mississippi, in whose muddy current they will lose their clear and limpid character, and become as opaque and impetuous as the waves of that mighty river of the West, we see a dark object making its way towards the spot on which we stand, through the swiftly dividing element. It has not observed us: we remain perfectly still, and presently it is distinctly visible; it is an Otter, and now within the range of our old gun "Tear Jacket," we take but one moment to raise our piece and fire; the water is agitated by a violent convulsive movement of the animal, our dog plunges into the river, and swimming eagerly to the Otter, seizes it, but the latter dives, dragging the dog with it beneath the surface, and when they reappear, the Otter has caught the dog by the nose and is struggling violently. The brave dog, however, does not give up, but in a few moments drags the wounded Otter to the shore, and we immediately despatch it. Being anxious to figure the animal, we smooth its disordered fur and proceed homewards with it, where, although at that time we had not drawn many quadrupeds, we soon select a position in which to figure the Otter, and accordingly draw it with one foot in a steel-trap, and endeavour to represent the pain and terror felt by the creature when its foot is caught by the sharp saw-like teeth of the trap.

Not far from the town of Henderson, (Kentucky), but on the opposite side of the Ohio river, in the State of Indiana, there is a pond nearly one mile in length, with a depth of water varying from twelve to fifteen feet. Its shores are thickly lined with cane, and on the edge of the water stand many large and lofty cypress trees. We often used to seat ourselves on a fallen trunk, and watch in this secluded spot the actions of the birds and animals which resorted to it, and here we several times observed Otters engaged in catching fishes and devouring them. When pursuing a fish, they dived expertly and occasionally remained for more than a minute below the surface. They generally held their prey when they came to the top of the water, by the head, and almost invariably swam with it to a half-sunken log, or to the margin of the pond, to eat the fish at their ease, having done which, they returned again to the deep water to obtain more.

One morning we observed that some of these animals resorted to the neighbourhood of the root of a large tree which stood on the side of the pond opposite to us, and with its overhanging branches shaded the water. After a fatiguing walk through the tangled cane-brake and thick underwood which bordered the sides of this lonely place, we reached the opposite side of the pond near the large tree, and moved cautiously through the mud and water towards its roots: but the hearing or sight of the Otters was attracted to us, and we saw several of them hastily make off at our approach. On sounding the tree with the butt of our gun, we discovered that it was hollow, and then having placed a large stick in a slanting position against the trunk, we succeeded in reaching the lowest bough, and thence climbed up to a broken branch from which an aperture into the upper part of the hollow enabled us to examine the interior. At the bottom there was quite a large space or chamber to which the Otters retired, but whether for security or to sleep we could not decide.

Next morning we returned to the spot, accompanied by one of our neighbours, and having approached, and stopped up the entrance under water as noiselessly as possible, we cut a hole in the side of the tree four or five feet from the ground, and as soon as it was large enough to admit our heads, we peeped in and discovered three Otters on a sort of bed composed of the inner bark of trees and other soft substances, such as water grasses. We continued cutting the hole we had made, larger, and when sufficiently widened, took some green saplings, split them at the but-end, and managed to fix the head of each animal firmly to the ground by passing one of these split pieces over his neck, and then pressing the stick forcibly downwards. Our companion then crept into the hollow, and soon killed the Otters, with which we returned home.

The American Otter frequents running streams, large ponds, and more sparingly the shores of some of our great lakes. It prefers those waters which are clear, and makes a hole or burrow in the banks, the entrance to which is under water.

This species has a singular habit of sliding off the wet sloping banks into the water, and the trappers take advantage of this habit to catch the animal by placing a steel-trap near the bottom of their sliding places, so that the Otters occasionally put their foot into it as they are swiftly gliding toward the water.

In Carolina, a very common mode of capturing the Otter is by tying a pretty large fish on the pan of a steel-trap, which is sunk in the water where it is from five to ten feet deep. The Otter dives to the bottom to seize the fish, is caught either by the nose or foot, and is generally found drowned. At other times the trap is set under the water, without bait, on a log, one end of which projects into the water, whilst the other rests on the banks of a pond or river ; the Otter, in endeavouring to mount the log, is caught in the trap.

Mr. Godman, in his account of these singular quadrupeds, states that "their favourite sport is sliding, and for this purpose in winter the highest ridge of snow is selected, to the top of which the Otters scramble, where, lying on the belly with the fore-feet bent backwards, they give themselves an impulse with their hind legs and swiftly glide head-foremost down the declivity, sometimes for the distance of twenty yards. This sport they continue apparently with the keenest enjoyment until fatigue or hunger induces them to desist."

This statement is confirmed by CARTWRIGHT, HEARNE, RICHARDSON, and more recent writers who have given the history of this species, and is in accordance with our own personal observations.

The Otters ascend the bank at a place suitable for their diversion, and sometimes where it is very steep, so that they are obliged to make quite an effort to gain the top; they slide down in rapid succession where there are many at a sliding place. On one occasion we were resting ourself on the bank of Canoe Creek, a small stream near Henderson, which empties into the Ohio, when a pair of Otters made their appearance, and not observing our proximity, began to enjoy their sliding pastime. They glided down the soap-like muddy surface of the slide with the rapidity of an arrow from a bow, and we counted each one making twenty-two slides before we disturbed their sportive occupation.

This habit of the Otter of sliding down from elevated places to the borders of streams, is not confined to cold countries, or to slides on the snow or ice, but is pursued in the Southern States, where the earth is seldom covered with snow, or the waters frozen over. Along the reserve-dams of the rice fields of Carolina and Georgia, these slides are very common. From the fact that this occurs in most cases during winter, about the period of the rutting season, we are inclined to the belief that this propensity may be traced to those instincts which lead the sexes to their periodical associations.

RICHARDSON says that this species has the habit of travelling to a great distance through the snow in search of some rapid that has resisted the severity of the winter frosts, and that if seen and pursued by hunters on these journeys, it will throw itself forward on its belly and slide through the snow for several yards, leaving a deep furrow behind it, which movement is repeated with so much rapidity, that even a swift runner on snow shoes has some difficulty in overtaking it. He also remarks that it doubles on its track with much cunning, and dives under the snow to elude its pursuers.

The Otter is a very expert swimmer, and can overtake almost any fish, and as it is a voracious animal, it doubtless destroys a great number of fresh water fishes annually. We are not aware of its having a preference for any particular species, although it is highly probable that it has. About twenty-five years ago we went early one autumnal morning to study the habits of the Otter at Gordon and Spring's Ferry, on the Cooper River, six miles above Charleston, where they were represented as being quite abundant. They came down with the receding tide in groups or families of five or six together. In the space of two hours we counted forty-six. They soon separated, ascended the different creeks in the salt marshes, and engaged in capturing mullets (*Mugil*). In most cases they came to the bank with a fish in their mouth, despatching it in a minute, and then hastened again after more prey. They returned up the river to their more secure retreats with the rising tide. In the small lakes and ponds of the interior of Carolina, there is found a favourite fish with the Otter, called the fresh-water trout (*Grystes salmoides*).

Although the food of the Otter in general is fish, yet when hard pressed by hunger, it will not reject animal food of any kind. Those we had in confinement, when no fish could be obtained were fed on beef, which they always preferred boiled. During the last winter we ascertained that the skeleton and feathers of a wild duck were taken from an Otter's nest on the banks of a rice field reserve-dam. It was conjectured that the duck had either been killed or wounded by the hunters, and was in this state seized by the Otter. This species can be kept in confinement easily in a pond surrounded by a proper fence where a good supply of fish is procurable.

On throwing some live fishes into a small pond in the Zoological Gardens in London, where an Otter was kept alive, it immediately plunged off the bank after them, and soon securing one, rose to the surface holding its prize in its teeth, and ascending the bank, rapidly ate it by large mouthfuls, and dived into the water again for another. This it repeated until it had caught and eaten all the fish which had been thrown into the water for its use. When thus engaged in devouring the luckless fishes the Otter bit through them, crushing the bones, which we could hear snapping under the pressure of its powerful jaws.

When an Otter is shot and killed in the water, it sinks from the weight of its skeleton, the bones being nearly solid and therefore heavy, and the hunter consequently is apt to lose the game if the water be deep; this animal is, however, usually caught in strong steel-traps placed and baited in its haunts; if caught by one of the fore-feet, it will sometimes gnaw the foot off, in order to make its escape.

Otters when caught young are easily tamed, and although their gait is ungainly, will follow their owner about, and at times are quite playful. We have on two occasions domesticated the Otter. The individuals had been captured when quite young, and in the space of two or three days became as tame and gentle as the young of the domestic dog. They preferred milk and boiled corn meal, and refused to eat fish or meat of any kind, until they were several months old. They became so attached to us, that at the moment of their entrance into our study they commenced crawling into our lap—mounting our table, romping among our books and writing materials, and not unfrequently upsetting our ink-stand and de-ranging our papers.

The American Otter has one litter annually, and the young, usually two and occasionally three in number, are brought forth about the middle of April, according to Dr. RICHARDSON, in high northern latitudes. In the Middle and Southern States they are about a month earlier, and probably litter in Texas and Mexico about the end of February.

The nest, in which the Otter spends a great portion of the day and in which the young are deposited, we have had opportunities of examining on several occasions. One we observed in an excavation three feet in diameter, in the bank of a rice field; one in the hollow of a fallen tree, and a third under the root of a cypress, on the banks of Cooper river, in South Carolina; the materials—sticks, grasses and leaves—were abundant; the nest was large, in all cases protected from the rains, and above and beyond the influence of high water or freshets.

J. W. AUDUBON procured a fine specimen of the Otter, near Lagrange in Texas, on the twenty-third of February, 1846. It was shot whilst playing or sporting in a piece of swampy and partially flooded ground, about sunset,—its dimensions we have already given.

Early writers have told us that the common Otter of Europe had long been taught to catch fish for its owners, and that in the houses of the great in Sweden, these animals were kept for that purpose, and would go out at a signal from the cook, catch fish and bring it into the kitchen in order to be dressed for dinner.

This, however improbable it may at first appear, is by no means unlikely, except that we doubt the fact of the animal's going by itself for the fish.

BEWICK relates some anecdotes of Otters which captured salmon and other fish for their owners, for particulars of which we must refer our readers to his History of Quadrupeds.

Our late relative and friend, N. BERTHOUD, Esq., of St. Louis, told us some time since, that while travelling through the interior of the State of Ohio, he stopped at a house where the landlord had four Otters alive which were so gentle that they never failed to come when he whistled for them, and that when they approached their master they crawled along slowly and with much apparent humility towards him, and looked somewhat like enormous thick and short snakes.

GEOGRAPHICAL DISTRIBUTION.

The geographical range of this species includes almost the whole continent of North America, and possibly a portion of South America. It has, however, been nearly extirpated in our Atlantic States east of Maryland, and is no longer found abundantly in many parts of the country in which it formerly was numerously distributed.

It is now procured most readily, in the western portions of the United States and on the Eastern shore of Maryland. It is still abundant on the rivers and the reserve-dams of the rice fields of Carolina, and is not rare in Georgia, Louisiana and Texas.

A considerable number are also annually obtained in the British provinces. We did not capture any Otters during our journey up the Missouri to the Yellow Stone River, but observed traces of them in the small water courses in that direction.

GENERAL REMARKS.

Much perplexity exists in regard to the number of species of American Otters, and consequently in determining their nomenclature. RAY, in 1693, described a specimen from Brazil under the name of *Braziliensis*. It was subsequently noticed by BRISSON, BLUMENBACH, D'AZARA, MARCGRAVE, SCHREBER, SHAW, and others. We have not had an opportunity of comparing our North American species with any specimen obtained from Brazil. The loose and unscientific descriptions we have met with of the Brazilian Otter, do not agree in several particulars with any variety of the species found in North America ; there is, however, a general resemblance in size and colour. Should it hereafter be ascertained by closer investigations that the species existing in these widely removed localities are mere varieties, then the previous name of *Braziliensis* (RAY) must be substituted for that of *L. Canadensis*, FR. CUVIER.

In addition to the yet undecided species of RAY, FR. CUVIER has separated the Canada from the Carolina species, bestowing on the former the name of *L. Canadensis*, and on the latter that of *L. Lataxina*. GRAY has published a specimen from the more northern portions of North America under the name *Lataxina Mollis ;* and a specimen which we obtained in Carolina, and presented to our friend Mr. WATERHOUSE of London, was, we believe, published by him under another name.

Notwithstanding these high authorities, we confess we have not been able to regard them in any other light than varieties, some more strongly marked than others, of the same species. The *L. Lataxina* of FR. CUVIER, and the specimen published by WATERHOUSE, do not present such distinctive characters as to justify us in separating the species from each other or from *L. Canadensis*. The specimen published by RICHARDSON under the name

of *L. Canadensis*, (Fauna Boreali Americana,) was that of a large animal;
and the *Mollis* of GRAY was, we think, a fine specimen of the Canada
Otter, with fur of a particular softness.   We have, after much deliberation,
come to the conclusion that all these must be regarded as varieties of one
species.   In dentition, in general form, in markings and in habits, they are
very similar.   The specimen from Texas, on account of its lighter colour
and somewhat coarser fur, differs most from the other varieties; but it does
not on the whole present greater differences than are often seen in the
common mink of the salt marshes of Carolina, when compared with specimens obtained from the streams and ponds in the interior of the Middle
States.   Indeed, in colour it much resembles the rusty brown of the Carolina mink.   In the many specimens we have examined, we have discovered shades of difference in colour as well as in the pelage among individuals obtained from the same neighbourhood.   In many individuals which
were obtained from the South and North, in localities removed a thousand
miles from each other, we could not discover that they were even varieties.
In other cases these differences may be accounted for from the known effects
of climate on other nearly allied species, as evidenced in the common mink.
On the whole we may observe, that the Otters of the North are of a darker
colour and have the fur longer and more dense than those of the South.   As
we proceed southward the hair gradually becomes a little lighter in colour
and the fur less dense, shorter, and coarser.   These changes, however, are
not peculiar to the Otter.   They are not only observed in the mink, but in
the raccoon, the common American rabbit, the Virginian deer, and nearly
all the species that exist both in the northern and southern portions of our
continent.

   We shall give a figure of *L. Mollis* of GRAY, in our third volume.

Plate LII

Drawn on Stone by Wᵐ E. Hitchcock

Swift Fox.

Drawn from Nature by J. J. Audubon. F.R.S. F.L.S.

Lith. Printed & Cold by J. T. Bowen, Philada

VULPES VELOX.—Say.

Swift Fox.   Kit Fox.

PLATE LII.—Male.

V. gracilis, supra cano fulvaque varices, infra albus; v. fulvo minor.

CHARACTERS.

Smaller than the American red fox, body slender, gray above, varied with fulvous; beneath, white.

SYNONYMES.

Kit Fox, or small burrowing fox of the plains.   Lewis and Clark, vol. i., p. 400.
    Vol. iii., pp. 28. 29.
Canis Velox, Say.   Long's Expedition, vol. ii., p. 339.
    "   "   Harlan's Fauna, 91.
    "   "   Godman's Nat. Hist., vol. i., p. 282.
Canis Cinereo Argentatus, Sabine, Franklin's Journey, p. 658.
    " (vulpes) Cinereo Argentatus, Richardson, Fa. B. Ame. p. 98.

DESCRIPTION.

This little species of Fox bears a great resemblance to our American red fox, in shape, but has a broader face and shorter nose than the latter species; in colour it approaches nearer to the gray fox.   Its form is light and slender, and gives indication of a considerable capacity for speed; the tail is long, cylindrical, bushy, and tapering at the end.

The entire length from the insertion of the superior incisors to the tip of the occipital crest, is rather more than four inches and three-tenths: the least distance between the orbital cavities nine-tenths of an inch; between the insertion of the lateral muscles at the junction of the frontal and parietal bones, half an inch.   The greatest breadth of this space on the parietal bones, thirteen-twentieths of an inch."—(Say.)   The hair is of two kinds, a soft dense and rather woolly fur beneath, intermixed with longer and stronger hairs.

COLOUR.

The fur on the back, when the hairs are separately examined, is from

the roots, for three-fourths of its length, of a light brownish gray colour,
then yellowish brown, then a narrow ring of black, then a larger ring
of pure white, slightly tipped at the apical part with black. The upper
part of the nose is pale yellowish brown, on each side of which there
is a patch of brownish, giving it a hoary appearance in consequence of
some of the hairs being tipped with white; moustaches black; upper
lip margined by a stripe of white hairs. There is a narrow blackish
brown line between the white of the posterior angle of the mouth, which
is prolonged around the margin of the lower lip. The upper part of
the head, the orbits of the eyes, the cheeks and superior surface of the
neck, back, and hips, covered with intermixed hairs, tipped with brown,
black, and white, giving those parts a grizzled colour. Towards the posterior parts of the back there are many long hairs interspersed, that
are black from the roots to the tip. The sides of the neck, the chest,
the shoulders and flanks, are of a dull reddish orange colour; the lower
jaw is white, with a tinge of blackish brown on its margins; the throat,
belly, inner surface of legs, aad upper surface of feet, are white. The
outside of the forelegs, and the posterior parts of the hindlegs, are brownish orange. The slight hairs between the callosities of the toes are
brownish. The tail is on the under surface yellowish gray with a mixture of black, and a few white hairs; the under surface is brownish
yellow and black at the end.

DIMENSIONS.

	Feet.	Inches.
From point of nose to root of tail, - - -	1	8
Tail, (vertebræ,) - - - - - -	0	$9\frac{3}{4}$
" to end of hair, - - - - - -	1	0
From tip of nose to end of head, - - -	0	$2\frac{1}{8}$
Between the eyes, - - - - - -	0	$1\frac{3}{4}$
Breadth between the ears, - - - - -	0	$2\frac{3}{8}$

Weight $8\frac{1}{2}$ lbs.

Measurement of a young animal killed at Fort Union.

	Feet.	Inches.
From point of nose to root of tail, - - -	1	$0\frac{1}{4}$
Tail, (vertebræ,) - - - - - -	0	$4\frac{7}{8}$
" to end of hair, - - - - - -	0	$5\frac{3}{4}$
Width at the shoulders, - - - - -	0	$7\frac{1}{4}$
Length of head, - - - - - -	0	$3\frac{1}{2}$
Between the eyes, - - - - -	0	$0\frac{7}{8}$
Breadth between the ears, - - - - -	0	$1\frac{3}{4}$

HABITS.

The First Swift Fox we ever saw alive was at Fort Clark on the upper Missouri river, at which place we arrived on the 7th of June, 1843. It had been caught in a steel-trap by one of its fore-feet, and belonged to Mr. CHARDON, the principal at the Fort, who with great kindness and politeness presented it to us ; assuring us that good care would be taken of it during our absence, (as we were then ascending the river to proceed to the base of the Rocky Mountains,) and that on our return to the Mandan village, we might easily take it with us to New-York.

Mr. CHARDON informed us that this Fox was a most expert rat catcher, and that it had been kept in a loft without any other food than the rats and mice that it caught there. It was a beautiful animal, and ran with great rapidity from one side of the loft to another, to avoid us. On our approaching, it showed its teeth and growled much like the common red fox.

Soon after we left Fort Clark, between the western shore of the Missouri river and the hills called the " Trois mamelles " by the Canadian and French trappers, on an open prairie, we saw the second Swift Fox we met with on this journey. Our party had been shooting several buffaloes, and our friend ED. HARRIS, Esq., and ourself, were approaching the hunters apace. We were on foot, and Mr. HARRIS was mounted on his buffalo horse, when a Swift Fox darted from a concealed hole in the prairie almost under the hoofs of my friend's steed. My gun was unfortunately loaded with ball, but the Fox was chased by Mr. HARRIS, who took aim at it several times but could not draw sight on the animal ; and the cunning fellow doubled and turned about and around in such a dexterous manner, that it finally escaped in a neighbouring ravine, and we suppose gained its burrow, or sheltered itself in the cleft of a rock, as we did not see it start again. This slight adventure with this (so called) Swift Fox convinced us that the accounts of the wonderful speed of this animal are considerably exaggerated ; and were we not disposed to retain its name as given by Mr. SAY, we should select that of Prairie Fox as being most appropriate for it. Mr. HARRIS, mounted on an Indian horse, had no difficulty in keeping up with it and overrunning it, which caused it to double as just mentioned. Had our guns been loaded with buck shot we should no doubt have killed it. It is necessary to say, perhaps, that all the authors who have written about this fox (most of whom appear to have copied Mr. SAY's account of it) assert that its extraordinary swiftness is one of the most remarkable characteristics of the animal. GODMAN observes that the fleetest antelope or deer,

when running at full speed, is passed by this little Fox with the greatest
ease, and such is the celerity of its motion, that it is compared by the celebrated travellers above quoted, Lewis and Clark and Mr. Say, "to the
flight of a bird along the ground rather than the course of a quadruped."

There is nothing in the conformation of this species, anatomically
viewed, indicating extraordinary speed. On the contrary, when we compare it with the red fox or even the gray, we find its body and legs
shorter in proportion than in those species, and its large head and
bushy tail give it rather a more heavy appearance than either of the
foxes just named.

Dr. Richardson informs us that the Saskachewan river is the most
northern limit of the range of the Kit Fox. Its burrows he says are
very deep and excavated in the open plains, at some distance from the
woody country. Lewis and Clark describe it as being extremely vigi-
lant, and say that it betakes itself on the slightest alarm to its burrow.

On our return to Fort Union after an excursion through a part of the
adjacent country, we found at some distance from the stockade a young
Swift Fox which we probably might easily have captured alive; but
fearing that its burrow was near at hand, and that it would soon reach
it and evade our pursuit, Mr. Harris shot it. This was the last specimen of this Fox that we were able to observe during our journey; we
have given its measurement in a former part of this article. On our return voyage, we found on arriving at Fort Clark that the living Swift
Fox given us by Mr. Chardon was in excellent condition. It was placed in a
strong wooden box lined in part with tin, and for greater security against
its escape, had a chain fastened to a collar around its neck. During
our homeward journey it was fed on birds, squirrels, and the flesh of
other animals, and finally safely reached our residence, near New-York,
where it was placed in a large cage box two-thirds sunk beneath the
surface of the ground, completely tinned inside, and half filled with earth.
When thus allowed a comparatively large space and plenty of earth to
burrow in, the Fox immediately began to make his way into the loose
ground, and soon had dug a hole large enough to conceal himself entirely.
While in this commodious prison he fed regularly and ate any kind of fresh
meat, growing fatter every day. He drank more water than foxes generally do, seemed anxious to play or wash in the cup which held his supply,
and would frequently turn it over, spilling the water on the floor of the
cage.

The cross fox which we described in our first volume does not appear to
require water, during the winter months at least, when fed on fresh meat;
as one that we have had in confinement during the past winter would not

drink any, and was not supplied with it for two or three months. Probably in a wild state all predatory animals drink more than when in confinement, for they are compelled to take so much exercise in the pursuit of their prey, that the evaporation of fluids, by perspiration, must go on rapidly; besides which, they would probably often try to appease the cravings of hunger by drinking freely, when unable to procure sufficient food.

GEOGRAPHICAL DISTRIBUTION.

The Swift Fox appears to be found on the plains of the Columbia river valley, as well as the open country of the region in which it has generally been observed, the extensive prairies of the eastern side of the Rocky Mountains.

It does not appear to be an inhabitant of New Mexico, Texas or California, as far as our information on the subject extends.

GENERAL REMARKS.

Our esteemed friend, Sir John Richardson, (Fauna Boreali Americana, p. 98,) has supposed that Schreber's description of *Canis cinereo argentatus*, applied to this species, and hence adopted his specific name, to the exclusion of Say's name of *C. Velox.* In our first volume, (p. 172,) we explained our views on this subject. In the descriptions of *C. Virginianus* of Schreber, and *C. Argenteus*, Erx., they evidently described mere varieties of the gray fox, (*V. Virginianus*); we have consequently restored Say's specific name, and awarded to him the credit of having been the first scientific describer of this animal.

MEPHITIS MESOLEUCA.—Licht.

Texan Skunk.

PLATE LIII.—Male.

M. Vitta solitaria media antice (in vertice) rotundata, acque lata ad basin caudae usque continuata, hac tota alba.

CHARACTERS.

The whole back, from the forehead to the tail, and the tail, white ; nose not covered with hair.

SYNONYMES.

Mephitis Mesoleuca, Lichtenstein. Darstellung neuer oder wenig bekannter Säugethiere. Berlin, 1827, 1834. Tab. 44, Fig. 2.
Mephitis Nasuta, Bennett. Proceedings of the Zoological Society, 1833, p. 39.
M. Mesoleuca, Licht. Ueber die Gattung Mephitis. Berlin, 1838, p. 23.

DESCRIPTION.

In form, this species bears a considerable resemblance to the common American skunk, (*Mephitis chinga.*) Like all the other species of skunk, this animal has a broad and fleshy body ; it is wider at the hips than at the shoulders, and when walking, the head is carried near the ground, whilst the back is obliquely raised six or seven inches higher ; it stands low on its legs, and progresses rather slowly. Forehead, slightly rounded ; eyes, small ; ears, short and rounded ; hair, coarse and long ; under fur, sparse, woolly, and not very fine ; tail, of moderate length and bushy ; nose, for three-fourths of an inch above the snout, naked. This is a characteristic mark, by which it may always be distinguished from the common American skunk, the latter being covered with short hair to the snout. Palms naked.

COLOUR.

The whole of the long hair, including the under fur on the back, and the tail on both surfaces, is white. This broad stripe commences on the forehead about two inches from the point of the nose, running near the ears,

Plate LIII

On Stone by Wm H. Radbrook.

Common Skunk.

Lith. Printed & Col. by J.T.Bowen, Phil.

and in a straight line along the sides and over the haunches, taking in the whole of the tail. The nails are white; the whole of the under surface of the body black, with here and there a white hair interspersed. On the forefeet around the palms and on the edges of the under surface, there are coarse whitish hairs.

The peculiarities in the colour of this species appear to be very uniform, as the specimens we examined in the Berlin Museum and in the collection of the Zoological Society in London, corresponded precisely with the specimen from which this description has been made.

DIMENSIONS.

	Ft.	Inches.
From point of nose to root of tail, - - -	1	$4\frac{1}{2}$
Tail (vertebræ), - - - - - - -	0	7
Do. to end of hair, - - - - - -	0	11
Breadth of head between the ears, - - -	0	3
Height of ear, - - - - - - -	0	$0\frac{3}{10}$
Length of heel to longest claw, - - - -	0	$2\frac{1}{4}$
Breadth of white stripe on the middle of the back,	0	5
Weight, 5 lbs.		

HABITS.

This odoriferous animal is found in Texas and Mexico, and is very similar in its habits to the common skunk of the Eastern, Middle and Southwestern States. A specimen procured by J. W. Audubon, who travelled through a portion of the State of Texas in 1845 and 6, for the purpose of obtaining a knowledge of the quadrupeds of that country, was caught alive in the neighbourhood of the San Jacinto; it was secured to the pack saddle of one of his baggage mules, but managed in some way to escape during the day's march, and as the scent was still strong on the saddle, it was not missed until the party arrived at the rancho of Mr. McFadden, who kept a house of entertainment for man and beast, which by this time was greatly needed by the travellers.

The almost endless varieties of the *Mephitis chinga*, the common skunk, many of which have been described as distinct species by naturalists, have, from our knowledge of their curious yet not specific differences, led us to admit any new species with doubt; but from the peculiar characteristics of this animal, there can be no hesitation in awarding to Prof. Lichtenstein the honour of having given to the world the first knowledge of this interesting quadruped.

The *Mephitis Mesoleuca* is found on the brown, broomy, sedgy plains, as

well as in the woods, and the cultivated districts of Texas and Mexico.
Its food consists in part of grubs, beetles, and other insects, and occasionally a small quadruped or bird, the eggs of birds, and in fact everything which this carnivorous but timid animal can appropriate to its
sustenance.

The retreats of this Skunk are hollows in the roots of trees or fallen
trunks, cavities under rocks, &c.; and it is, like the northern species, easily
caught when seen, (if any one has the resolution to venture on the experiment,) as it will not endeavour to escape unless it be very near its hiding
place, in which case it will avoid its pursuer by retreating into its burrow,
and there remaining for some time motionless, if not annoyed by a dog, or
by digging after it.

The stomach of the specimen from which our drawing was made, contained a number of worms, in some degree resembling the tape worm at
times found in the human subject. Notwithstanding this circumstance,
the individual appeared to be healthy and was fat. The rainy season
having set in (or at least the weather being invariably stormy for some
time) after it was killed, it became necessary to dry its skin in a chimney.
When first taken, the white streak along the back was as pure and free
from any stain or tinge of darkness or soiled colour as new fallen snow.
The two glands containing the fetid matter, discharged from time to time
by the animal for its defence, somewhat resembled in appearance a
soft egg.

This species apparently takes the place of the common American skunk,
(*Mephitis chinga*,) in the vicinity of the ranchos and plantations of the
Mexicans, and is quite as destructive to poultry, eggs, &c., as its northern
relative. We have not ascertained anything about its season of breeding,
or the time the female goes with young; we have no doubt, however, that
in these characteristics it resembles the other and closely allied species.

The long and beautiful tail of this Skunk makes it conspicuous among
the thickets or in the musquit bushes of Texas, and it most frequently
keeps this part elevated so that in high grass or weeds it is first seen by
the hunters who may be looking for the animal in such places.

GEOGRAPHICAL DISTRIBUTION.

The Mephitis Mesoleuca is not met with in any portion of the United
States eastward and northward of Texas. It is found in the latter State
and in most parts of Mexico. We have, however, not seen any skunk from
South America which corresponds with it.

GENERAL REMARKS.

Naturalists have been somewhat at a loss to decide on the name by which this species should be designated, and to what author the credit is due of having been the first describer.

The specimens obtained by LICHTENSTEIN were procured by Mr. DEPPE, in the vicinity of Chico, in Mexico, in 1825, and deposited in the museum of Berlin. In occasional papers published by Dr. LICHTENSTEIN, from 1827 to 1834, this species with many others was first published. In 1833, BENNETT published in the proceedings of the Zoological Society, the same species under the name of *M. Nasuta*. The papers of LICHTENSTEIN, although printed and circulated at Berlin, were not reprinted and collected into a volume till 1834. Having seen the original papers as well as the specimens at Berlin, and being satisfied of their earlier publication, we have no hesitation in adopting the name of LICHTENSTEIN as the first describer and publisher.

MUS DECUMANUS.—Pall.

Brown or Norway Rat.

PLATE LIV.—Males, Female, and Young.

Mus, cauda longissima squamata, corpore setoso griseo, subtus albido.

CHARACTERS.

Grayish-brown above, dull white beneath, tail nearly as long as the body, feet not webbed ; of a dingy white colour.

SYNONYMES.

Mus Decumanus, Pallas, Glir., p. 91–40.
   "   "   Schreber, Saugthiere, p. 645.
   "   "   Linn., Syst. Nat. ed. Gmel., t. p. 127.
Mus Aquaticus, Gesner's Quadr., p. 732.
Mus Decumanus, Shaw's Genl. Zool., ii., p. 50 t. 130.
Surmulat, Buff., Hist. Nat. viii., p. 206 t. 27.
Mus Decumanus, Cuv., Regne Animal, 1, p. 197.
   "   "   Godman, vol. ii., p. 78.
   "   "   Dekay, p. 79.
Mus Americanus, Dekay, American Black Rat, p. 81.

DESCRIPTION.

Body, robust ; head, long ; muzzle, long, but less acute than that of the black rat ; eyes, large and prominent ; moustaches, long, reaching to the ears ; ears, rounded and nearly naked ; tail, generally a little shorter than the body, (although occasionally a specimen may be found where it is of equal length,) slightly covered with short rigid hairs. There are four toes on each of the fore-feet, with a scarcely visible rudimental thumb, protected by a small blunt nail ; five toes on each of the hind feet ; the feet are clothed with short adpressed hairs. The fur seldom lies smooth, and the animal has a rough and not an inviting appearance.

COLOUR.

Outer surface of the incisors, reddish-brown ; moustaches, white and black ; the former colour preponderating ; the few short scattered hairs along the outer edges of the ear, yellowish brown ; eyes, black ; hair on

Plate LXV

Brown or Norway Rat

the back, from the roots, bluish-gray, then reddish-brown, broadly tipped with dark brown and black. On the under surface, the softer and shorter hair is from the roots ashy-gray broadly tipped with white.

VARIETIES.

1st. We have on several occasions, through the kindness of friends, received specimens of white rats which were supposed to be new species. They proved to be albinos of the present species. Their colour was white throughout, presenting the usual characteristics of the albino, with red eyes. One of this variety was preserved for many months in a cage with the brown rat, producing young, that in this instance all proved to be brown.

2d. We have at different times been able to procure specimens of a singular variety of this species that seems to have originated in this country. For the first specimen we were indebted to our friend Dr. Samuel Wilson of Charleston. Two others were sent to us from the interior of South Carolina. One was presented to us by a cat, and another was caught in a trap. In form, in size, and in dentition, they are precisely like the brown rat. The colour, however, is on both surfaces quite black. In some specimens there is under the chest and on the abdomen, a longitudinal white stripe similar to those of the mink. The specimens, after being preserved for a year or two, lose their intense black colour, which gradually assumes a more brownish hue. We examined a nest of the common brown rat containing 8 young, 5 of which were of the usual colour, and 3 black. The specimen obtained by Mr. Bell of New-York and published by Dr. Dekay, New-York Fauna, p. 81, under the name of *Mus Americanus*, undoubtedly belonged to this variety, which appears to have of late years become more common in the Southern than in the Northern States. This is evidently not a hybrid produced between *Mus Decumanus* and *Mus Rattus*, as those we have seen present the shape and size of the former, only differing in colour.

DIMENSIONS.

	Inches.
From point of nose to root of tail, - - - -	10
Tail, - - - - - - - - -	9
From point of nose to ear, - - - - -	$2\frac{1}{2}$
Height of ear, - - - - - - - -	$\frac{3}{8}$

HABITS.

The brown rat is unfortunately but too well known almost in every portion of our country, and in fact throughout the world, to require an elaborate account of its habits, but we will give such particulars as may we hope be interesting.   It is one of the most prolific and destructive little quadrupeds about the residences of man, and is as fierce as voracious. Some cases are on record where this rat has attacked a man when he was asleep, and we have seen both adults and children who, by their wanting a piece of the ear, or a bit of the end of the nose, bore painful testimony to its having attacked them while they were in bed; it has been known to nibble at an exposed toe or finger, and sometimes to have bitten even the remains of the shrouded dead who may have been exposed to its attacks.

The Norway Rat is very pugnacious, and several individuals may often be seen fighting together, squealing, biting, and inflicting severe wounds on each other.   On one occasion, we saw two of these rats in furious combat, and so enraged were they, that one of them whose tail was turned towards us, allowed us to seize him, which we did, giving him at the same time such a swing against a gate post which was near, that the blow killed him instantly—his antagonist making his escape.

During the great floods or freshets which almost annually submerge the flat bottom-lands on the Ohio river at various places, the rats are driven out from their holes and seek shelter under the barns, stables, and houses in the vicinity, and as the increasing waters cover the low grounds, may be seen taking to pieces of drift wood and floating logs, &c., on which they sometimes remain driving along with the currents for some distance. They also at such times climb up into the lofts of barns, smokehouses, &c., or betake themselves to the trees in the orchards or gardens.   We once, at Shippingport, near the foot of the falls of the Ohio river, whilst residing with our brother-in-law, the late N. BERTHOUD, went out in a skiff, during a freshet which had exceeded those of many previous years in its altitude, and after rowing about over the tops of fences that were secured from rising with the waters by being anchored by large cross-timbers placed when they were put up, under the ground, to which the posts were dove-tailed, and occasionally rowing through floating worm-fences which had broken away from their proper locations and were lying flat upon the surface of the flowing tide, we came to the orchard attached to the garden, and found the peach and apple trees full of rats, which seemed almost as active in running among the branches as squirrels.   We had our gun with us and tried to shoot some of them, but the cunning rogues dived into the water whenever we approached, and swam off in various directions, some to one tree and some to another, so that we were puzzled which to follow. The rats swam and dived with equal facility and made rapid progress through the water. Many of them remained in the orchard until the freshet subsided, which was in the course of a few days. Whether they caught any fish or not during this time we cannot say, but most of them found food enough to keep them alive until they were able once more to occupy their customary holes and burrows. During these occasional floods on our western rivers, immense numbers of spiders and other insects take refuge in the upper stories of the houses, and the inhabitants find themselves much incommoded by them as well as by the turbulent waters around their dwellings. Such times are, however, quite holidays to the young folks, and skiffs and batteaux of every description are in requisition, while some go about on a couple of boards, or paddle from street to street on large square pine logs. When the flats are thus covered, there is generally but little current running on them, although the main channel of the river flows majestically onward, covered with floating logs and the fragments of sheds, haystacks, &c., which have left their quiet homes on the sides of the river many miles above, to float on a voyage of discovery down to the great Mississippi, unless stopped by the way by the exertions of some fortunate discoverer of their value, who rowing out among the drifting logs, roots and branches, ties a rope to the frail floating tenement, and tows it to the trunk of a tree, where he makes it fast, for the water to leave it ready for his service, when the river has again returned to ,its quiet and customary channel. Stray flat boats loaded with produce, flour, corn and tobacco, &c., are often thus taken up, and are generally found and claimed afterwards by their owners. The sight of the beautiful Ohio thus swelling proudly along, and sometimes embracing the country with its watery margin extended for miles beyond its ordinary limits, is well worth a trip to the West in February or March. But these high freshets do not occur every year, and depend on the melting of the snows, which are generally dissolved so gradually that the channel of the river is sufficient to carry them off.

In a former work, (Ornithological Biography, vol. 1, p. 155,) we have given a more detailed account of one of the booming floods of the Ohio and Mississippi rivers, to which we beg now to refer such of our readers as have never witnessed one of those remarkable periodical inundations.

Mr. Ogden Hammond, formerly of Throg's Neck, near New-York, furnished us with the following account of the mode in which the Norway Rat captures and feeds upon the small sand clams which abound on the sandy places along the East river below high water mark. He repaired to a wharf on his farm with one of his men at low water: in a few moments a rat was seen issuing from the lower part of the wharf, peeping cautiously around before he ventured from his hiding place. Presently one of the small clams buried in the soft mud and sand which they inhabit, threw up a thin jet of water about a foot above the surface of the ground, upon seeing which, the rat leaped quickly to the spot, and digging with its fore-paws, in a few moments was seen bringing the clam towards his retreat, where he immediately devoured it.

When any of these clams lie too deep to be dug up by the rats, they continue on the watch and dig after the next which may make known its whereabouts by the customary jet of water. These clams are about ¾ of an inch long and not more than ⅝ of an inch wide ; their shells are slight, and they are sometimes used as bait by fishermen.

The Brown or Norway Rat was first introduced in the neighbourhood of Henderson, Kentucky, our old and happy residence for several years, within our recollection.

One day a barge arrived from New-Orleans (we think in 1811) loaded with sugar and other goods ; some of the cargo belonged to us. During the landing of the packages we saw several of these rats make their escape from the vessel to the shore, and run off in different directions. In a year from this time they had become quite a nuisance ; whether they had been re-inforced by other importations, or had multiplied to an incredible extent, we know not. Shortly after this period we had our smokehouse floor taken up on account of their having burrowed under it in nearly every direction. We killed at that time a great many of them with the aid of our dogs, but they continued to annoy us, and the readers of our Ornithological Biography are aware, that ere we left Henderson some rats destroyed many of our valued drawings.

This species migrates either in troops or singly, and for this purpose takes passage in any conveyance that may offer, or it plods along on foot. It swims and dives well, as we have already remarked, so that rivers or water-courses do not obstruct its progress. We once knew a female to secrete herself in a wagon, loaded with bale rope, sent from Lexington, (Ky.) to Louisville, and on the wagon reaching its destination, when the coils of rope were turned out, it was discovered that the animal had a litter of several young ones : she darted into the warehouse through the iron bars which were placed like a grating in front of the cellar windows. Some of the young escaped also, but several of them were killed by the wagoner. How this rat was fed during the journey we do not know, but as the wagons stop every evening at some tavern, the probability is that she procured food for herself by getting out during the night and picking up corn, &c.

The Norway Rat frequently deserts a locality in which it has for some time remained and proved a great pest. When this is the case, the whole tribe journey to other quarters, keeping together and generally appearing in numbers in their new locality without any previous warning to the unlucky farmer or housekeeper to whose premises they have taken a fancy.

When we first moved to our retreat, nine miles above the city of New-York, we had no rats to annoy us, and we hoped it would be some time before they discovered the spot where we had located ourselves. But in the course of a few months a great many of them appeared, and we have occasionally had eggs, chickens and ducklings carried off by them to the number of six or more in a night. We have never been able to get rid of this colony of rats, and they have even made large burrows in the banks on the water side, where they can hardly be extirpated.

The Norway Rat is quite abundant in New-York and most other maritime cities, along the wharves and docks, and becomes very large. These animals are frequently destroyed in great numbers, while a ship is in port, after her cargo has been discharged, by smoking them; the fumes of sulphur and other suffocating materials, being confined to the hold by closing all parts, windows and hatches. After a thorough cleaning out, a large ship has been known to have had many thousands on board. Our old friend, Capt. Cumings, who in early life made many voyages to the East Indies, relates to us, that one of his captains used to have rats caught, when on long voyages, and had them cooked and served up at his table as a luxury. He allowed his sailors a glass of grog for every rat they caught, and as the supply was generally ample, he used to invite his mates and passengers to partake of them with due hospitality. Our friend, who was a mate, had a great horror of the captain's invitations, for it was sometimes difficult to ascertain in what form the delicate animals would appear, and to avoid eating them. Not having ourselves eaten rats, (as far as we know,) we cannot say whether the old India captain's fondness for them was justified by their possessing a fine flavour, but we do think prejudices are entertained against many animals and reptiles that are, after all, pretty good eating.

In the account of the black rat in our first volume, (*Mus rattus*,) pp. 190, 191, and 192, we gave some details of the habits of the present species, and stated our opinion in regard to its destroying the black rat. Dr. Godman considered the Norway Rat so thorough an enemy of the black rat, that he says, (vol. 2, p. 83,) in speaking of the latter, that it is now found only in situations to which the Brown Rat has not extended its migrations.

According to the same author, who quotes R. SMITH, Rat Catcher, p. 5, 1768, (see GODMAN, vol. 2, p. 77,) the Brown Rat was not known even in Europe prior to the year 1750. RICHARDSON says, (probably quoting from HARLAN, Fauna, p. 149,) that it was brought from Asia to Europe, according to the accounts of historians of the seventeenth century, and was unknown in England before 1730. PENNANT, writing in 1785, says he has no authority for considering it an inhabitant of the new continent (America). HARLAN states that the Norwegian rat did not, as he was credibly informed, make its appearance in the United States any length of time previous to the year 1775. HARLAN does not give the Brown Rat as an American species, giving only what he considered indigenous species.

The Brown Rat brings forth from 10 to 15 young at a litter, and breeds several times in a year. Fortunately for mankind, it has many enemies: weasels, skunks, owls, hawks, &c., as well as cats and dogs. We have never known the latter to eat them, but they may at times do so. Rats are also killed by each other, and the weak ones devoured by the stronger.

This species becomes very fat and clumsy when living a long time in mills or warehouses. We have often seen old ones so fat and inactive that they would fall back when attempting to ascend a staircase.

We will take our leave of this disagreeable pest, by saying, that it is omnivorous, devouring with equal voracity meat of all kinds, eggs, poultry, fish, reptiles, vegetables, &c. &c. It prefers eels to other kinds of fish, having been known to select an eel out of a large bucket of fresh fish, and drag it off to its hole. In vegetable gardens it devours melons, cucumbers, &c., and will eat into a melon, entering through a hole large enough to admit its body, consuming the tender sweet fruit, seeds and all, and leaving the rind almost perfect. Where rats have gained access to a field or vegetable garden, they generally dig holes near the fruits or vegetables, into which they can make an easy retreat at the approach of an enemy.

We have represented several of these rats in our plate about to devour muskmelons, for which they have a strong predilection.

GEOGRAPHICAL DISTRIBUTION.

The *Mus Decumanus* is found in all the temperate parts of the world where man has been able to carry it in ships. It has not as yet penetrated into the fur countries, to the Rocky Mountains and California. The *Neotoma Drummondi* would probably be able to destroy it, being quite as fierce and much larger, should its wanderings lead it into the territory occupied by the latter. The Brown or Norway Rat is met with almost every where from Nova Scotia to and beyond our southern range, except in the western and northern regions above mentioned, and there even it will soon be found in California, at the mouth of the Columbia river, and among the settlements in Oregon.

GENERAL REMARKS.

We had assigned to LINNÆUS the credit of having been the first describer of the Brown Rat. On turning however to his 12th edition, we find no notice of this species. In a subsequent edition published by GMELIN in 1778, a description is added. It had however been previously described by PALLAS in 1767 under the name which it still retains. He is therefore entitled to the priority.

SCIURUS RUBICAUDATUS.—Aud. and Bach.

Red-Tailed Squirrel.

PLATE LV.

S. supra sub rufus cano mistus, subtus sordide flavus, magnitudine inter s. cinereum et s. migratorium ; cauda auriculisque rufis.

CHARACTERS.

Intermediate in size between the cat squirrel (S. Cinereus) and the Northern gray squirrel (S. Migratorius) ; ears and tail, red ; body, light-brown mixed with gray above, soiled buff beneath.

DESCRIPTION.

In form this species resembles the northern gray squirrel, possessing evidently all its activity ; its proportions are more delicate, and it weighs less, than the cat squirrel. It is considerably smaller than the great-tailed squirrel of Say, (S. Sayi). Although a little larger than the northern gray squirrel, its tail is shorter, and its fur a little coarser. The only specimen in which we were enabled to examine the dentition, had but twenty teeth ; the small front molars which appear to be permanent in the northern gray squirrel, and deciduous in several other species, were here entirely wanting.

COLOUR.

The fur on the back is in half its length from the roots, plumbeous, succeeded by a narrow marking of light brown, then black, tipped with whitish, a few interspersed hairs are black at the apical portion ; on the under surface the hairs are yellowish-white at the roots, and reddish-buff at the tips. The long hairs on the under surface of the tail are red through their whole extent. On the upper surface of the tail the hairs are reddish with three black annulations, tipped with red. Moustaches, black ; ears, around the eye, sides of face, throat and neck, inner surface of legs, upper surface of feet and belly, dull buff : tail, rufous.

On Stone by W.E. Hitchcock.

Red-tailed Squirrel.

Drawn from Nature by J J Audubon.F R S FL S   Lith. Printed & Col.d by J. T. Bowen Philad.

DIMENSIONS.

			Inches
Length, from point of nose to root of tail,	-	-	13
Do. vertebræ,	-	-	10
Do. to end of hair,	-	-	12½
Height of ear,	-	-	½
Heel to end of longest nail,	-	-	2¼

HABITS.

We have obtained no information in regard to the habits of this species, but have no doubt it possesses all the sprightliness and activity of other squirrels, particularly the Northern gray and cat squirrels, as well as the great tailed squirrel, to which in form and size it is allied.

GEOGRAPHICAL DISTRIBUTION.

The specimen from which our drawing was made, was procured in the State of Illinois.   This squirrel is also found in the barrens of Kentucky: we possess a skin sent to us by our good friend Dr. CROGHAN, procured we believe near the celebrated Mammoth cave, of which he is proprietor.

Mr. CABOT, of Boston, likewise has one, as well as we can recollect, in his collection.   We sought in vain, while on our journey in the wilds of the Upper Missouri country, for this species, which apparently does not extend its range west of the well-wooded districts lying to the east of the great prairies.   It will probably be found abundant in Indiana, although it has been hitherto most frequently observed in Illinois.   Of its northern and southern limits, we know nothing, and it may have a much more extended distribution than is at present supposed.

GENUS BISON.—Pliny.

DENTAL FORMULA.

$$Incisive \; \frac{0}{8}; \quad Canine \; \frac{0-0}{0-0}; \quad Molar \; \frac{6-6}{6-6} = 32.$$

Head, large and broad ; forehead, slightly arched ; horns, placed before the salient line of the frontal crest ; tail, short ; shoulders, elevated ; hair, soft and woolly.

The generic name is derived from Pliny, who applied the word Bison, wild ox, to one of the species on the Eastern continent.

There are five species of Buffalo that may be conveniently arranged under this genus : one existing in the forests of Southern Russia in Asia, in the Circassian mountains, and the desert of Kobi ; one in Ethiopia and the forests of India, one on the mountains of Central Asia, one in Ceylon, and one in America. In addition to this, the genus Bos, which formerly included the present, contains five well determined species, one inhabiting the country near the Cape of Good Hope, one in Central Africa, one in the Himalaya mountains and the Birman Empire, one in India, and one in the forests of Middle Europe.

BISON AMERICANUS.—Gmel.

AMERICAN BISON.—Buffalo.

PLATE LVI. Male.
PLATE LVII. Female, Male and Young.

B. capite magno, lato, fronte leviter arcuata ; cornibus parvis, brevibus, teretibus, extrorsum dein sursum versis ; cauda breve, cruribus gracilibus armis excelsis, villo molli, lanoso.

CHARACTERS.

Forehead, broad, slightly arched ; horns, small, short, directed laterally and upwards ; tail, short ; legs, slender ; shoulders, elevated · hair, soft and woolly.

Plate LVI

American Bison or Buffalo.

On Stone by Wᵐ E. Hitchcock

Lith. Printed & Col'd by J. T. Bowen. Phil.

Drawn from Nature by J. J. Audubon F.R.S. F.L.S.

Plate LVII.

Drawn from Nature by I.J. Audubon F.R.S.F.L.S.

On stone by W.M.E. Hitchcock

American Bison or Buffalo

Printed & Col.d by J.T. Bowen, Phila.a

TAURUS MEXICANUS, Hernandez, Mex., p. 587, Fig. male, 1651.
TAUREAU SAUVAGE, Hennepin, Nouv. Discov., vol. i., p. 186, 1699.
THE BUFFALO, Lawson's Carolina, p. 115, Fig.
    "   "   Catesby's Carolina, Appendix xxxii., tab. 20.
    "   "   Hearne's Journey, p. 412.
    "   "   Franklin's First Voy., p. 113.
    "   "   Pennant's Arctic Zool., vol. i., p. 1.
    "   "   Long's Expedition, vol. iii., p. 68.
    "   "   Warden's U. S., vol. i., p. 248.
BOS AMERICANUS, Linn., S. N., ed Gmel. 1, p. 204.
    "   "   Cuv., Regne an 1, p. 270.
BOS AMERICANUS, Harlan, 268.
    "   "   Godman, vol. iii., 4.
    "   "   Richardson, Fa., p. 79.
BUFFALO, Hudson's Bay Traders, Le Boeuf, Canadian Voyagers.
AMERICAN OX, Dobs, Hudson's Bay, 41.

DESCRIPTION.

Male, killed on the Yellow Stone river, July 16th, 1843.

The form bears a considerable resemblance to that of an overgrown domestic bull, the top of the hump on the shoulders being considerably higher than the rump, although the fore-legs are very short; horns, short, stout, curved upward and inward, one foot one inch and a half around the curve; ears, short and slightly triangular towards the point; nose, bare; nostrils, covered internally with hairs; eyes, rather small in proportion to the size of the animal, sunk into the prominent projection of the skull; neck, and forehead to near the nose, covered with a dense mass of shaggy hair fourteen inches long between the horns, which, as well as the eyes and ears, are thereby partially concealed; these hairs become gradually shorter and more woolly towards the muzzle. Under the chin and lower jaw there is an immense beard, a foot or upwards in length.

Neck, short; hairs along the shoulder and fore-legs about four inches long. The beard around the muzzle resembles that of the common bull. A mass of hair rises on the hind part of the fore-leg, considerably below the knee. A ridge of hairs commences on the back and runs to a point near the insertion of the tail. On the flanks, rump and fore-legs the hairs are very short and fine.

On the hind-legs there are straggling long hairs extending to the knee, and a few tufts extending six inches below the knee; hind-legs, and tail, covered with short hairs; within a few inches of the tip of the tail there is a tuft of hair nearly a foot in length. The pelage on the head has scarcely any of the soft woolly hair which covers other parts of the body, and approaches nearer to hair than to wool.

A winter killed specimen.

From the neck, around the shoulder and sides, the body is covered with a dense heavy coat of woolly hair, with much longer and coarser hairs intermixed. There is a fleshy membrane between the forelegs, like that in the common domestic bull, but not so pendulous.

Female.

In form and colour the female bears a strong resemblance to the male; she is, however, considerably smaller, and of a more delicate structure. Her horns are of the same length and shape as those of the male, but are thinner and more perfect, in consequence of the cows engaging less in combat than the bulls. The hump is less elevated; the hair on the forehead shorter and less bushy; the rings on the horns are more corrugated than on those of our domestic cattle.

Spinous processes rising from the back bone or vertebræ of the bull, and forming the hump: they are flat, with sharp edges both anteriorly and posteriorly; the two longest are eighteen and a quarter inches long, three inches at the end which is the widest, and two inches at the narrowest; the first, fifteen inches; second, (largest,) eighteen and a quarter inches in length; third, sixteen and a half; fourth, sixteen; the fifth, fifteen inches, and the rest gradually diminishing in size; the fifteenth spinous process being three and a half inches long; the remainder are wanting in our specimen. The whole of the processes are placed almost touching each other at the insertion and at the end, and their breadth is parallel to the course of the back-bone. In the centre or about half the distance from the insertion to the outer end of them, they are (the bone being narrower in that part) from a quarter to one inch apart. The ribs originate and incline outward backward and downward from between these upright spinous bones.

COLOUR.

A summer specimen.

Head, neck, throat, fore-legs, tail and beard, dark brownish-black; hoofs, brown; rump, flanks, line on the back, blackish brown; horns nearly black. Upper surface of body light-brown; the hairs uniform in colour from the roots, the whole under surface blackish-brown.

The colour of the female is similar to that of the male.

At the close of the summer when the new coat of hair has been obtained, the Buffalo is in colour between a dark umber, and liver-shining brown; as the hair lengthens during winter, the tips become paler.

Young male, twelve months old.

A uniform dingy brown colour, with a dark brown stripe of twisted woolly upright hairs, extending from the head over the neck shoulders and back to the insertion of the tail. The hairs on the forehead, which form the enormous mass on the head of the adult, are just beginning to be developed.

Under the throat and along the chest, the hairs extend in a narrow line of about three inches in length; the bush at the end of the tail is tolerably well developed. Hairs on the whole body short and woolly.

A calf, six weeks old, presents the same general appearance, but is more woolly. The legs, especially near the hoofs, are of a lighter colour than the adult.

A calf taken from the body of a cow, in September, was covered with woolly hair; the uniform brownish, or dim yellow, strongly resembling the young of a domesticated cow.

HABITS.

Whether we consider this noble animal as an object of the chase, or as an article of food for man, it is decidedly the most important of all our contemporary American quadrupeds; and as we can no longer see the gigantic mastodon passing over the broad savannas, or laving his enormous sides in the deep rivers of our wide-spread land, we will consider the Buffalo as a link, (perhaps sooner to be forever lost than is generally supposed,) which to a slight degree yet connects us with larger American animals, belonging to extinct creations.

But ere we endeavour to place before you the living and breathing herds of Buffaloes, you must journey with us in imagination to the vast western prairies, the secluded and almost inaccessible valleys of the Rocky Mountain chain, and the arid and nearly impassable deserts of the western table lands of our country; and here we may be allowed to express our deep, though unavailing regret, that the world now contains only few and imperfect remains of the lost races, of which we have our sole knowledge through the researches and profound deductions of geologists; and even though our knowledge of the osteology of the more recently exterminated species be sufficient to place them before our " mind's eye," we have no description and no figures of the once living and moving, but now departed possessors of these woods, plains, mountains and waters, in which, ages ago, they are supposed to have dwelt. Let us however hope, that our humble efforts may at least enable us to perpetuate a knowledge of such species as the Giver of all good has allowed to remain with us to the present day. And now we will endeavour to give a good account of the majestic Bison.

In the days of our boyhood and youth, Buffaloes roamed over the small and beautiful prairies of Indiana and Illinois, and herds of them stalked through the open woods of Kentucky and Tennessee ; but they had dwindled down to a few stragglers, which resorted chiefly to the " Barrens," towards the years 1808 and 1809, and soon after entirely disappeared. Their range has since that period gradually tended westward, and now you must direct your steps " to the Indian country," and travel many hundred miles beyond the fair valleys of the Ohio, towards the great rocky chain of mountains which forms the backbone of North-America, before you can reach the Buffalo, and see him roving in his sturdy independence upon the vast elevated plains, which extend to the base of the Rocky Mountains.

Hie with us then to the West ! let us quit the busy streets of St. Louis, once considered the outpost of civilization, but now a flourishing city, in the midst of a fertile and rapidly growing country, with towns and villages scattered for hundreds of miles beyond it ; let us leave the busy haunts of men, and on good horses take the course that will lead us into the Buffalo region, and when we have arrived at the sterile and extended plains which we desire to reach, we shall be recompensed for our toilsome and tedious journey : for there we may find thousands of these noble animals, and be enabled to study their habits, as they graze and ramble over the prairies, or migrate from one range of country to another, crossing on their route water-courses, or swimming rivers at places where they often plunge from the muddy bank into the stream, to gain a sand-bar or shoal, midway in the river, that affords them a resting place, from which, after a little time, they can direct their course to the opposite shore, when, having reached it, they must scramble up the bank, ere they can gain the open prairie beyond.

There we may also witness severe combats between the valiant bulls, in the rutting season, hear their angry bellowing, and observe their sagacity, as well as courage, when disturbed by the approach of man.

The American Bison is much addicted to wandering, and the various herds annually remove from the North, at the approach of winter, although many may be found, during that season, remaining in high latitudes, their thick woolly coats enabling them to resist a low temperature, without suffering greatly. During a severe winter, however, numbers of them perish, especially the old, and the very young ones. The breeding season is generally the months of June and July, and the calves are brought forth in April and May ; although occasionally they are produced as early as March or as late as July. The Buffalo most frequently has but one calf at a time, but instances occur of their having two. The females usually retire from the herd either singly or several in company, select as solitary a spot as can be found, remote from the haunt of wolves, bears, or other enemies that would be most likely to molest them, and there produce their young.

Occasionally, however, they bring forth their offspring when the herd is migrating, and at such times they are left by the main body, which they rejoin as soon as possible. The young usually follow the mother until she is nearly ready to have a calf again. The Buffalo seldom produces young until the third year, but will continue breeding until very old. When a cow and her very young calf are attacked by wolves, the cow bellows and sometimes runs at the enemy, and not unfrequently frightens him away ; this, however, is more generally the case when several cows are together, as the wolf, ever on the watch, is sometimes able to secure a calf when it is only protected by its mother.

The Buffalo begins to shed its hair as early as February. This falling of the winter coat shows first between the fore-legs and around the udder in the female on the inner surface of the thighs, &c. Next, the entire pelage of long hairs drop gradually but irregularly, leaving almost naked patches in some places, whilst other portions are covered with loosely hanging wool and hair. At this period these animals have an extremely ragged and miserable appearance. The last part of the shedding process takes place on the hump. During the time of shedding, the Bison searches for trees, bushes, &c., against which to rub himself, and thereby facilitate the speedy falling off of his old hair. It is not until the end of September, or later, that he gains his new coat of hair. The skin of a Buffalo, killed in October, the hunters generally consider, makes a good Buffalo robe ; and who is there, that has driven in an open sleigh or wagon, that will not be ready to admit this covering to be the cheapest and the best, as a protection from the cold, rain, sleet, and the drifting snows of winter ? for it is not only a warm covering, but impervious to water.

The Bison bulls generally select a mate from among a herd of cows and do not leave their chosen one until she is about to calve.

When two or more males fancy the same female, furious battles ensue and the conqueror leads off the fair cause of the contest in triumph. Should the cow be alone, the defeated lovers follow the happy pair at such a respectful distance, as will ensure to them a chance to make their escape, if they should again become obnoxious to the victor, and at the same time enable them to take advantage of any accident that might happen in their favour. But should the fight have been caused by a female who is in a large herd of cows, the discomfited bull soon finds a substitute for his first passion. It frequently happens, that a bull leads off a cow, and remains with her separated during the season from all others, either male or female.

When the Buffalo bull is working himself up to a belligerent state, he paws the ground, bellows loudly, and goes through nearly all the actions we may see performed by the domesticated bull under similar circumstances, and finally rushes at his foe head foremost, with all his speed and strength. Notwithstanding the violent shock with which two bulls thus meet in mad career, these encounters have never been known to result fatally, probably owing to the strength of the spinous process commonly called the hump, the shortness of their horns, and the quantity of hair about all their fore-parts.

When congregated together in fair weather, calm or nearly so, the bellowing of a large herd (which sometimes contains a thousand) may be heard at the extraordinary distance of ten miles at least.

During the rutting season, or while fighting, (we are not sure which,) the bulls scrape or paw up the grass in a circle, sometimes ten feet in diameter, and these places being resorted to, from time to time, by other fighting bulls, become larger and deeper, and are easily recognised even after rains have filled them with water.

In winter, when the ice has become strong enough to bear the weight of many tons, Buffaloes are often drowned in great numbers, for they are in the habit of crossing rivers on the ice, and should any alarm occur, rush in a dense crowd to one place ; the ice gives way beneath the pressure of hundreds of these huge animals, they are precipitated into the water, and if it is deep enough to reach over their backs, soon perish. Should the water, however, be shallow, they scuffle through the broken and breaking ice, in the greatest disorder, to the shore.

From time to time small herds, crossing rivers on the ice in the spring, are set adrift, in consequence of the sudden breaking of the ice after a rise in the river. They have been seen floating on such occasions in groups of three, four, and sometimes eight or ten together, although on separate cakes of ice. A few stragglers have been known to reach the shore in an almost exhausted state, but the majority perish from cold and want of food rather than trust themselves boldly to the turbulent waters.

Buffalo calves are often drowned, from being unable to ascend the steep banks of the rivers across which they have just swam, as the cows cannot help them, although they stand near the bank, and will not leave them to their fate unless something alarms them.

On one occasion Mr. KIPP, of the American Fur Company, caught eleven calves, their dams all the time standing near the top of the bank. Frequently, however, the cows leave the young to their fate, when most of them perish. In connection with this part of the subject, we may add, that we were informed when on the Upper Missouri river, that when the banks of that river were practicable for cows, and their calves could not follow them, they went down again, after having gained the top, and would remain by them until forced away by the cravings of hunger. When thus forced by the necessity of saving themselves to quit their young, they seldom, if ever, returned to them.

When a large herd of these wild animals are crossing a river, the calves or yearlings manage to get on the backs of the cows, and are thus conveyed safely over; but when the heavy animals, old and young, reach the shore, they sometimes find it muddy or even deeply miry; the strength of the old ones struggling in such cases to gain a solid footing, enables them to work their way out of danger in a wonderfully short time. Old bulls, indeed, have been known to extricate themselves when they had got into the mire so deep that but little more than their heads and backs could be seen. On one occasion we saw an unfortunate cow that had fallen into, or rather sank into a quicksand only seven or eight feet wide; she was quite dead, and we walked on her still fresh carcase safely across the ravine which had buried her in its treacherous and shifting sands.

The gaits of the Bison are walking, cantering, and galloping, and when at full speed, he can get over the ground nearly as fast as the best horses found in the Indian country. In lying down, this species bends the forelegs first, and its movements are almost exactly the same as those of the common cow. It also rises with the same kind of action as cattle.

When surprised in a recumbent posture by the sudden approach of a hunter, who has succeeded in nearing it under the cover of a hill, clump of trees or other interposing object, the Bison springs from the ground and is in full race almost as quick as thought, and is so very alert, that one can scarcely perceive his manner of rising on such occasions.

The bulls never grow as fat as the cows, the latter having been occasionally killed with as much as two inches of fat on the boss or hump and along the back to the tail. The fat rarely exceeds half an inch on the sides or ribs, but is thicker on the belly. The males have only one inch of fat, and their flesh is never considered equal to that of the females in delicacy or flavour. In a herd of Buffaloes many are poor, and even at the best season it is not likely that all will be found in good condition; and we have occasionally known a hunting party, when Buffalo was scarce, compelled to feed on a straggling old bull as tough as leather. For ourselves, this was rather uncomfortable, as we had unfortunately lost our molars long ago.

The Bison is sometimes more abundant in particular districts one year than another, and is probably influenced in its wanderings by the mildness or severity of the weather, as well as by the choice it makes of the best pasturage and most quiet portions of the prairies. While we were at Fort Union, the hunters were during the month of June obliged to go out twenty-five or thirty miles to procure Buffalo meat, although at other times, the animal was quite abundant in sight of the fort. The tramping of a large herd, in wet weather, cuts up the soft clayey soil of the river bottoms, (we do not not mean the bottom of rivers,) into a complete mush. One day, when on our journey up the Missouri river, we landed on one of the narrow strips of land called bottoms, which formed the margin of the river and was backed by hills of considerable height at a short distance. At this spot the tracks of these animals were literally innumerable, as far as the eye could reach in every direction, the plain was covered with them; and in some places the soil had been so trampled as to resemble mud or clay, when prepared for making bricks. The trees in the vicinity were rubbed by these buffaloes, and their hair and wool were hanging on the rough bark or lying at their roots. We collected some of this wool, we think it might be usefully worked up into coarse cloth, and consider it worth attention. The roads that are made by these animals, so much resemble the tracks left by a large wagon-train, that the inexperienced traveller may occasionally imagine himself following the course of an ordinary wagon-road. These great tracks run for hundreds of miles across the prairies, and are usually found to lead to some salt-spring, or some river or creek, where the animals can allay their thirst.

The captain of the steamboat on which we ascended the Missouri, informed us, that on his last annual voyage up that river, he had caught several Buffaloes, that were swimming the river. The boat was run close upon them, they were lassoed by a Spaniard, who happened to be on board, and then hoisted on the deck, where they were butchered secundum artem. One day we saw several that had taken to the water, and were coming towards our boat. We passed so near them, that we fired at them, but did not procure a single one. On another occasion, one was killed from the shore, and brought on board, when it was immediately divided among the men. We were greatly surprised to see some of the Indians, that were going up with us, ask for certain portions of the entrails, which they devoured with the greatest voracity. This gluttony excited our curiosity, and being always willing to ascertain the quality of any sort of meat, we tasted some of this sort of tripe, and found it very good, although at first its appearance was rather revolting.

The Indians sometimes eat the carcasses of Buffaloes that have been drowned, and some of those on board the Omega one day asked the captain most earnestly to allow them to land and get at the bodies of three Buffaloes which we passed, that had lodged among the drift-logs and were probably half putrid. In this extraordinary request some of the squaws joined. That, when stimulated by the gnawings of hunger, Indians, or even Whites, should feed upon carrion, is not to be wondered at, since we have many instances of cannibalism and other horrors, when men are in a state of starvation, but these Indians were in the midst of plenty of wholesome food and we are inclined to think their hankering after this disgusting flesh must be attributed to a natural taste for it, probably acquired when young, as they are no doubt sometimes obliged in their wanderings over the prairies in winter, to devour carrion and even bones and hides, to preserve their lives. In the height of the rutting-season, the flesh of the Buffalo bull is quite rank, and unfit to be eaten, except from necessity, and at this time the animal can be scented at a considerable distance.

When a herd of Bisons is chased, although the bulls run with great swiftness their speed cannot be compared with that of the cows and yearling calves. These, in a few moments leave the bulls behind them, but as they are greatly preferred by the hunter, he always (if well mounted) pursues them and allows the bulls to escape. During the winter of 1842 and 43, as we were told, Buffaloes were abundant around Fort Union, and during the night picked up straggling handfuls of hay that happened to be scattered about the place. An attempt was made to secure some of them alive, by strewing hay as a bait, from the interior of the old fort, which is about two hundred yards off, to some distance from the gateway, hoping the animals would feed along into the enclosure. They ate the hay to the very gate; but as the hogs and common cattle were regularly placed there, for security, during the night, the Buffaloes would not enter, probably on account of the various odours issuing from the interior. As the Buffaloes generally found some hay scattered around, they soon became accustomed to sleep in the vicinity of the fort, but went off every morning, and disappeared behind the hills, about a mile off.

One night they were fired at, from a four-pounder loaded with musket-balls. Three were killed, and several were wounded, but this disaster did not prevent them from returning frequently to the fort at night, and they were occasionally shot, during the whole winter, quite near the fort.

As various accounts of Buffalo-hunts have been already written, we will pass over our earliest adventures in that way, which occurred many years ago, and give you merely a sketch of the mode in which we killed them during our journey to the West, in 1843.

One morning in July, our party and several persons attached to Fort Union, (for we were then located there,) crossed the river, landed opposite the fort, and passing through the rich alluvial belt of woodland which margins the river, were early on our way to the adjacent prairie, beyond the hills. Our equipment consisted of an old Jersey wagon, to which we had two horses attached, tandem, driven by Mr. CULBERTSON, principal at the fort. This wagon carried Mr. HARRIS, BELL, and ourselves, and we were followed by two carts, which contained the rest of the party, while behind came the running horses or hunters, led carefully along. After crossing the lower prairie, we ascended between the steep banks of the rugged ravines, until we reached the high undulating plains above. On turning to take a retrospective view, we beheld the fort and a considerable expanse of broken and prairie-land behind us, and the course of the river was seen as it wound along, for some distance. Resuming our advance we soon saw a number of antelopes, some of which had young ones with them. After travelling about ten miles farther we approached the Fox river, and at this point one of the party espied a small herd of Bisons at a considerable distance off. Mr. CULBERTSON, after searching for them with the telescope, handed it to us and showed us where they were. They were all lying down and appeared perfectly unconscious of the existence of our party. Our vehicles and horses were now turned towards them and we travelled cautiously to within about a quarter of a mile of the herd, covered by a high ridge of land which concealed us from their view. The wind was favourable, (blowing towards us,) and now the hunters threw aside their coats, tied handkerchiefs around their heads, looked to their guns, mounted their steeds, and moved slowly and cautiously towards the game. The rest of the party crawled carefully to the top of the ridge to see the chase. At the word of command, given by Mr. CULBERTSON, the hunters dashed forward after the bulls, which already began to run off in a line nearly parallel with the ridge we were upon. The swift horses, urged on by their eager riders and their own impetuosity, soon began to overtake the affrighted animals; two of them separated from the others and were pursued by Mr. CULBERTSON and Mr. BELL; presently the former fired, and we could see that he had wounded one of the bulls. It stopped after going a little way and stood with its head hanging down and its nose near the ground. The blood appeared to be pouring from its mouth and nostrils, and its drooping tail showed the agony of the poor beast. Yet it stood firm, and its sturdy legs upheld its ponderous body as if nought had happened. We hastened toward it but ere we approached the spot, the wounded animal fell, rolled on its side, and expired. It was quite dead when we reached it. In the mean time Mr. Bell had continued in hot haste after the other, and Mr. Harris and Mr. Squire had each selected, and were following one of the main party. Mr. Bell shot, and his ball took effect in the buttocks of the animal. At this moment Mr. Squire's horse threw him over his head fully ten feet: he fell on his powder-horn and was severely bruised; he called to some one to stop his horse and was soon on his legs, but felt sick for a few moments. Friend Harris, who was perfectly cool, neared his bull, shot it through the lungs, and it fell dead on the spot. Mr. Bell was still in pursuit of his wounded animal and Mr. Harris and Mr. Squire joined and followed the fourth, which, however, was soon out of sight. We saw Mr. Bell shoot two or three times, and heard guns fired, either by Mr. Harris or Mr. Squire, but the weather was so hot that fearful of injuring their horses they were obliged to allow the bull they pursued to escape. The one shot by Mr. Bell, tumbled upon his knees, got up again, and rushed on one of the hunters, who shot it once more, when it paused, and almost immediately fell dead.

The flesh of the Buffaloes thus killed was sent to the fort in the cart, and we continued our route and passed the night on the prairie, at a spot about half way between the Yellow-Stone and the Missouri rivers. Here, just before sundown, seven more bulls were discovered by the hunters, and Mr. Harris, Mr. Bell and Mr. Culbertson each killed one. In this part of the prairie we observed several burrows made by the swift fox, but could not see any of those animals although we watched for some time in hopes of doing so. They probably scented our party and would not approach. The hunters on the prairies, either from hunger or because they have not a very delicate appetite, sometimes break in the skull of a buffalo and eat the brains raw. At sunrise we were all up, and soon had our coffee, after which a mulatto man called Lafleur, an excellent hunter attached to the American Fur-Company, accompanied Mr. Harris and Mr. Bell on a hunt for antelopes, as we wanted no more Buffaloes. After waiting the return of the party, who came back unsuccessful, we broke up our camp and turned our steps homeward.

The Buffalo bulls which have been with their fair ones are at this season wretchedly poor, but some of them, which appear not to have much fondness for the latter, or may have been driven off by their rivals, are in pretty good condition. The prairies are in some places whitened with the skulls of the Buffalo, dried and bleached by the summer's sun and the frosts and snows of those severe latitudes in winter. Thousands are killed merely for their tongues, and their large carcasses remain to feed the wolves and other rapacious prowlers on the grassy wastes.

A large Bison bull will generally weigh nearly two thousand pounds, and a fat cow, about twelve hundred. We weighed one of the bulls killed by our party and found it to reach seventeen hundred and twenty seven pounds, although it had already lost a good deal of blood. This was an old bull and was not fat; it had probably weighed more at some previous period. We were told that at this season a great many half-breed Indians were engaged in killing Buffaloes and curing their flesh for winter-use, on Moose river, about 200 miles north of us.

When these animals are shot at a distance of fifty or sixty yards, they rarely, if ever, charge on the hunters. Mr. CULBERTSON told us he had killed as many as nine bulls from the same spot, unseen by these terrible animals. There are times, however, when they have been known to gore both horse and rider, after being severely wounded, and have dropped down dead but a few minutes afterwards. There are indeed instances of bulls receiving many balls without being immediately killed, and we saw one which during one of our hunts was shot no less than twenty-four times before it dropped.

A bull that our party had wounded in the shoulder, and which was thought too badly hurt to do much harm to any one, was found rather dangerous when we approached him, as he would dart forward at the nearest of his foes, and but that his wound prevented him from wheeling and turning rapidly, he would certainly have done some mischief. We fired at him from our six-barrelled revolving pistol, which, however, seemed to have little other effect than to render him more savage and furious. His appearance was well calculated to appal the bravest, had we not felt assured that his strength was fast diminishing. We ourselves were a little too confident, and narrowly escaped being overtaken by him through our imprudence. We placed ourselves directly in his front, and as he advanced, fired at his head and ran back, not supposing that he could overtake us; but he soon got within a few feet of our rear, with head lowered, and every preparation made for giving us a hoist; the next instant, however, we had jumped aside, and the animal was unable to alter his headlong course quick enough to avenge himself on us. Mr. BELL now put a ball directly through his lungs, and with a gush of blood from the mouth and nostrils, he fell upon his knees and gave up the ghost, falling (as usual) on the side, quite dead.

On another occasion, when the same party were hunting near the end of the month of July, Mr. SQUIRE wounded a bull twice, but no blood flowing from the mouth, it was concluded the wounds were only in the flesh;

and the animal was shot by Mr. CULBERTSON, OWEN McKENZIE, and Mr. SQUIRE, again. This renewed fire only seemed to enrage him the more, and he made a dash at the hunters so sudden and unexpected, that Mr. SQUIRE, attempting to escape, rode between the beast and a ravine which was near, when the bull turned upon him, his horse became frightened and leaped down the bank, the Buffalo following him so closely that he was nearly unhorsed; he lost his presence of mind and dropped his gun; he, however, fortunately hung on by the mane and recovered his seat. The horse was the fleetest, and saved his life. He told us subsequently that he had never been so terrified before. This bull was fired at several times after SQUIRE's adventure, and was found to have twelve balls lodged in him when he was killed. He was in very bad condition, and being in the rutting season we found the flesh too rank for our dainty palates and only took the tongue with us.

Soon afterwards we killed a cow in company with many bulls and were at first afraid that they would charge upon us, which in similar cases they frequently do, but our party was too large and they did not venture near, although their angry bellowings and their unwillingness to leave the spot showed their rage at parting with her. As the sun was now sinking fast towards the horizon on the extended prairie, we soon began to make our way toward the camping ground and passed within a moderate distance of a large herd of Buffaloes, which we did not stop to molest but increasing our speed reached our quarters for the night, just as the shadows of the western plain indicated that we should not behold the orb of day until the morrow.

Our camp was near three conical hills called the Mamelles, only about thirty miles from Fort Union, although we had travelled nearly fifty by the time we reached the spot. After unloading and unsaddling our tired beasts, all hands assisted in getting wood and bringing water, and we were soon quietly enjoying a cup of coffee. The time of refreshment to the weary hunter is always one of interest: the group of stalwart frames stretched in various attitudes around or near the blazing watch-fires, recalls to our minds the masterpieces of the great delineators of night scenes; and we have often at such times beheld living pictures, far surpassing any of those contained in the galleries of Europe.

There were signs of grizzly bears around us, and during the night we heard a number of wolves howling among the bushes in the vicinity. The service berry was abundant and we ate a good many of them, and after a hasty preparation in the morning, started again after the Buffaloes we had seen the previous evening. Having rode for some time, one of our party who was in advance as a scout, made the customary signal from the top of a high hill, that Buffaloes were in sight ; this is done by walking the hunter's horse backward and forward several times. We hurried on and found our scout lying close to his horse's neck, as if asleep on the back of the animal. He pointed out where he had discovered the game, but they had gone out of sight, and (as he said) were travelling fast, the herd being composed of both bulls and cows. The hunters mounted at once, and galloped on in rapid pursuit, while we followed more leisurely over hills and plains and across ravines and broken ground, at the risk of our necks. Now and then we could see the hunters, and occasionally the Buffaloes, which had taken a direction toward the Fort. At last we reached an eminence from which we saw the hunters approaching the Buffaloes in order to begin the chase in earnest. It seems that there is no etiquette among Buffalo hunters, and this not being understood beforehand by our friend HARRIS, he was disappointed in his wish to kill a cow. The country was not as favourable to the hunters as it was to the flying herd. The females separated from the males, and the latter turned in our direction and passed within a few hundred yards of us without our being able to fire at them. Indeed we willingly suffered them to pass unmolested, as they are always very dangerous when they have been parted from the cows. Only one female was killed on this occasion. On our way homeward we made towards the coupee, an opening in the hills, where we expected to find water for our horses and mules, as our supply of Missouri water was only enough for ourselves.

The water found on these prairies is generally unfit to drink, (unless as a matter of necessity,) and we most frequently carried eight or ten gallons from the river, on our journey through the plains. We did not find water where we expected, and were obliged to proceed about two miles to the eastward, where we luckily found a puddle sufficient for the wants of our horses and mules. There was not a bush in sight at this place, and we collected Buffalo dung to make a fire to cook with. In the winter this prairie fuel is often too wet to burn, and the hunters and Indians have to eat their meat raw. It can however hardly be new to our readers to hear that they are often glad to get any thing, either raw or cooked, when in this desolate region.

Young Buffalo bulls are sometimes castrated by the Indians, as we were told, for the purpose of rendering them larger and fatter ; and we were informed, that when full grown they have been shot, and found to be far superior to others in the herd, in size as well as flavour. During severe winters the Buffaloes become very poor, and when the snow has covered the ground for several months to the depth of two or three feet, they are wretched objects to behold. They frequently in this emaciated state lose their hair and become covered with scabs ; and the magpies alight on their backs and pick the sores.   The poor animals in these dreadful seasons die in great numbers.

A singular trait in the Buffalo when caught young, was related to us, as follows : When a calf is taken, if the person who captures it places one of his fingers in its mouth, it will follow him afterwards, whether on foot or on horseback, for several miles.

We now give a few notes from our journal kept at Fort Union, which may interest our readers.

August 7th, 1843, a Buffalo cow was killed and brought into the fort, and to the astonishment of all, was found to be near her time of calving. This was an extraordinary circumstance at that season of the year.

August 8th, The young Buffaloes have commenced shedding their first (or red) coat of hair, which drops off in patches about the size of the palm of a man's hand.   The new hair is dark brownish black.   We caught one of these calves with a lasso, and had several men to hold him, but on approaching to pull off some of the old hair, he kicked and bounced about in such a furious manner that we could not get near him.   Mr. CULBERTSON had it however taken to the press post, and there it was drawn up and held so closely that we could handle it, and we tore off some pieces of its old pelage, which hung to the side with surprising tenacity.

The process of butchering or cutting up the carcass of the Buffalo is generally performed in a slovenly and disgusting manner by the hunters, and the choicest parts only are saved, unless food is scarce.   The liver and brains are eagerly sought for, and the hump is excellent when broiled. The pieces of flesh from the sides are called by the French, fillets, or the depouille ; the marrow bones are sometimes cut out, and the paunch is stripped of its covering of fat.

Some idea of the immense number of Bisons to be still seen on the wild prairies, may be formed from the following account, given to us by Mr. KIPP, one of the principals of the American Fur Company. " While he was travelling from Travers' Bay to the Mandan nation in the month of August, in a cart heavily laden, he passed through herds of Buffalo for six days in succession.   At another time he saw the great prairie near Fort Clark on the Missouri river, almost blackened by these animals, which covered the plain to the hills that bounded the view in all directions, and probably extended farther.

When the Bisons first see a person, whether white or red, they trot or canter off forty or fifty yards, and then stop suddenly, turn their heads and gaze on their foe for a few moments, then take a course and go off at full speed until out of sight, and beyond the scent of man.

Although large, heavy, and comparatively clumsy, the Bison is at times brisk and frolicksome, and these huge animals often play and gambol about, kicking their heels in the air with surprising agility, and throwing their hinder parts to the right and left alternately, or from one side to the other, their heels the while flying about and their tails whisking in the air. They are very impatient in the fly and mosquito season, and are often seen kicking and running against the wind to rid themselves of these tormentors.

The different Indian tribes hunt the Buffalo in various ways: some pursue them on horseback and shoot them with arrows, which they point with old bits of iron, or old knife blades. They are rarely expert in loading or reloading guns, (even if they have them,) but in the closely contested race between their horse and the animal, they prefer the rifle to the bow and arrow. Other tribes follow them with patient perseverance on foot, until they come within shooting distance, or kill them by stratagem.

The Mandan Indians chase the Buffalo in parties of from twenty to fifty, and each man is provided with two horses, one of which he rides, and the other being trained expressly for the chase, is led to the place where the Buffaloes are started. The hunters are armed with bows and arrows, their quivers containing from thirty to fifty arrows according to the wealth of the owner. When they come in sight of their game, they quit the horses on which they have ridden, mount those led for them, ply the whip, soon gain the flank or even the centre of the herd, and shoot their arrows into the fattest, according to their fancy. When a Buffalo has been shot, if the blood flows from the nose or mouth, he is considered mortally wounded; if not, they shoot a second or a third arrow into the wounded animal.

The Buffalo, when first started by the hunters, carries his tail close down between the legs; but when wounded, he switches his tail about, especially if intending to fight his pursuer, and it behooves the hunter to watch these movements closely, as the horse will often shy, and without due care the rider may be thrown, which when in a herd of Buffalo is almost certain death. An arrow will kill a Buffalo instantly if it takes effect in the heart, but if it does not reach the right spot, a dozen arrows will not even arrest one in his course, and of the wounded, many run out of sight and are lost to the hunter.

At times the wounded Bison turns so quickly and makes such a sudden rush upon the hunter, that if the steed is not a good one and the rider perfectly cool, they are overtaken, the horse gored and knocked down, and the hunter thrown off and either gored or trampled to death. But if the horse is a fleet one, and the hunter expert, the Bison is easily outrun and they escape. At best it may be said that this mode of Buffalo hunting is

dangerous sport, and one requires both skill and nerve to come off successfully.

The Gros Ventres, Blackfeet and Assinaboines often take the Buffalo in
large pens, usually called parks, constructed in the following manner.

Two converging fences built of sticks logs and brushwood are made,
leading to the mouth of a pen somewhat in the shape of a funnel. The
pen itself is either square or round, according to the nature of the ground
where it is to be placed, at the narrow end of the funnel, which is always
on the verge of a sudden break or precipice in the prairie ten or fifteen feet
deep, and is made as strong as possible. When this trap is completed, a
young man very swift of foot starts at daylight, provided with a Bison's
hide and head, to cover his body and head when he approaches the herd
that is to be taken, on nearing which he bleats like a young Buffalo calf,
and makes his way slowly towards the mouth of the converging fences
leading to the pen. He repeats this cry at intervals, the Buffaloes follow
the decoy, and a dozen or more of mounted Indians at some distance behind
the herd gallop from one side to the other on both their flanks, urging them
by this means to enter the funnel, which having done, a crowd of men women and children come and assist in frightening them, and as soon as they
have fairly entered the road to the pen beneath the precipice, the disguised
Indian, still bleating occasionally, runs to the edge of the precipice, quickly
descends, and makes his escape, climbing over the barricade or fence of the
pen beneath, while the herd follow on till the leader (probably an old bull)
is forced to leap down into the pen, and is followed by the whole herd, which
is thus ensnared, and easily destroyed even by the women and children,
as there is no means of escape for them.

This method of capturing the Bison is especially resorted to in October
and November, as the hide is at that season in good condition and saleable,
and the meat can be preserved for the winter supply. When the Indians
have thus driven a herd of Buffalo into a pen, the warriors all assemble
by the side of the enclosure, the pipe is lighted, and the chiefs smoke to the
honour of the Great Spirit, to the four points of the compass, and to the
herd of Bisons. As soon as this ceremony has ended, the destruction commences, guns are fired and arrows shot from every direction at the devoted animals, and the whole herd is slaughtered before the Indians enter the
space where the Buffaloes have become their victims. Even the children
shoot tiny arrows at them when thus captured, and try the strength of their
young arms upon them.

It sometimes happens, however, that the leader of the herd becomes alarmed and restless while driving to the precipice, and should the fence be weak,
breaks through, and the whole drove follow and escape. It also some

times occurs, that after the Bisons are in the pen, which is often so filled that they touch each other, the terrified crowd swaying to and
fro, their weight against the fence breaks it down, and if the smallest
gap is made, it is immediately widened, when they dash through and
scamper off, leaving the Indians in dismay and disappointment. The side
fences for the purpose of leading the Buffaloes to the pens extend at
times nearly half a mile, and some of the pens cover two or three hundred yards of ground. It takes much time and labour to construct one
of these great traps or snares, as the Indians sometimes have to bring
timber from a considerable distance to make the fences and render
them strong and efficient.

The Bison has several enemies : the worst is, of course, man ; then comes
the grizzly bear ; and next, the wolf. The bear follows them and succeeds
in destroying a good many ; the wolf hunts them in packs, and commits
great havoc among them, especially among the calves and the cows
when calving. Many Buffaloes are killed when they are struggling in the
mire on the shores of rivers where they sometimes stick fast, so that the
wolves or bears can attack them to advantage ; eating out their eyes and
devouring the unresisting animals by piecemeal.

When we were ascending the Missouri river, the first Buffaloes were
heard of near Fort Leavenworth, some having a short time before been
killed within forty miles of that place. We did not, however, see any of
these animals until we had passed Fort Croghan, but above this point we
met with them almost daily, either floating dead on the river, or gazing at
our steamboat from the shore.

Every part of the Bison is useful to the Indians, and their method of
making boats, by stretching the raw hide over a sort of bowl-shaped frame
work, is well known. These boats are generally made by the women, and we saw some of them at the Mandan village. The horns are
made into drinking vessels, ladles, and spoons. The skins form a good
bed, or admirable covering from the cold, and the flesh is excellent food,
whether fresh or dried or made into pemmican ; the fat is reduced and
put up in bladders, and in some cases used for frying fish, &c.

The hide of the Buffalo is tanned or dressed altogether by the women,
or squaws, and the children ; the process is as follows : The skin is first
hung on a post, and all the adhering flesh taken off with a bone, toothed
somewhat like a saw ; this is performed by scraping the skin downwards, and requires considerable labour. The hide is then stretched on
the ground and fastened down with pegs ; it is then allowed to remain
till dry, which is usually the case in a day or two. After it is dry-
the flesh side is pared down with the blade of a knife fastened in a

bone, called a grate, which renders the skin even and takes off about a quarter of its thickness. The hair is taken off with the same instrument and these operations being performed, and the skin reduced to a proper thickness, it is covered over either with brains, liver or grease, and left for a night. The next day the skin is rubbed and scraped either in the sun or by a fire, until the greasy matter has been worked into it, and it is nearly dry; then a cord is fastened to two poles and over this the skin is thrown, and pulled, rubbed and worked until quite dry; after which it is sewed together around the edges excepting at one end; a smoke is made with rotten wood in a hole dug in the earth, and the skin is suspended over it, on sticks set up like a tripod, and thoroughly smoked, which completes the tanning and renders the skin able to bear wet without losing its softness or pliability afterwards.

Buffalo robes are dressed in the same manner, only that the hair is not removed and they are not smoked. They are generally divided into two parts: a strip is taken from each half on the back of the skin where the hump was, and the two halves, or sides, are sewed together after they are dressed, with thread made of the sinews of the animal; which process being finished, the robe is complete and ready for market.

The scrapings of the skins, we were informed, are sometimes boiled with berries, and make a kind of jelly which is considered good food in some cases by the Indians. The strips cut off from the skins are sewed together and make robes for the children, or caps, mittens, shoes, &c. The bones are pounded fine with a large stone and boiled, the grease which rises to the top is skimmed off and put into bladders. This is the favourite and famous marrow grease, which is equal to butter. The sinews are used for stringing their bows, and are a substitute for thread; the intestines are eaten, the shoulder-blades made into hoes, and in fact (as we have already stated) nothing is lost or wasted, but every portion of the animal, by the skill and industry of the Indians, is rendered useful.

Balls are found in the stomach of the Buffalo, as in our common domestic cattle.

Having heard frequent discussions respecting the breeding of the Bison in a domesticated state, and knowing that ROBERT WICKLIFFE, Esq., of Kentucky, had raised some of these animals, we requested his son, then on his way to Europe, to ask that gentleman to give us some account of their habits under his care, and shortly afterwards received a letter from him, dated Lexington Nov. 6th, 1843, in which he gives an interesting account of the Bison breeding with the common cow, and other particulars connected with this animal. After expressing his desire to comply with our request intimated to him by his son, he proceeds to give us the following information : "as far," he writes, "as his limited knowledge of natural history and his attention to these animals will permit him to do." He proceeds: "The herd of Buffalo I now possess have descended from one or two cows that I purchased from a man who brought them from the country called the Upper Missouri; I have had them for about thirty years, but from giving them away and the occasional killing of them by mischievous persons, as well as other causes, my whole stock at this time does not exceed ten or twelve. I have sometimes confined them in separate parks from other cattle, but generally they herd and feed with my stock of farm cattle, They graze in company with them as gently as the others. The Buffalo cows, I think, go with young about the same time the common cow does, and produce once a year ; none of mine have ever had more than one at a birth. The approach of the sexes is similar to that of the common bull and cow under similar circumstances at all times when the cow is in heat, a period which seems, as with the common cow, confined neither to day, nor night, nor any particular season, and the cows bring forth their young of course at different times and seasons of the year, the same as our domesticated cattle. I do not find my Buffaloes more furious or wild than the common cattle of the same age that graze with them.

"Although the Buffalo, like the domestic cow, brings forth its young at different seasons of the year, this I attribute to the effect of domestication, as it is different with all animals in a state of nature. I have always heard their time for calving in our latitude was from March until July, and it is very obviously the season which nature assigns for the increase of both races, as most of my calves were from the Buffaloes and common cows at this season. On getting possession of the tame Buffalo, I endeavoured to cross them as much as I could with my common cows, to which experiment I found the tame or common bull unwilling to accede, and he was always shy of a Buffalo cow, but the Buffalo bull was willing to breed with the common cow.

"From the domestic cow I have several half breeds, one of which was a heifer ; this I put with a domestic bull, and it produced a bull calf. This I castrated, and it made a very fine steer, and when killed produced very fine beef. I bred from the same heifer several calves, and then, that the experiment might be perfect, I put one of them to the Buffalo bull, and she brought me a bull calf which I raised to be a very fine large animal, perhaps the only one to be met with in the world of his blood, viz., a three quarter, half quarter, and half quarter of the common blood. After making these experiments, I have left them to propagate their breed themselves, so that I have only had a few half breeds, and they always prove the same, even by a Buffalo bull. The full blood is not as large as the improved stock, but as large as the ordinary cattle of the country. The crossed or half blood are larger than either the Buffalo or common cow. The hump, brisket, ribs and tongue of the full and half blooded are preferable to those of the common beef, but the round and other parts are much inferior. The udder or bag of the Buffalo is smaller than that of the common cow, but I have allowed the calves of both to run with their dams upon the same pasture, and those of the Buffalo were always the fattest; and old hunters have told me, that when a young Buffalo calf is taken, it requires the milk of two common cows to raise it. Of this I have no doubt, having received the same information from hunters of the greatest veracity. The bag or udder of the half breed is larger than that of full blooded animals, and they would, I have no doubt, make good milkers.

"The wool of the wild Buffalo grows on their descendants when domesticated, but I think they have less of wool than their progenitors. The domesticated Buffalo still retains the grunt of the wild animal, and is incapable of making any other noise, and they still observe the habit of having select places within their feeding grounds to wallow in.

"The Buffalo has a much deeper shoulder than the tame ox, but is lighter behind. He walks more actively than the latter, and I think has more strength than a common ox of the same weight. I have broke them to the yoke, and found them capable of making excellent oxen ; and for drawing wagons, carts, or other heavily laden vehicles on long journeys, they would, I think, be greatly preferable to the common ox. I have as yet had no opportunity of testing the longevity of the Buffalo, as all mine that have died, did so from accident or were killed because they became aged. I have some cows that are nearly twenty years old, that are healthy and vigorous, and one of them has now a sucking calf.

"The young Buffalo calf is of a sandy red or rufous colour, and commences changing to a dark brown at about six months old, which last colour it always retains. The mixed breeds are of various colours ; I have had them striped with black, on a gray ground like the zebra, some of them brindled red, some pure red with white faces, and others red without any markings of white. The mixed bloods have not only produced in my stock from the tame and the Buffalo bull, but I have seen the half bloods reproducing ; viz. : those that were the product of the common cow and wild Buffalo bull. I was informed that at the first settlement of the country, cows that were considered the best for milking, were from the half blood, down to the quarter, and even eighth of the Buffalo blood. But my experiments have not satisfied me that the half Buffalo bull will produce again. That the half breed heifer will be productive from either race, as I have before stated, I have tested beyond the possibility of a doubt.

"The domesticated Buffalo retains the same haughty bearing that distinguishes him in his natural state. He will, however, feed or fatten on whatever suits the tame cow, and requires about the same amount of food. I have never milked either the full blood or mixed breed, but have no doubt they might be made good milkers, although their bags or udders are less than those of the common cow; yet from the strength of the calf, the dam must yield as much or even more milk than the common cow."

Since reading the above letter, we recollect that the Buffalo calves that were kept at Fort Union, though well fed every day, were in the habit of sucking each other's ears for hours together.

There exists a singular variety of the Bison, which is however very scarce, and the skin of which is called by both the hunters and fur traders a "beaver robe." These are valued so highly that some have sold for more than three hundred dollars. Of this variety Mr. CULBERTSON had the goodness to present us with a superb specimen, which we had lined with cloth, and find a most excellent defence against the cold, whilst driving in our wagon during the severity of our northern winters.

GEOGRAPHICAL DISTRIBUTION.

The range of the Bison is still very extensive; but although it was once met with on the Atlantic coast, it has, like many others, receded and gone west and south, driven onward by the march of civilization and the advance of the axe and plough. His habits, as we have seen, are migratory, and the extreme northern and southern limits of the wandering herds not exactly defined. Authors state, that at the time of the first settlement of Canada it was not known in that country, and SAGARD THEODAT mentions having heard that bulls existed in the far west, but saw none himself. According to Dr. RICHARDSON, Great Slave Lake, latitude 60°, was at one time the northern boundary of their range; but of late years, according to the testimony of the natives, they have taken possession of the flat limestone district of Slave Point on the north side of that lake, and have wandered to the vicinity of Great Marten Lake, in latitude 63° or 64°. The Bison was not known formerly to the north of the Columbia river on the Pacific coast, and LEWIS and CLARK found Buffalo robes were an important article of traffic between the inhabitants of the east side and those west of the Rocky mountains.

The Bison is spoken of by HERNANDEZ as being found in New Spain or Mexico, and it probably extended farther south. LAWSON speaks of two Buffaloes that were killed in one season on Cape Fear river, in North Carolina. The Bison formerly existed in South Carolina on the seaboard, and we were informed that from the last herd seen in that State, two were killed in the vicinity of Columbia. It thus appears that at one period this animal ranged over nearly the whole of North America.

At the present time, the Buffalo is found in vast herds in some of the great prairies, and scattered more sparsely nearly over the whole length and breadth of the valleys east and west that adjoin the Rocky Mountain chain

PUTORIUS ERMINEA.—Linn.

White Weasel.—Stoat.

PLATE LIX—Male and Female in summer pelage.

P. Hyeme alba; æstate supra rutila, infra alba caudae apice nigro.

CHARACTERS.

White, in winter; in summer, brown above, white beneath; tip of the tail, black.

SYNONYMES.

Mustela Erminea, Briss. Règne An., p. 243, 2.
    "   "   Linn., Syst. Nat., 12. i., p. 68. 7.
    "   "   Schreb., Säugth., p. 496, 11 t. 137.
    "   "   Erxleben Syst., p. 474, 13.
Vivera Erminea, Shaw, Gen. Zool., i., 2 p. 426 t. 99.
    "   "   Pennant, Arctic Zoology, i., p. 75.
Hermine, Buffon, C. C., p. 240, t.
Mustela Erminea, Parry's First Voyage, Sup. 135.
    "   "   Parry's Second Voy., App. 294.
    "   "   Franklin's First Journey, p. 652.
    "   "   Godman, Ame. Nat. Hist., vol. i., p. 193, fig. 1.
    "   "   Harlan, p.62.
Putorius Noveboracensis, Dekay, Nat. Hist. New-York, p. 36.

DESCRIPTION.

Body, long and slender, with a convex nose and forehead; limbs, short, and rather stout; tail, long and cylindrical; moustaches, long, extending beyond the ears; ears, low, broad and round, do not entirely surround the auditory opening, sparingly covered with short hairs on both surfaces. There are five toes on each foot, the inner toe much the shortest; the toes are clothed with hairs, covering the nails; fur, soft and short; tail, hairy, and bushy at the end. There are two glands situated on each side of the under surface of the tail, which contain an offensive white musky fluid.

Plate LIX

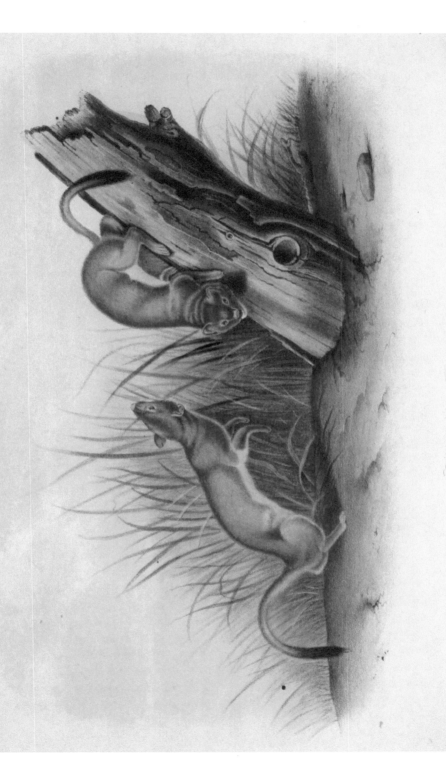

Drawn from Nature by J.J. Audubon, FRS. F.L.S.

White Weasel. (Stoat.)

Lith Printed & Col.d by J.T. Bowen, Philad.a

COLOUR.

In winter, in the latitude of Pennsylvania and New-York, all the hairs are snowy white from the roots, except those on the end of the tail, which for about one and three-fourth inches is black. We received specimens from Virginia obtained in January, in which the colours on the back had undergone no change, and remained brown ; and from the upper and middle districts of South Carolina killed at the same period, when no change had taken place, and it was stated that this, the only species of Weasel found there, remained brown through the whole year. These specimens are now in our possession, and we have arrived at the conclusion that the farther South we advance, the less perfect is the change from brown to white. We have specimens from Long Island, obtained in winter, which retain shades of brown on the head and dorsal line. Those from the valleys of the Virginia mountains have broad stripes of brown on the back, and specimens from Abbeville and Lexington, S. Carolina, have not undergone the slightest change. We were informed by our friend Mr. Bromfield an eminent botanist of England, that in the Isle of Wight, the place of his residence, the Ermine underwent only a partial change in winter.

In summer, the upper surface of the body is of a chesnut-brown colour, a little darker on the dorsal line ; under surface, the upper lips to the nose, chin, throat, inner surfaces of legs, and belly, white ; the line separating the colour of the back from that on the under surface, is very distinct, but irregular, and in some specimens, the white on the belly extends further up along the sides than in others. Whiskers white and black ; the former preponderating ; end of tail, as in winter, black.

DIMENSIONS.

Inches.

Old male.

	Inches.
Nose to root of tail,	$10\frac{1}{2}$
Tail (vertebræ),	$5\frac{1}{2}$
" to end of hair,	7
Breadth between the ears,	$1\frac{1}{4}$
Length of head,	2
Stretch of legs from end, to end of claws,	14
Length of hind foot, to end of nails,	$1\frac{3}{4}$
" fore-foot, to " "	$1\frac{1}{4}$
Black tip of tail,	3

HABITS.

The name of Ermine is associated with the pride of state and luxury, its fur having from time immemorial been the favourite ornament of the robes of princes, judges and prelates. From its snowy whiteness it is emblematic of the purity which they *ought* to possess.

To us the Ermine, in its winter dress, has always appeared strikingly beautiful. On a wintry day, when the earth was covered with a broad sheet of snow, our attention has sometimes been arrested by this little animal peering out from a log heap, or the crevices of a stone fence ; its eyes in certain shades of light appearing like sapphires, its colour vieing in whiteness and brilliancy with the snowy mantle of the surrounding landscape.

Graceful in form, rapid in his movements, and of untiring industry, he is withal a brave and fearless little fellow; conscious of security within the windings of his retreat among the logs, or heap of stones, he permits us to approach him to within a few feet, then suddenly withdraws his head; we remain still for a moment, and he once more returns to his post of observation, watching curiously our every motion, seeming willing to claim association so long as we abstain from becoming his persecutor.

Yet with all these external attractions, this little Weasel is fierce and bloodthirsty, possessing an intuitive propensity to destroy every animal and bird within its reach, some of which, such as the American rabbit, the ruffed grouse, and domestic fowl, are ten times its own size. It is a notorious and hated depredator of the poultry house, and we have known forty well grown fowls to have been killed in one night by a single Ermine. Satiated with the blood of probably a single fowl, the rest, like the flock slaughtered by the wolf in the sheepfold, were destroyed in obedience to a law of nature, an instinctive propensity to kill. We have traced the footsteps of this bloodsucking little animal on the snow, pursuing the trail of the American rabbit, and although it could not overtake its prey by superior speed, yet the timid hare soon took refuge in the hollow of a tree, or in a hole dug by the marmot, or skunk. Thither it was pursued by the Ermine, and destroyed, the skin and other remains at the mouth of the burrow bearing evidence of the fact. We observed an Ermine, after having captured a hare of the above species, first behead it and then drag the body some twenty yards over the fresh fallen snow, beneath which it was concealed, and the snow tightly pressed over it ; the little prowler displaying thereby a habit of which we became aware for the first time on that occasion. To avoid a dog that was in close pursuit, it mounted a tree and laid itself flat on a limb about twenty feet from the ground, from which it was finally shot. We have ascertained by successful experiments, repeated more than a hundred times, that the Ermine can be employed, in the manner of the ferret of Europe, in driving our American rabbit from the burrow into which it has retreated. In one instance, the Ermine employed had been captured only a few days before, and its canine teeth were filed in order to prevent its destroying the rabbit; a cord was placed around its neck to secure its return. It pursued the hare through all the windings of its burrow and forced it to the mouth, where it could be taken in a net, or by the hand. In winter, after a snow storm, the ruffed grouse has a habit of plunging into the loose snow, where it remains at times for one or two days. In this passive state the Ermine sometimes detects and destroys it. In an unsuccessful attempt at domesticating this grouse by fastening its feet to a board in the mode adopted with the stool pigeon, and placing it high on a shelf, an Ermine which we had kept as a pet, found its way by the curtains of the window and put an end to our experiment by eating off the head of our grouse.

Notwithstanding all these mischievous and destructive habits, it is doubtful whether the Ermine is not rather a benefactor than an enemy to the farmer, ridding his granaries and fields of many depredators on the product of his labour, that would devour ten times the value of the poultry and eggs which, at long and uncertain intervals, it occasionally destroys. A mission appears to have been assigned it by Providence to lessen the rapidly multiplying number of mice of various species and the smaller rodentia.

The white-footed mouse is destructive to the grains in the wheat fields and in the stacks, as well as the nurseries of fruit trees. LE CONTE's pine-mouse is injurious to the Irish and sweet potato crops, causing more to rot by nibbling holes into them than it consumes, and WILSON's meadow-mouse lessens our annual product of hay by feeding on the grasses, and by its long and tortuous galleries among their roots.

Wherever an Ermine has taken up its residence, the mice in its vicinity for half a mile round have been found rapidly to diminish in number. Their active little enemy is able to force its thin vermiform body into the burrows, it follows them to the end of their galleries, and destroys whole families. We have on several occasions, after a light snow, followed the trail of this weasel through fields and meadows, and witnessed the immense destruction which it occasioned in a single night. It enters every hole under stumps, logs, stone heaps and fences, and evidences of its bloody deeds are seen in the mutilated remains of the mice scattered on the snow. The little chipping or ground squirrel, *Tamias Lysteri*, takes up its residence in the vicinity of the grain fields, and is known to carry off in its cheek pouches vast quantities of wheat and buckwheat, to serve as winter stores. The Ermine instinctively discovers these snug retreats, and in the space of a few minutes destroys a whole family of these beautiful little *Tamiæ*; without even resting awhile until it has consumed its now abundant food its appetite craving for more blood, as if impelled by an irresistible destiny it proceeds in search of other objects on which it may glut its insatiable vampire-like thirst. The Norway rat and the common house-mouse take possession of our barns, wheat stacks, and granaries, and destroy vast quantities of grain. In some instances the farmer is reluctantly compelled to pay even more than a tithe in contributions towards the support of these pests. Let however an Ermine find its way into these barns and granaries, and there take up its winter residence, and the havoc which is made among the rats and mice will soon be observable. The Ermine pursues them to their farthest retreats, and in a few weeks the premises are entirely free from their depredations. We once placed a half domesticated Ermine in an outhouse infested with rats, shutting up the holes on the outside to prevent their escape. The little animal soon commenced his work of destruction. The squeaking of the rats was heard throughout the day. In the evening, it came out licking its mouth, and seeming like a hound after a long chase, much fatigued. A board of the floor was raised to enable us to ascertain the result of our experiment, and an immense number of rats were observed, which, although they had been killed on different parts of the building, had been dragged together, forming a compact heap.

The Ermine is then of immense benefit to the farmer. We are of the opinion that it has been over-hated and too indiscriminately persecuted. If detected in the poultry house, there is some excuse for destroying it, as, like the dog that has once been caught in the sheepfold, it may return to commit farther depredations; but when it has taken up its residence under stone heaps and fences, in his fields, or his barns, the farmer would consult his interest by suffering it to remain, as by thus inviting it to a home, it will probably destroy more formidable enemies, relieve him from many petty annoyances, and save him many a bushel of grain.

Let us not too hastily condemn the little Ermine for its bloodthirsty propensities. It possesses well-developed canine teeth, and obeys an instinct of nature. Man, with organs not so decidedly carnivorous, and possessed of the restraining powers of reason and conscience, often commits a wanton havoc on the inferior animals, not so much from want of food, as from a mere love of sport. The buffalo and the elk he has driven across the Mississippi, and their haunts are now restricted to the prairies of the far West. Even now thousands are slaughtered for amusement, and their tongues only are used, whilst their carcasses are left to the wolves. He fills his game bag with more woodcock, partridges and snipe, than he requires; his fishing-rod does not remain idle even after he has provided a full meal for his whole family; and our youngsters are taught to shoot the little warbler and the sparrow as a preparatory training for the destruction of larger game.

The Ermine is far from being shy in its habits. It is not easily alarmed, and becomes tolerably tame when taken young, for we have on several occasions succeeded in our attempts at domesticating it, but it appeared to us that these pets were not quite as gentle as many ferrets that we have seen in Europe. When not kept in confinement, they were apt to stray off into the fields and woods, and finally became wild. The tracks of this species on the snow are peculiar, exhibiting only two footprints, placed near each other, the succeeding tracks being far removed, giving evidences of long leaps. We have frequently observed where it had made long galleries in the deep snow for twenty or thirty yards, and thus in going from one burrow to another, instead of travelling over the surface, it had constructed for itself a kind of tunnel beneath.

The Ermine is easily taken in any kind of trap. We have on several occasions, when observing one peeping at us from its secure hole in the wall, kept it gazing until a servant brought a box trap baited with a bird or piece of meat, which was placed within a few feet of its retreat. The Ermine, after eyeing the trap for a few moments, gradually approached it, then after two or three hasty springs backwards returned stealthily into the trap, seized the bait, and was caught. We find in our note-book the following memorandum: "On the 19th June, 1846, we baited a large wire trap with maize: on visiting the trap on the following day we found it had caught seven young rats and a Weasel; the throats of the former had all been cut by the Weasel, and their blood sucked; but what appeared strange to us, the Weasel itself was also dead. The rats had been attracted by the bait: the Weasel went into the trap and killed them; and whether it met its death by excessive gluttony, or from a wound inflicted by its host of enemies, we are unable to determine.

This species does not appear to be very abundant any where. We have seldom found more than two or three on any farm in the Northern or Eastern States. We have ascertained that the immense number of tracks often seen in the snow in particular localities were made by a single ani-

mal, as by capturing one, no signs of other individuals were afterwards
seen.   We have observed it most abundant in stony regions : in Dutchess
and Ontario counties in New-York, on the hills of Connecticut and Vermont, and at the foot of the Alleghanies in Pennsylvania and Virginia.   It
is solitary in its habits, as we have seldom seen a pair together except
in the rutting season.   A family of young, however, are apt to remain in
the same locality till autumn.   In winter they separate, and we are inclined to think that they do not hunt in couples or in packs like the wolf,
but that, like the bat and the mink, each individual pursues its prey without copartnership, and hunts for its own benefit.

The only note we have ever heard uttered by the Ermine is a shrill
querulous cry : this was heard only when it was suddenly alarmed, or
received a hurt, when its sharp scream was always attended with an
emission of the offensive odour with which nature has furnished it as
a means of defence.   Although nocturnal in its habits, the Ermine is frequently met with at all hours of the day, and we have seen it in pursuit of the common rabbit under a bright shining sun at noon-day.

We doubt whether the Ermine ever digs its own burrows, and although
when fastened to a chain in a state of confinement we observed it digging shallow holes in the ground, its attempts at burrowing were as
awkward as those of the rat ; the nests we have seen were placed under roots of trees, in stone heaps, or in the burrows of the ground squirrel, from which the original occupants had been expelled.   The rutting season is in winter, from the middle of February to the beginning
of March.   The young, from four to seven, are born in May, in the
latitude of New-York.   We were informed by a close observer, that in
the upper country of Carolina, the young had been seen as early as
the 25th of March.   The colour of the young when a week old, is
pale yellow on the upper surface.

The Ermine avoids water, and if forcibly thrown into it, swims awk-
wardly like a cat.   It does not, like the fisher and pine marten, pursue
its prey on trees, and seems never to ascend them from choice ; but from
dire necessity, when closely pursued by its implacable enemy, the dog.

One of the most singular characteristics of this species, viz., its change
of colour from brown in summer to pure white in winter, and from
white in spring to its summer colour, remains to be considered.   It is
well known that about the middle of October the Ermine gradually
loses its brown summer-coat and assumes its white winter-pelage, which
about the middle of March is replaced by the usual summer colour·

As far as our observations have enabled us to form an opinion on
this subject, we have arrived at the conclusion, that the animal sheds

its coat twice a year, i. e., at the periods when these semi-annual changes take place. In autumn, the summer hair gradually and almost imperceptibly drops out, and is succeeded by a fresh coat of hair, which in the course of two or three weeks becomes pure white; while in the spring the animal undergoes its change from white to brown in consequence of shedding its winter coat, the new hairs then coming out brown. We have in our possession a specimen captured in November, in which the change of colour has considerably advanced, but is not completed. The whole of the under surface, the sides, neck and body to within half an inch of the back, together with the legs, are white, as well as the edges of the ears. On the upper surface, the nose, forehead, neck, and an irregular line on the back, together with a spot on the outer surface of the fore-leg, are brown, showing that these parts change colour last.

In reference to the change of pelage and colour as exhibited in spring, we add some notes made by the senior author of this work, in March, 1842, on a specimen sent to him alive by OGDEN HAMMOND, Esq.

The Weasel this evening, the 6th of March, began to show a change of colour; we were surprised to see that all around its nose, the white hair of its winter dress had changed suddenly to a silky black hue, and this extended to nearly between the ears. Here and there also were seen small spots of black about its rump, becoming more apparent toward the shoulders, and forming as it were a ridge along the back of the animal.

March 10th. By noon the change was wonderfully manifested. The whole upper surface of the head had become black to the eye, as well as the ridge of the back, the latter part having become quite clouded, and showing an indescribable motley mixture of closely-blended white, black, and blackish brown.

18th. This day the change of colour reached the root of the tail, where it formed a ring of about one inch, of the same reddish black colour. All other parts remained white, slightly tinged with pale lemon colour. It fed, as we perceived, more voraciously than ever since we have had it in our possession. No less than three or four mice were devoured to-day, and what is very strange, it left no remains of either hair, skull, feet, or any other part of these animals; and on this day, the 18th of March, it ate a very large piece of fresh beef, weighing nearly half a pound.

19th. Last night our Weasel made great progress, for this morning we found the coloured ridge on the back broader and less mottled. The posterior coloured part of the head had joined the ridge of the back. The posterior part of the hind legs had become brown, and we observed a small spot the size of a sixpence on each upper part of the thighs. At this juncture we think the animal is beautiful.

22d. This morning we found all the white hair on the outward ridge of the back had fallen, and portions of the thighs and shoulders had become broader; the coloured parts were of a rich brown to the very nose, and there existed indications of small dark spots coming from the sides of the belly, somewhat like so many beads strung on a thread, separated from the lower edge of the back ridge by a line of white of about half an inch. The weasel continues as lively as ever. When asleep, it curls its body around, and the tail encircles the whole animal, the end covering the nose. The eyes appear to be kept carefully uncovered. The general tints of the coloured parts of this Weasel were very much darker than in any other specimen which we have in our collection. When angry, it emitted a sharp shrill cry, and snapped with all its might at the objects presented to it. It was very cleanly in its habits, never rendering its sleeping apartment disagreeable.

28th. Our Weasel got out of its cage by pushing the wires apart, passing through an aperture not exceeding five-eighths of an inch, as we suppose by putting its head diagonally through the bars. The door and windows of our room were closed, however, and, when we entered, our little fellow looked at us as if well acquainted, but soon ran behind a box. It devoured last night at least half a pound of beef, kept in the room for its day's ration. We placed the cage, with the door open, on the floor, and by walking round the box that concealed it, the animal was induced to run towards the cage, and was again secured in it.

We have often observed this species whilst retreating; if near its place of concealment, it does so backwards, and we observed the same movement when it passed from one section of its cage to the other, dragging its food and concealing it among the straw. While we were sitting at a distance from its retreat, it proceeded by leaps very swiftly to within two or three feet of us, when it suddenly threw itself round and retreated backward, as mentioned before.

The purplish brown was now augmented on the thighs and shoulders to the knee joints, no white hairs remaining mixed with those that were coloured. Beneath the jaws, separate small brown spots appeared at equal distances, leaving an intermediate space of white, as was the case along the flanks. The root of the tail had acquired no farther change. Since last week our animal has diffused a very strong disagreeable odour, musky and fetid, which may be attributable to this being its breeding season; we observed that the smell was more disagreeable in the mornings and evenings, than at mid-day.

April.—On paying our accustomed visit to our Weasel this evening, we found it dead, which put a stop to any further observation of its habits. Its measurements are as follows:

	Inches.
From point of nose to end of tail, - - - -	10¼
Tail (vertebræ), - - - - - -	5
Tail to end of hair, - - - - - -	6
Height of ear, - - - - - - -	¾
Breadth of ear, - - - - - - -	⅘
Fore claws and hind claws stretching out to the black hair of the tail, - - - - - - - -	14¼

GEOGRAPHICAL DISTRIBUTION

If, as we feel confident after having examined more than a hundred specimens from both continents, the American Ermine is identical with that of Europe, it will be found to have the widest range of any quadruped at present known. It exists in the colder portions of Asia, and in the temperate, as well as in all the Northern States of Europe. We have seen specimens from England and Scotland, from France, Germany, Switzerland, Denmark, Sweden, and Russia.

In America, its geographical range is also very extensive. Dr. DEKAY (see Fauna, N. Y., p. 37) supposes it to be a northern animal, found as far south as Pennsylvania. We agree with him in his supposition that it is a northern animal, as it is only found in the Southern States where the country is mountainous or considerably elevated. It exists in the polar regions of America as far north as FRANKLIN, PARRY, RICHARDSON, LYON and other explorers were able to penetrate. It is found in Nova Scotia and Canada, and in all the Eastern and Northern States. We observed it along the whole chain of mountains in Virginia and North Carolina. We obtained a specimen from Abbeville in South Carolina, from our friend Dr. BARRETT, a close observer and a good naturalist; and another from Mr. FISHER, from Orangeburg District. We have ascertained that it exists in the mountains of Georgia, where we are penning this article. We saw a specimen procured by TOWNSEND in Oregon, and have heard of its existence in North California. It is, however, not found in the maritime districts of any of the Southern States, and in Carolina and Georgia does not approach within fifty miles of the seaboard; and even when it exists on the most elevated portions of country, it is, like the ruffed grouse in similar localities, a rare species.

GENERAL REMARKS.

Writers on Natural History, up to the time of HARLAN, GODMAN and RICHARDSON, without having instituted very close comparisons, considered the species existing in Asia, Europe and America, to be identical. At a somewhat later period, however, naturalists, discovering on patient and close investigation that nearly all our species of quadrupeds as well as birds differed from the closely allied species on the eastern continent, began to doubt the identity of the Ermine existing in Europe and America. We have been unable to ascertain whether these doubts originated from any difference in specimens from these countries, or from a belief that so small an animal could scarcely be found on both continents, and thus prove an exception to a general rule. We admit that were an animal restricted to the temperate climates on either continent, and not found in the polar regions, there would be a strong presumptive argument against the identity of closely allied species existing in Europe and America. The Ermine of the eastern continent is known to exist where the two continents nearly approach each other, perhaps occasionally have been united by a solid bridge of ice, and probably may be so again during some of the coldest seasons of the polar winters and being capable of travelling on the snow, and resisting the severest cold, this animal is fully able to cross from one continent to the other, like the white bear, or Arctic fox, species which are admitted as identical on both continents. Our species, moreover, is known to exist equally far north, and has been traced nearer to the poles than even the musk-ox.

We observed, in the Museum of the Zoological Society, that the specimen brought by RICHARDSON was regarded as a new species by C. L. BONAPARTE, Esq., (now Prince of Musignano.)

In the recent work of Dr. DEKAY, we perceive it has been described as a new species, under the name of *Putorius Noveboracensis*. In a spirit of great fairness and candour, however, he states: " I have never seen the true Ermine in its summer dress, and only know it from PENNANT's description : ears edged with white ; head, back, sides and legs, pale tawny brown ; under side of the body white ; lower part of the tail brown, end black." The only point of difference, then, is in the ears edged with white. PENNANT's specimen unquestionably was obtained at the period of time when the animal had only partially changed colour, as in all these cases the specimens before us, both from Europe and America, have their ears edged with white. We have compared a great number of specimens from both continents, and have several of each lying before us ; the edges of the ears in summer colour are all brown, and neither in size, dentition, nor colour, can we observe a shade of difference.

Plate LVIII.

On Stone by W.E. Hitchcock

Orange bellied Squirrel.

Drawn from Nature by J. J. Audubon, F.R.S. F.L.S.   Lith. Printed & Col.d by J. T. Bowen, Philad.a

SCIURUS SUB-AURATUS.—Bach.

ORANGE - BELLIED SQUIRREL. — GOLDEN - BELLIED SQUIRREL.

PLATE LVIII.—MALE AND FEMALE

S. Magnitudine, S. migratorium superæns, S. Carolinensi cedens ; supra cinereus flavido-undutus, subtus saturate aureus, cauda corpore longiore.

CHARACTERS.

Size intermediate between the Northern gray and the little Carolina squirrel ; tail longer than the body ; colour, above, gray, with a wash of yellow ; beneath, deep golden yellow.

SYNONYME.

GOLDEN-BELLIED SQUIRREL, Sciurus Sub-auratus.—Bachman, Mon. Genus Sciurus, p. 12.

DESCRIPTION.

In the two specimens now before us, which are very similar in size and markings, there is no appearance of the small anterior upper molar found in several other species of this genus. We conclude, therefore, that it either does not exist at all, or drops out at a very early period ; and accordingly set down this species as having only twenty teeth, viz. :

$$Incisive\ \frac{2}{2};\ Canine\ \frac{0-0}{0-0};\ Molar\ \frac{4-4}{4-4}=20.$$

The upper incisors are of moderate size ; their colour is deep orange brown ; the lower incisors are a little paler ; head, of medium size ; ears short and pointed, clothed with hair on both surfaces. The body seems more formed for sprightliness and agility than that of the small Carolina Squirrel, and in this respect comes nearest to the northern gray squirrel. The tail is long, and nearly as broad as that of the last named species.

COLOUR.

The whole upper surface gray, with a distinct yellow wash. The hairs which give this outward appearance are grayish slate colour at their base, then broadly annulated with yellowish, then black, and near the tips annulated with yellowish-white; sides of the face and neck, the whole of the inner side of the limbs, feet, and the under parts, deep golden yellow; on the cheeks and sides of the neck, however, the hairs are obscurely annulated with black and whitish; the ears are well clothed on both surfaces with tolerably long hair of the same deep golden hue as the sides of the face; hairs of the feet mostly blackish at the root, some obscurely tipped with black; hairs of the tail, black at the root, and the remaining portion bright rusty yellow; each hair annulated with black three times; the under surface of the tail is chiefly bright rusty yellow; whiskers, longer than the head, black.

DIMENSIONS.

	Inches.	Lines.
Length of head and body, - - - -	10	6
" of tail, (vertebræ,) - - - - -	9	2
" including fur, - - - - - -	12	0
" of palm to end of middle fore-claw, - -	1	7
" of heel to point of middle nail, - - -	2	7
" of fur on the back, - - - - -	0	7
Height of ear posteriorly, - - - -	0	5
Breadth of tail with hair extended, - - -	8	6

Weight 1¼ lbs.

HABITS.

During the winter season the city of New-Orleans is thronged by na tives of almost every land, and the Levee (which is an embankment extending along the margin of the river) presents a scene so unlike anything American, that as we walk along its smooth surface we may imagine ourselves in some twenty different countries, as our eyes fall upon many a strange costume, whose wearer has come from afar, and is, like ourselves, perchance, intent on seeing the curiosities of this Salmagundi city. Here a Spanish gentleman from Cuba, or a Mexican, next a pirate or thief, perhaps, from the same countries; all Europe is here represented, and the languages of many parts of the world can be heard whilst walking even half a mile; the descendants of Africa are here metamorphosed into French folks, and the gay bandanna that turbans the heads of the coloured women, is always adjusted with good taste, and is their favourite head-dress.

But the most interesting figures are the few straggling Choctaw and Chickasaw Indians, who bring a variety of game to the markets, and in their blankets, red flannel leggings, moccasins and bead finery, form a sort of dirty picturesque feature in the motley scene, and generally attract the artist's eye : many of these Indians have well formed legs and bodies, and their half-covered shoulders display a strength and symmetry indicating almost a perfect development of the manly form—their sinews and muscles being as large as is compatible with activity and grace. Whilst conversing with one of these remnants of a once numerous race, it was our good fortune to see for the first time the singular and beautiful little Orange-bellied Squirrel which the Indian hunter had brought with him along with other animals for sale, having procured it in the recesses of the forest on the borders of an extensive swamp.

Rarely indeed does the Orange-bellied Squirrel leave its solitary haunts and quit the cypress or sweet-gum shades, except to feed upon pecannuts, berries, persimmons, or other delicacies growing in the uplands ; and it does not hoard up the small acorn from the swamp-oak until late in the autumn, knowing that the mild winters of Louisiana are seldom cold enough to prevent it from catching an unlucky beetle from time to time during the middle of the day, or interfere with searches for food among the dry leaves and decaying vegetable substances in the woods. Besides, early in the year the red-maple buds will afford a treat to which this little squirrel turns with as much eagerness as the horse that has been kept all winter upon hay and corn, dashes into a fine field of grass in the month of May.

The hole inhabited by the present species is generally in some tall tree growing in the swamp, and perhaps sixty or one hundred yards from the dry land, and the animal passes to it from tree to tree, or along some fallen monarch of the woods, over the shallow water, keeping his large eye bent upon the surrounding lands in fear of some enemy ; and, in faith, he runs no little risk, for should the red-shouldered hawk, or the sharped-shinned, dart upon him, he is an easy prey; or, on a warm day, a snake, called the " water moccasin," curled up in his way, might swallow him, " tail and all." But good fun it must be to see the sportsman following in pursuit, splashing and floundering through the water, sometimes half-leg deep, and at others only up to the ankles, but stumbling occasionally, and making the " water fly ;"

so that when he *has* a chance to pull trigger, he is certain to snap *both barrels !*

Of the breeding of this species we know nothing, nor can we say more of its habits, which are yet to be farther investigated.

GEOGRAPHICAL DISTRIBUTION.

We have not heard of the occurrence of this species farther north than Louisiana, and think it probable its range will be found to extend west and south of that state into Texas, and perhaps Mexico.

Plate LX

Bridled Weasel.

PUTORIUS FRENATA.—Licht.

BRIDLED WEASEL.

PLATE LX.—MALES.

P. magnitudine P. ermineæ, supra fulvus, infra ex flavicante albus; naso, dorso, majore capitis parte, auribusque nigris; macula inter aures et vitta frontali albis.

CHARACTERS.

Size of the ermine; nose, back part of the head, and ears, black; a white spot between the ears, and a band over the forehead, white; yellowish-brown above, yellowish-white beneath.

SYNONYME.

MUSTELA FRENATA, Lichtenstein. Darstellung neuer oder wenig bekannter Säugethiere XLII., Tafel. Berlin, 1827-1834.

DESCRIPTION.

This species in form bears a considerable resemblance to the Ermine of the more northern parts of America. It is however rather stouter, the neck shorter, the ears narrower and higher, and the tail a little longer. In its dentition it is also similar to the common weasel, being a true *putorius*, with thirty-four teeth, having only four molars on each side of the upper jaw, and five beneath, whilst the genus *Mustela* is characterized by having thirty-eight teeth, five on each side of the upper jaw, and six beneath. The ears and tail are clothed with hair, the fur is a little shorter and slightly coarser than that of the Ermine.

COLOUR.

Moustaches, ears on both surfaces, nose, and around the eyes, black; a broad band of white rises in the forehead above the nose, extending around the head between the eyes and ears, reaching the neck and throat including the chin, the colours of which as well as the inner surfaces of the fore-legs are white; there is also a white spot on the back of the head between the ears. The colour is dark brownish black from the neck, reaching the white band on the forehead, where the lines of separation are distinctly but irregularly preserved. On the under surface from the chest to the tail including the inner surface of the thighs, a light fawn colour; tail, the colour of the back till within an inch of the tip, where it gradually darkens into black. The black at the end of the tail is not only shorter but less distinct than the corresponding parts on the ermine in summer colour.

The colour of the back and outer surfaces of the legs is light yellowish brown, gradually darkening on the neck till it reaches and blends with the dark brown colours on the hind head.

DIMENSIONS.

	Inches.
From point of nose to root of tail, - - - -	11
Tail (vertebræ), - - - - - - -	5
Do. to end of hair, - - - - - - -	6
Height of ear, - - - - - - -	$0\frac{1}{2}$
Breadth of skull, - - - - - - -	$1\frac{1}{4}$
From heel to end of longest nail, - - - -	$1\frac{5}{8}$

HABITS.

We have personally no knowledge of the habits of this rare and comparatively new species. The specimen from which Dr. LICHTENSTEIN made his description and figure, was obtained by F. DEPPE, Esq., in the vicinity of the city of Mexico, where the animal was indiscriminately called *Comadreja*, *Oronzito* and *Onzito*. He was unable to collect any information in regard to its habits. The specimen from which our description and figure were made, was captured by Mr. JOHN K. TOWNSEND. We conversed with an American officer, who informed us that he had occasionally seen it near Monterey in Mexico, that it there bore no better character than its congener the Ermine in the more northern parts of America; that it was destructive to poultry and eggs, and very commonly took up its residence in the outhouses on plantations, and under such circumstances was regarded as a great nuisance. Fortunately for them, the species was considered as quite rare in the northern parts of Mexico, as the Mexican who pointed out this animal to our officer stated, this was the first *Comadreja* he had seen in five years.

GEOGRAPHICAL DISTRIBUTION.

As we have not heard of the existence of our Ermine in Mexico, we are inclined to the belief that this species takes the place of the

Ermine in the South, and that with similar roving and predacious habits it has a more extended geographical range than is at present known. The field of natural history in Texas, California, and Mexico, has been as yet very imperfectly explored. We have only heard of the Bridled Weasel as being found in four widely separated localities—in Texas between the Colorado and Rio Grande, in Mexico in the vicinity of the capital, and in the northern parts near Monterey, and in the valleys of the mountains south-west of that city.

GENERAL REMARKS.

In comparing this singularly marked species with others from the Eastern and Western hemispheres, we have been struck with the uniformity existing on both continents in the nearly equal distribution of predacious animals, and in their close resemblance to each other, in size, form and habits. The badger in Europe (*Meles vulgaris*) is in America replaced by *M. Labradoria*. The European Otter (*Lutra vulgaris*) has its representative in America in our Canada otter (*Lutra Canadensis*). The European mink (*P. lutreola*) is replaced by our nearly similar (*P. vison*). The European ferret (*P. furo*) by our western black-footed ferret (*P. nigripes*). The ermine and common weasel of the north of Europe (*P. erminea*) and (*P. vulgaris*) by our ermine and brown weasel (*P. erminea*) and (*P. fusca*) in the Northern and Middle States of America, and the Java ferret (*P. nudipes*) has its representative near the tropics in America in our (*P. frenata*), nearly of the same size, and with similar habits. There is evidently great wisdom in this arrangement of Providence. Countries under similar latitudes producing large numbers of the smaller rodentia, require a certain number of carnivorous animals to prevent their too rapid multiplication, which in the absence of such a provision of nature would be destructive of the interests of the husbandman.

GENUS PROCYON.—Storr.

DENTAL FORMULA.

$$Incisive \frac{6}{6}; \quad Canine \frac{1-1}{1-1}; \quad Molar \frac{6-6}{6-6} = 40.$$

Muzzle, pointed and projecting beyond the lower jaw ; ears, short and oval ; tail, bushy, and long. Feet, five toed, with strong nails not retractile ; soles of feet, (posterior,) naked ; the species rest on the heel, but walk on the toes. Mammæ, six ventral ; there is a gland on each side of the anus which secretes a slightly offensive fluid.

The generic name is derived from the Greek προ, before, and κυων, a dog.

Two species only have been noticed : one in the northern, and the other in the southern parts of North America.

PROCYON LOTOR.—Linn.

Raccoon.

PLATE LXI.—Male and Young.

P. corpore supra canescente plus minus in nigrum vergente, infra, auriculis pedibusque albicantibus ; facie albida, fascia sub oculari obliqua nigra, cauda rufescente annulis 4–5 nigris.

CHARACTERS.

Body above, grayish mixed with black ; ears, and beneath, whitish ; a black patch across the eye. Tail with 4 or 5 annulations of black and gray.

SYNONYMES.

Arecon, Smith's Voyages, xiii., p. 31.
Ursus Lotor, Linn., 12th ed., p. 70.
    "   "   Erxleben, Syst., p. 165–4.
    "   "   Schreber Säugth., p. 521, 5 t. 143.
Le Raton, Buffon, vol. viii., p. p. 337, t. xliii.

on Stone by N J Hitchcock

Raccoon.

Drawn from Nature by J W Audubon   Lith Printed & Col.d by J T Bowen, Phil

RACCOON BEAR, Pennant's Arct. Zool., vol. i., p. 69.
PROCYON LOTOR, Cuv., Règne Animal, vol. i., p. 143.
"   "   Sabine, Journal, p. 649.
"   "   Harlan, p. 53.
"   "   Godman, vol. i., p. 53.
"   "   Dekay, New-York Fauna, p. 26.
PROCYON NIVEA, Gray, Magazine of Nat. Hist., vol. i., 1837, p. 580.

DESCRIPTION.

The body is rather stout, the legs of moderate length, and the appearance of the animal would indicate that although he is not intended for great speed, he is still by his compact and well organized structure, his strong and muscular limbs and short and stout claws, capable of a tolerably rapid race, and is able to climb, although not with the agility of the squirrel, still with greater alacrity than his near relative the bear.

Head, rather round ; nose, tapering, sharp, and the snout moveable ; point of the nose, naked ; eyes, round, and of moderate size ; moustaches, few, very rigid, resembling bristles, extending to the chin ; ears, low, erect, elliptical, with their tips much rounded, clothed with hair on both sides ; on the inner surface the hairs are longer and less dense ; tail, of moderate length and bushy. In its feet the Raccoon is partially plantigrade, hence it was classed by LINNÆUS among the bears, under the genus *Ursus ;* soles of feet, naked. When it sits, it often brings the whole hind sole to the ground, resting in the manner of the bear. The canine teeth are large and extend beyond the lips. The nails are strong, hooked and sharp, not covered with hair. The body is densely clothed with two kinds of hair ; the outer and longer, long and coarse ; the inner, softer and more like wool.

COLOUR.

Point of nose, and soles of feet, black ; nails, dark brown ; moustaches, nearly all white ; ears, lips, above the snout and chin, dingy white ; above the eyes, and around the forehead, light gray. A dark brown patch extends from each side of the neck and passes the eyes, over the nose, nearly reaching the snout, and gradually fading on the forehead into the colours of the back ; eyes, black ; the longer hairs on the back are dark brown at the roots, then yellowish-white for half their length, and are broadly tipped with black ; the softer fur beneath, pale brown throughout the whole body ; on the sides and belly, the longer hairs are dingy white from the roots ; the tail has about six distinct black rings, and is tipped with black ; these rings alternate with five light yellowish-brown annulations.

DIMENSIONS.

Old male, received from Dr. John Wright.

Inches.

Nose to anterior canthus,	-	-	-	-	-	-	-	$1\frac{3}{8}$
" " corner of mouth,	-	-	-	-	-	-	-	$2\frac{1}{4}$
" " root of ear,	-	-	-	-	-	-	-	$7\frac{3}{8}$
" " " of tail,	-	-	-	-	-	-	-	$26\frac{1}{2}$
Tail, (vertebræ), ·	-	-	-	-	-	-	-	8
" to end of hair,	-	-	-	-	-	-	-	$9\frac{1}{2}$
Length of head,	-	-	-	-	-	-	-	$7\frac{1}{2}$
Breadth of head,	-	-	-	-	-	-	-	$4\frac{1}{8}$

Weight, 22 lbs.

HABITS.

The Raccoon is a cunning animal, is easily tamed, and makes a pleasant monkey-like pet. It is quite dexterous in the use of its fore-feet, and will amble after its master in the manner of a bear, and even follow him into the streets. It is fond of eggs, and devours them raw or cooked with avidity, but prefers them raw of course, and if it finds a nest will feast on them morning, noon and night without being satiated. It will adroitly pick its keeper's pockets of anything it likes to eat, and is always on the watch for dainties. The habits of the muscles (*unios*) that inhabit our fresh water rivers are better known to the Raccoon than to most conchologists, and their flavour is as highly relished by this animal as is that of the best bowl of clam soup by the epicure in that condiment.

Being an expert climber, the Raccoon ascends trees with facility and frequently invades the nest of the woodpecker, although it may be secure against ordinary thieves, by means of his fore-feet getting hold of the eggs or the young birds. He watches too the soft-shelled turtle when she is about to deposit her eggs, for which purpose she leaves the water and crawling on to the white sand-bar, digs a hole and places them underneath the heated surface. Quickly does the rogue dig up the elastic ova, although ever so carefully covered, and appropriate them to his own use, notwithstanding the efforts of the luckless turtle to conceal them.

Sometimes, by the margin of a pond, shrouded, or crouched among tall reeds and grasses, Grimalkin-like, the Raccoon lies still as death, waiting with patience for some ill-fated duck that may come within his reach. No negro on a plantation knows with more accuracy when the corn (maize) is juicy and ready for the connoisseur in roasting ears, and he does not require the aid of fire to improve its flavour, but attacks it more

voraciously than the squirrel or the blackbird, and is the last to quit the cornfield.

The favourite resorts of the Raccoon are retired swampy lands well covered with lofty trees, and through which are small water-courses. In such places its tracks may be seen following the margins of the bayous and creeks, which it occasionally crosses in search of frogs and muscles which are found on their banks. It also follows the margins of rivers for the same purpose, and is dexterous in getting at the shell-fish, notwithstanding the hardness of the siliceous covering with which nature has provided them. In dry seasons, the receding waters sometimes leave the muscles exposed to the heat of the sun, which destroys their life and causes their shells to open, leaving them accessible to the first animal or bird that approaches.

In the dreary months of winter should you be encamped in any of the great Western forests, obliged by the pitiless storm to remain for some days, as we have been, you will not be unthankful if you have a fat Raccoon suspended on a tree above your camp, for when kept awhile, the flesh of this species is both tender and well-flavoured.

The Raccoon when full grown and in good condition we consider quite a handsome animal. We have often watched him with interest, cautiously moving from one trunk to another to escape his view. His bright eye, however, almost invariably detected us ere we could take aim at him, and he adroitly fled into a hollow tree and escaped from us.

We once met with one of these animals whilst we were travelling on horseback from Henderson to Vincennes, on the edge of a large prairie in a copse, and on approaching it ran up a small sapling from which we shook it off with ease; but as soon as it reached the ground it opened its mouth and made directly towards us, and looked so fierce, that drawing a pistol from our holsters, we shot it dead when it was only a few feet from us.

The young are at their birth quite small; (about the size of a half-grown rat;) some that we saw in Texas were not more than two days old and were kept in a barrel. They uttered a plaintive cry not unlike the wail of an infant.

The Raccoon usually produces from four to six young at a time, which are generally brought forth early in May, although the period of their littering varies in different latitudes.

When the Indian corn is ripening, the Raccoons invade the fields to feast on the rich milky grain, as we have just stated, and as the stalks are too weak to bear the weight of these marauders, they generally break them down with their fore-paws, tear off the husks from the ears, and

then munch them at their leisure.   During this inviting season, the
Raccoon is not the only trespasser on the corn fields, but various animals
are attracted thither to receive their portion, and even the merry school-
boy shares the feast with them, at the risk of paying for his indulgence
by incurring the necessity of a physician's prescription the next day.   The
havoc committed in the Western States by squirrels and other animals
is almost incalculable, and no vigilance of the farmer can guard against
the depredations of these hungry intruders, which extend from farm to
farm, and even penetrate to those embosomed in the forests, where settlements are few and far between.

The Raccoon is not strictly a nocturnal animal ; and although it generally visits the corn fields at night, sometimes feeds on the green corn
during the day ; we have seen it thus employed during the heat of summer, and it will occasionally enter a poultry house at mid-day, and
destroy many of the feathered inhabitants, contenting itself with the head
and blood of the fowls it kills.

The nest or lair of the Raccoon is usually made in the hollow of some
broken branch of a tree.   When tamed, these animals are seldom induced
to lie or sleep on a layer of straw.

There exists a species of oyster in the Southern States of inferior
quality which bears the name of Raccoon Oyster : it lies imbedded in
masses in the shallow waters of the rivers.   These oysters are covered by
high tides, but are exposed at low water.   On these the Raccoons are
fond of feeding, and we have on several occasions seen them on the oyster
banks.   We have however never had an opportunity of ascertaining by
personal observation the accuracy of a statement which we have frequently heard made with great confidence, viz., that the Raccoon at low
tide in endeavouring to extricate these oysters from the shell, is occasionally caught by the foot in consequence of the closing of the valve of the
shell fish, when numbers of these being clustered and imbedded together,
the Raccoon cannot drag them from their bed, and the returning tide
drowns him.

The naturalist has many difficulties to encounter when inquiring into
facts connected with his pursuit : every one acquainted with the habits
of even our common species must know, that the information gained
from most of those who reside near their localities, from their want of
particular observation, is generally very limited, and probably the most
interesting knowledge gained by such queries, would be the result of a
comparison of the accounts given at different places.   From the Alleghany mountains, the swamps of Louisiana, and the marshes of Carolina,

we have received nearly the same history of the cunning manœuvres and sly tricks of the Raccoon in procuring food.

We add the following notes on a Raccoon kept for a considerable time in a tame state or partially domesticated.

When it first came into our possession it was about one-third grown. By kind treatment it soon became very docile, but from its well known mischievous propensities we always kept it chained.

It was truly omnivorous : never refusing any thing eatable, vegetable or animal, cooked or uncooked, all was devoured with equal avidity. Of some articles however it seemed particularly fond : as sugar, honey, chestnuts, fish and poultry. The animal would become almost frantic when either of the two first was placed near it, but beyond its reach. No means would be left untried to obtain the dainty morsel. It would rush forward as far as the chain permitted, and stretch out a fore-paw toward the object of its wishes to its utmost extent, which failing to reach it, the other was extended ; again disappointed, the hind limbs were tried in succession, by which there was a nearer approach to the food, on account of the animal being chained by the neck.

On being offered food when hungry, or roused up suddenly from any cause, or when in active play, the eye was of a lustrous green, changing apparently the whole countenance.

It had a strong propensity to roll food and other things under its paws ; segars in particular, especially when lighted. We have observed a similar propensity in young bears.

On placing a pail of water within its reach, it ran to it, and after drinking would examine the contents to the bottom with the fore-paws, seemingly expecting to find some fish or frog. If any thing was found it was speedily brought to the surface and scrutinized. We have seen it throw chips, bits of china and pebbles, &c., into the pail, and then fish them out for amusement, but never saw it put a particle of its food in to soak, except in a few instances when it threw in hard corn, but we do not think it was for this purpose.

After playing for a short time in the water it would commonly urinate in it and then upset the pail.

We gave it a fish weighing two pounds. The Raccoon turned it in all directions in search of a convenient point of attack. The mouth, nose, fins, vent, &c., were tried. At length an opening was made at the vent, into which a paw was deeply inserted ; the intestines were withdrawn and eaten with avidity. At the same time an attempt was made to insert the other paw into the mouth of the fish to meet its fellow. This disposition to use the paws in concert, was shown in almost every action, sometimes in a very ludicrous manner. On giving the animal a jug, one paw would be inserted in the aperture, and a hundred twists and turns would be made to join its fellow on the outside.

After devouring as much of the fish as it wished, it placed the paws on the remainder and lay down to doze, until hunger returned, watching the favourite food, and growling at any animal which happened to pass near it. By degrees this propensity to defend its food passed off, and it would allow the dog or fox to partake of it freely. We placed a half-grown fox within its reach : the Raccoon instantly grasped it with its legs and paws and commenced a close examination. It thrust its pointed nose in the ear of the fox to the very bottom, smelling and snuffing as if determined to find out the nature of the animal. During this time it showed no disposition to injure the fox.

The Raccoon can scent an object for some distance with accuracy. We suffered ours to go loose on one occasion, when it made directly for some small marmots confined in a cage in another room.

Our pet Raccoon whose habits we are relating evinced a singular propensity to listen to things at a distance, however many persons were around him, even though he might be at the moment eating a frog, of which food he was very fond. He would apparently hear some distant noise, then raise his head and continue listening, seeming every moment more absorbed ; at last he would suddenly run and hide himself in his burrow. This seems to be connected with some instinct of the animal in his wild state, probably whilst sitting on a tree sunning himself, when he is in the habit of listening to hear the approach of an enemy, and then hurrying to his hole in the tree.

Enjoying the hospitality of a friend one night at his plantation, the conversation turned on the habits of animals : and in speaking of the Raccoon he mentioned that it fed on birds and rabbits generally, but in winter robbed the poultry houses. The negroes on his plantation he said kept good dogs, and relied on them for hunting the Raccoon.

Whenever a Raccoon was about to attack the poultry house, the dogs scenting him give a shrill cry, which is the signal for his owner to commence the hunt. He comes out armed with an axe, with a companion or two, resolved on a Raccoon hunt. The dog soon gives chase with such rapidity, that the Raccoon, hard pressed, takes to a tree. The dog, close at his heels, changes his whining cry while running to a shrill short sharp bark. If the tree is small or has limbs near the ground so that it can be easily ascended, the eager hunters climb up after the " coon." He perceives his danger, endeavours to avoid his pursuers by ascending to the farthest topmost branch, or the extremity of a limb ; but all his efforts are in vain, his relentless pursuers shake the limb until he is compelled to let go his hold, and he comes toppling heavily to the ground, and is instantly seized by the dogs. It frequently happens however that the trees are tall and destitute of lower branches so that they cannot be climbed without the risk of life or limb. The negroes survey for a few moments in the bright moonlight the tall and formidable tree that shelters the coon, grumble a little at the beast for not having saved them trouble by mounting an easier tree, and then the ringing of their axes resounds through the still woods, awakening echoes of the solitude previously disturbed only by the hooting of the owl, or the impatient barking of the dogs. In half an hour the tree, is brought to the ground and with it the Raccoon, stunned by the fall: his foes give him no time to define his position, and after a short and bloody contest with the dogs, he is despatched, and the sable hunters remunerated, —for his skin they will sell to the hatters in the nearest town, and his flesh they will hang up in a tree to freeze and furnish them with many a savoury meal.

The greatest number of Raccoons, however, are killed by log-traps set with a figure of 4 trigger, and baited with a bird or squirrel, an ear of corn, or a fish: either the appetite or curiosity of these animals will entice them into a trap or entangle them in a snare.

Another mode of destroying this species is by fire-hunting, which requires good shooting, as the animal only shows one eye from behind the branch of a tree, which reflecting the light of the fire-hunter's torch, shines like a ball of phosphorus, and is generally knocked out at twenty-five or thirty yards by a good marksman.

The Raccoon, like the bear, hibernates for several months during winter in the latitude of New-York, and only occasionally and in a warm day leaves its retreat, which is found in the hollow of some large tree. We once however tracked in deep snow the footsteps of a pair of this species in the northern parts of New-York, and obtained them by having the tree in which they lay concealed cut down. They had made a circle in company of about a mile, and then returned to their winter domicil.

The specimen from which the large figure on our plate was taken was a remarkably fine male, and was sent to us alive by our friend, the late Dr. John Wright of Troy, New-York.

GEOGRAPHICAL DISTRIBUTION.

The Raccoon has a very extensive geographical range. Captain Cook saw skins at Nootka Sound which were supposed to be those of the Raccoon. Dixon and Parltock obtained Raccoon skins from the natives of Cook's River in latitude 60°. It is supposed by Richardson that this animal extends farther north on the shores of the Pacific, than it does on the eastern side of the Rocky Mountains. He farther states, that the Hudson's Bay Company procured about one hundred skins from the southern parts of the fur districts as far north as Red River, latitude 50°. We have not been able to trace it on the Atlantic coast farther north than Newfoundland. It is found in the Eastern, Northern and Middle States, and seems to become more abundant as we proceed southwardly. In some of the older States its numbers have greatly diminished, in consequence of the clearing of the forests, and the incessant wars waged against it by the hunters. In South Carolina, Georgia, Alabama, Mississippi and Louisiana, it is still found in great numbers, is regarded as a nuisance to the corn fields, and is at particular seasons hunted at night by sportsmen and negroes. We have been informed by our friend Daniel Morrison, Esq., of Madison Springs in Georgia, that in his frequent visits to Arkansas between the Washita and Red Rivers, the Raccoons are very plentiful and are frequently seen travelling about in open day, and that many corn fields are nearly destroyed by the Raccoon and the bear.

It was seen by Lewis and Clark at the mouth of the Columbia river. We possess several specimens obtained in Texas, and were informed by a friend, that although he had not seen it in California, he had heard of its existence in the northern parts of that State.

GENERAL REMARKS.

As might be expected, an occasional variety is found in this species.

We possess a specimen nearly black ; another yellowish white, with the annulations in the tail faint and indistinct. A nest of young was found in Christ Church parish in South Carolina, two of which were of the usual colour, the other two were white ; one of them was sent to us ; it was an albino, with red eyes, and all the hairs were perfectly white with the exception of faint traces of rings on the tail. We have no doubt that a similar variety was described by Gray, under the name of *Procyon nivea.*

We have accordingly added his name as a synonyme. Our friend Dr. Samuel George Morton of Philadelphia kept one for some time alive which was of a yellowish cream colour, and was also an albino.

Plate LXII

American Elk.— Wapiti Deer.

Drawn from Nature by J.J.Audubon, F.R.S.S.

On stone by W.E. Hitchcock.

Lith. Printed & Col⁴ by J. T. Bowen, Phil⁴

GENUS ELAPHUS. — Griffith.

$$Incisive \frac{0}{8}; \quad Canine \frac{1-1}{0-0}; \quad Molar \frac{6-6}{6-6} = 34.$$

Horns, (existing only in the male,) round; very large; antlers terminating in a fork or in snags from a common centre, suborbital sinus; canine teeth in the male, in the upper jaw; a muzzle.

The generic name is derived from the Greek Ελαφος, a Stag, or Elk; the name was applied by Pliny, Linnæus, and other naturalists, to designate a particular species existing in Europe, *Cervus Elaphus.*

Three well-determined species may be arranged under this genus; one existing in Europe, one in Walhihii, (the Nepaul Stag,) and one in America.

ELAPHUS CANADENSIS. — Ray.

American Elk. — Wapite Deer.

PLATE LXII. — Male and Female.

E. Cervus Virginianus robustior cornibus amplissimis ramosis teretibus, frontalibus amplis; cauda brevissima. Color rufescens, hieme fuscescens, uropygio flavicante stria nigra circumscripto.

CHARACTERS.

Larger than the Virginian deer. Horns, large, not palmated, with brow antlers; a naked space round the lachrymal opening. Tail, short. Colour, yellowish-brown above, a black mark extending from the angle of the mouth along the sides of the lower jaw. A broad pale yellowish spot on the buttocks.

SYNONYMES.

STAG, Pennant, Arctic Zool., vol. i., p. 27.
WEWASKISS, Hearne, Journal, p. 360.
RED DEER, Umfreville.
Do.   do.   Ray, Synops. Quad., p. 84.
C. STRONGYLOCEROS, Schreber, Säugethiere, vol. ii., p. 1074, pl. 247, F. q. G.
ALCES AMERICANUS, Jefferson's Notes on Virginia, p. 77.
THE ELK, Lewis and Clark, vol. ii., p. 167.
C. WAPTITE, Barton, Med. and Phys. Journal, vol. i., p. 36.
ELK, Smith, Med. Reports, vol. ii., p. 157, fig. Male, Female, and Young.
CERVUS (ELAPHUS) CANADENSIS, (The Wapite,) Synopsis of the Species of Mammalia.   Griffith's Cuvier, p. 776.
C. CANADENSIS, Harlan, p. 236.
Do.   do.   Godman, vol. ii., p. 294, fig. Male.
CERVUS STRONGYLOCEROS, Richardson, (The Wapite.) p. 251.
ELAPHUS CANADENSIS, Dekay, New-York Fauna, p. 118, plate 28, fig. 2.

DESCRIPTION

The Elk is of an elegant, stately and majestic form, and the whole animal is in admirable proportion. It bears so strong a resemblance to the red deer of Europe, that it was for a long time regarded as a mere variety of the same species. It is, however, much larger in size, and on closer examination differs from it in many particulars.

Head, of moderate size; muzzle, broad and long, rather small, not very prominent; ears, large; legs, rather stout, finely proportioned; hoofs, rather small.

From between the horns to the end of the frontal bone, beyond the nasal opening sixteen inches, length of horns following the curvature of the main branch four feet; with all the roots three and a quarter inches, by two and a quarter thick. There are six points on each horn, irregularly disposed, varying in length from nine to sixteen inches, excepting one which is two and a half inches only in length. At their points the horns curve backward and upward, and are about three feet five inches apart, at about half the distance from their roots to the extreme tip of the longest point or main branch. The horns at the insertion are three and three-quarter inches apart from the ring or crown at their roots.

In examining a number of elk horns we find a very remarkable variety, no two antlers being exactly alike on the same animal. We possess one pair which has a blunt prong extending downward on the right side of the face about nine inches, whilst the corresponding prong on the opposite side is turned upwards. The horns of this individual have five prongs on one horn **and seven on** the other. The horns are longitudinally channelled, most of the prongs inclining forward and upward, especially those nearest the roots of the main horn. All the horns are large and round, with brow antlers. The weight of the horns on full grown animals, as we have ascertained by weighing about a dozen of large size, is from thirty to forty-five pounds.

The three hindermost teeth in the upper jaw are double ; the remainder single. There are in the upper jaw of the male two very small canine teeth inclining forward almost on a line with the jaw. There is a short rudimentary mane on the fore-shoulder, and under the throat during the winter there are long black hairs.

There is a space on the outer side of the hind legs covered by a tuft, which is of an irregular oval shape, of about one and a half inch in length, the hairs which cover it being an inch long, lying flat and backwards, with shorter hairs extending down the leg several inches below the space.

The hairs on the body generally are very coarse, rather short ; longest on the back of the ham, where the whitish patch and the black line on the latter unite.

The tail, which in summer is not bushy, is thinly clothed with hair running to a point. A young male has its horns which are in velvet, nearly perpendicular, running but slightly backwards to the length of fourteen inches, where they divide into three short prongs.

COLOUR.

Male.

Muzzle, nostrils, and hoofs, black ; head, dark brown ; neck, rather darker, being nearly black ; on each side of the under jaw there is a longitudinal white patch, between which there is a large black stripe extending along the lines of the under jaw, dividing about four inches from the mouth, and continuing downward to the throat, where it unites again and is diffused in the general black colour of the throat and neck, leaving in its course a white space between the bone of the lower jaw, nearly as large as a man's hand.

There is no light-coloured ring, or space, around the eyes as in the European red deer, but in the present species the space around the socket of the eye is scarcely a shade lighter than the surrounding parts of the head.

Under surface of the ear, yellowish white, with a hue of dark brown on the margin ; on the outer surface of the ear, there is a white patch about four inches in length and nearly two inches wide, covering about a third of the ear, and running from near the root of the ear upwards at the lower edge

In the younger males the head, face and back of the neck are not nearly as dark as in specimens of old animals; the under jaw and throat however as well as a space above the nostrils are black as in the latter. The
upper and under surfaces of body and legs are light brownish gray, the
legs being rather darker than the body.

On the rump there is a broad patch of light grayish white commencing
nine inches above the root of the tail, spreading downward on each side
to a point in the ham, ten inches below the tail. It is fourteen inches
across opposite the root of the tail, (from one ham to the other,) and
twenty-two inches in length from the back to the termination on the thigh
or ham below the tail. This grayish white patch is bordered on the thighs
by a strongly marked black space which also separates it all around, although less conspicuously from the general colour of the body. We
have observed that in young specimens this pale mark on the rump is less
conspicuous, and in one specimen is not even perceptible, and this peculiarity has most probably misled some of our authors in regard to the species.

In specimens of about two years old the light but scarcely perceptible
markings on the rump gradually change to grayish brown between the hind
legs. In a still younger specimen of a male about eighteen months old
which has the horns three inches in height, (which are completely clothed
with soft brownish hairs to their summits,) there is scarcely any black on
the neck, and the white on the rump is not visible.

Female in summer colour.

We possess this animal in a state of confinement: she has like all the
females of this species no horns. She bears a strong resemblance in form
and colour to the male. Her neck is rather thinner and longer, and her
legs and body more slender. Her eyes are mild, and she is in her disposition very gentle and docile. The hair in summer is like that of the
male, uniform in colour from the roots to the surface.

Winter colour.

Both males and females in winter assume a very heavy coat of dark
gray hair all over the body. These hairs are about two and a half inches
to three long and are moderately coarse and strong.

When examined separately they have a wavy or crimped appearance.
The white patch on the rump is strongly developed in contrast with
the dark iron-gray colour of the winter coat. At this season the male
has a remarkable growth of hairs on the throat as well as on the back
of the neck, which increase considerably in length, so that the latter
might easily be mistaken for the rudiment of a mane.

DIMENSIONS.

Adult male (killed on the Upper Missouri River).

	Feet.	Inches.
From nose to root of tail, - - - - -	7	8¾
Length of tail, - - - - - - -	0	1¾
" of eye, - - - - - - - -	0	1¾
From tip of nose to root of ear, - - -	1	8
Length of ear, - - - - - - -	0	9¼
Height to shoulders, - - - - - -	4	10
Rump, - - - - - - - - -	5	2
Girth back of fore-legs, - - - - -	5	6¼

The females we measured were rather smaller than the above: one killed on the Yellow Stone River measured seven feet six and a half inches from nose to root of tail, and four feet seven inches from top of shoulder to the ground.

HABITS.

On our plate we have represented a pair of Elks in the foreground of a prairie scene, with a group of small figures in the distance; it gives but a faint idea of this animal in its wild and glorious prairie home: Observe the splendid buck, as he walks lightly, proudly, and gracefully along. It is the season of love: his head is raised above the willows bordering the large sand-bar on the shores of the Missouri, his spreading antlers have acquired their full growth, the velvet has been rubbed off, and they are hard and polished. His large amber-coloured eyes are brightened by the sun, his neck is arched, and every vein is distended. He looks around and snuffs the morning air with dilated nostrils: anon he stamps the earth with his fore-feet and utters a shrill cry somewhat like the noise made by the loon. When he discovers a group of females he raises his head, inclines it backwards, and giving another trumpet-like whistle, dashes off to meet them, making the willows and other small trees yield and crack as he rushes by. He soon reaches the group, but probably finds as large and brave a buck as himself gallanting the fair objects of his pursuit, and now his eyes glow with rage and jealousy, his teeth are fiercely champed together making a loud harsh noise, his hair stands erect, and with the points of his immense horns lowered like the lance of a doughty knight in times of yore, he leaps towards his rival and im-

mediately a desperate battle ensues. The furious combatants sway backwards and forwards, sideways or in circles, each struggling to get within the other's point, twisting their brawny necks, and writhing as they
endeavour to throw their opponent off the ground. At length our valorous
Elk triumphs and gores the other, so that he is worsted in the fight, and
turns ingloriously and flies, leaving the field and the females in possession of the victor : for should there be any young Elks present during such
a combat, they generally run off.

The victorious buck now ranges the tangled woods or leads the does
to the sand-bars or the willow-covered points along the broad stream.
After a certain period, however, he leaves them to other bucks, and towards the latter part of February his antlers drop off, his body is much
emaciated, and he retires to some secluded spot, where he hopes no enemies will discover him, as he is no longer vigorous and bold, and would
dread to encounter even a single wolf.

When we first settled (as it is termed) in the State of Kentucky, some
of these animals were still to be met with ; but at present we believe none
are to be found within hundreds of miles of our then residence. During
a journey we made through the lower part of the State, armed as usual
with our double-barrelled gun, whilst passing through a heavy-timbered
tract not far from Smithland at the mouth of the Cumberland River, we
espied two Elks, a male and female, which started out of a thicket not
more than forty or fifty yards from us. Our gun being loaded with balls,
we fired successfully and brought down the buck. The tavern keeper at
Smithland went after the animal with a wagon and brought him into the
little village. The hunters in the neighbourhood said they had not seen
or heard of Elks in that part of the State for several years, although
some were to be found across the Ohio, in the state of Illinois.

At the time we are writing (1847) the Elk is not seen in any numbers
until you ascend the Missouri River for a great distance. In that part of
the country, where the points in the river are well covered with wood
and under-brush, they are to be found at times in considerable numbers.
These animals however do not confine themselves to the neighbourhood of
the water-courses, but roam over the prairies in large herds. Unless
disturbed or chased, they seldom leave a secluded retreat in a thickly-
wooded dell, except to go to the river to drink, or sun themselves on the
sand-bars. They are partial to the islands covered with willow, cotton
wood, &c., and fringed with long grass, upon which they make a bed
during the hot sultry hours of the day. They also form a bed occasionally in the top of a fallen tree.

During hot weather, when mosquitoes abound in the woods, they retire to ponds or proceed to the rivers and immerse their bodies and heads, leaving merely enough of their noses above the water to allow them to breathe.

Whilst ascending the Missouri river in the steamer Omega, we observed a fawn of this species one morning running along the shore under a high bank. It was covered with yellowish white spots, was as nimble and active as a kitten, and soon reached a place where it could ascend the bank, when it scampered off amid the tall grass. We had on board a servant of Mr. CHARDON named ALEXIS LABOMBARDE who was a most expert hunter. We soon saw another fawn, and ALEXIS went after it, the boat having stopped to wood. He climbed the bank and soon overtook the little animal, but having no rope or cord with him, was at a loss how to secure his captive. He took off his suspenders and with these and his pocket-handkerchief managed to fasten the fawn around the neck, but on attempting to drag it toward the boat the suspenders gave way and the fawn dropped into the stream, and swam a few yards lower down, where it again landed; one of our party witnessed from the steamboat the ineffectual efforts of LABOMBARDE and ran up to his assistance, but also without a rope or cord, and after much ado the animal again swam off and escaped.

The food of the Elk consists generally of the grass found in the woods, the wild pea-vines, the branches of willows, lichens, and the buds of roses, &c. During the winter they scrape the snow from the ground with their fore-feet, and eat the tender roots and bark of shrubs and small trees.

On our reaching Fort Pierre we were presented by Mr. PICOT with a most splendidly prepared skin of a superb male Elk, and a pair of horns. The latter measured four feet six and a half inches in length; breadth between the points twenty-seven and a half inches. The circumference of the skull or base ten inches, the knob twelve inches, between the knobs three inches. This animal, one of the largest ever seen by Mr. PICOT, was killed in the month of November, 1832.

HEARNE says that the Elk is the most stupid of all the deer kind; but our experience has led us widely to differ from that traveller, as we have always found these animals as wary and cunning as any of the deer tribe with which we are acquainted. We strongly suspect HEARNE had reference to another species, the American reindeer.

We chanced one day to land on a sand-bar covered with the broad deep tracks of apparently some dozen Elks: all the hunters we had in our boat prepared to join in the chase, and we among the rest, with our old trusty double-barrelled gun, sallied forth, and while passing through a large patch of willows, came suddenly upon a very large buck; the noble animal was not more than a few steps from where we stood: our gun was levelled in an instant, and we pulled trigger, but the cap did not explode. The Elk was startled by the noise of the falling hammer, and wheeling round, throwing up the loose soil with his hoofs, galloped off among the willows towards the river, making a clear path through the small trees and grass. We ran to intercept him, but were too late, and on reaching the bank the Elk was already far out in the stream, swimming rapidly with its shoulders and part of its back above water. On the opposite shore there was a narrow beach, and the moment the Elk touched the bottom, it sprang forward and in a bound or two was out of sight behind the fringing margin of trees on the shore. This, we are sorry to say, was the only Elk we had an opportunity of firing at whilst on our last western expedition.

The pair from which the figures on our plate were taken we purchased at Philadelphia: they had been caught when young in the western part of Pennsylvania; the male was supposed to be four or five years old, and the female also was full grown. These Elks were transported from Philadelphia to our place near New-York, and we had a capacious and high enclosure made for them. The male retained much of its savage habits when at liberty, but the female was quite gentle. When she was first put in the pen, where the buck was already pacing round seeking for a weak point in the enclosure, he rushed towards her, and so terrified her that she made violent exertions to escape, and ran at full speed with her head up and her nostrils distended, round and round, until we had the large box in which she had been brought up from Philadelphia placed in the enclosure, when she entered it as a place of refuge, and with her head towards the opening stood on her defence, on which the male gave up the pursuit, and this box was afterwards resorted to whenever she wished to be undisturbed.

We had some difficulty in taking the bridle off from the head of the buck, as he kicked and pranced furiously whenever any one approached for that purpose, and we were forced to secure his head by means of a lasso over his horns, and drawing him by main force to a strong post, when one of our men cut the leather with a knife.

While these two Elks were kept by us they were fed on green oats, hay, Indian corn, and all such food as generally is given to the cow, excepting turnips, which they would not touch.

We found that the pair daily ate as much food as would have sufficed for two horses. They often whistled (as the hunters call this remarkable noise, which in calm weather can be heard nearly a mile); this shrill sound appears to be produced by an almost spasmodic effort, during which the animal turns its head upwards and then backwards. While we were outlining the male, we often observed him to dilate the lachrymal spaces or openings adjoining the eyes, so that they were almost as wide as long. When we drew near he would incline his head sideways, curl back his upper lip, and show a portion of his tongue and fine teeth, which last he ground or grated together, turning his head the while from side to side, and eyeing us with a look of angry suspicion. His eyes enlarged and his whole figure partook of the excitement he felt.

The process of rubbing off the velvet from the horns was soon accomplished by this animal; he began the moment he had been taken out of his box, to rub against the small dog-wood and other trees that stood within the enclosure. At a later period of the year we have observed the Elk rubbing his antlers against small trees, and acting as if engaged in fight; whether this manœuvre be performed for the purpose of loosening the horns, towards the period when they annually drop off, we, in parliamentary language, are not prepared to say.

Elks at times congregate from the number of fifty to several hundreds, and in these cases the whole herd follow the movements of their leader, which is generally the largest and the strongest male of the party. They all stop when he stops, and at times they will all turn about with as much order and with far greater celerity than a troop of horse, of which, when thus seen in array, they forcibly remind us.

From accident or otherwise great differences exist in the formation of the antlers of the Elk, although the horns of all the American *Cervii* are so specifically distinct as to enable the close observer to tell almost at a glance to what species any shown to him belonged. The ease with which these animals pass, encumbered with their ponderous and wide-spreading antlers, through the heavy-timbered lands of the West, is truly marvellous; and we can hardly help wondering that they are not oftener caught and entangled by their horns. Instances there doubtless are of their perishing from getting fastened between vines, or thick growing trees, but such cases are rare.

The male Elk drops his horns in February or March. The one we had dropped one on the ninth of March, and as the other horn held on for a day or two longer, the animal in this situation had quite an awkward appearance. After the horns fall, the head looks sore, and sometimes the places from which they have been detached are tinged with blood. As soon as the huge antlers drop off, the Elks lose their fierce and pugnacious character, and the females are no longer afraid of them; while on the other hand, the males show them no farther attentions whatever.

The young, sometimes one, but usually two in number, are brought forth in the latter end of May or June. It is stated by GODMAN, we know not on what authority, that when twins are produced they are generally male and female.

A friend of ours related to us some time ago the following anecdote. A gentleman in the interior of Pennsylvania who kept a pair of Elks in a large woodland pasture, was in the habit of taking pieces of bread or a few handfuls of corn with him when he walked in the enclosure, to feed these animals, calling them up for the amusement of his friends. Having occasion to pass through his park one day, and not having provided himself with bread or corn for his pets, he was followed by the buck, who expected his usual gratification : the gentleman, irritated by the pertinacity with which he was accompanied, turned round, and picking up a small stick, hit the animal a smart blow, upon which, to his astonishment and alarm, the buck, lowering his head, rushed at him and made a furious pass with his horns : luckily the gentleman stumbled as he attempted to fly, and fell over the prostrate trunk of a tree, near which lay another log, and being able to throw his body between the two trunks, the Elk was unable to injure him, although it butted at him repeatedly and kept him prisoner for more than an hour. Not relishing this proceeding, the gentleman, as soon as he escaped, gave orders to have the unruly animal destroyed.

The teeth of the Elk are much prized by the Indians to ornament their dresses ; a " queen's robe " presented to us is decorated with the teeth of fifty-six Elks. This splendid garment, which is made of antelope skins, was valued at no less than thirty horses !

The droppings of the Elk resemble those of other deer, but are much larger.

The Elk, like other deer, lie down during the middle of the day, and feed principally at early morning, and late in the evening. They drink a good deal of water.

This species can be easily domesticated, as we have observed it in menageries and in parks both of Europe and America. The males, like those of the Virginian deer, as they advance in age, by their pugnacious habits are apt to become troublesome and dangerous. The Elk lives to a great age, one having been kept in the possession of the elder PEALE of Philadelphia for thirteen years; we observed one in the Park of a nobleman in Austria that had been received from America twenty-five years before.

GEOGRAPHICAL DISTRIBUTION.

We have every reason to believe, that the Elk once was found on nearly every portion of the temperate latitudes of North America. It has never advanced as far north as the moose deer, but it ranges much farther to the south. The earliest explorers of America nearly all speak of the existence of the stag, which they supposed was identical with the stag or red deer of Europe. It differs from the Virginian deer, which continues to range in the vicinity of settlements and is not driven from its favourite haunts by the cry of the hounds or the crack of the rifle. On the contrary the Elk, like the buffalo, takes up its line of march, crosses broad rivers and flies to the yet unexplored forests, as soon as it catches the scent and hears the report of the gun of the white man. At present there is only a narrow range on the Alleghany mountains where the Elk still exists, in small and decreasing numbers, east of the Missouri, and these remnants probably of large herds would undoubtedly migrate elsewhere were they not restricted to their present wild mountainous and hardly accessible range, by the extensive settlements on the west and south.

Mr. PEALE of Philadelphia mentioned to us some fifteen years ago, that the only region in the Atlantic States where he could procure specimens of the Elk was the highest and most sterile mountains in the northwest of Pennsylvania, where he had on several occasions gone to hunt them.

Dr. DEKAY (New-York Fauna, p. 119) mentions, on the authority of BEACH and VAUGHAN, two hunters in whose statements confidence could be placed, that as late as 1826, Elks were seen and killed on the north branch of the Saranac. On a visit to Western Virginia in 1847, we heard of the existence of a small herd of Elk that had been known for many years to range along the high and sterile mountains about forty miles to the west of the Red Sulphur Springs. The herd was composed of eight males, whose number was ascertained by their tracks in the snow. One of these had been killed by a hunter, and the number was reduced to seven. Our informant, a friend in whom the highest confidence could be placed, supposed, as all the individuals in the herd had horns, the race would soon disappear from the mountains. As, however, the males at certain seasons keep in separate groups, we have no doubt there was a similar or larger herd of females in the same range ; but the number is doubtless annually lessening, and in all probability it will not be many years before the Elk will be entirely extirpated, to beyond several hundred miles west of the Mississippi.

This animal, according to RICHARDSON, does not extend its range farther to the north than the 56th or 57th parallel of latitude, nor is it found to the eastward of a line drawn from the south end of Lake Winnepeg to the Saskatchewan in the 103d degree of longitude, and from thence till it strikes the Elk river in the 111th degree. It is found on the western prairies, and ranges along the eastern sides of the mountains in Texas and New Mexico. It is also found in Oregon and California. Its most southern geographical range still remains undetermined.

GENERAL REMARKS.

The family of Elks was by all our old authors placed in the same genus with the true deer, (*Cervus*,) to which they are very closely allied in their character and habits. As that genus however has been greatly enlarged in consequence of the discovery of new species, the deer have been conveniently divided into several sub-genera, of which our species is the largest and most interesting among the true Elks (*Elaphus*).

The American Elk, Wappite, or Stag, was for a long period considered identical with the European red deer, (*C. Elaphus*,) and was, we believe, first treated as a distinct species by RAY. It was subsequently noticed by JEFFERSON and described and figured in the Medical Repository. The difference between these two species is so great that they may be distinguished at a glance. Our Elk is fully a foot higher at the shoulders than the European red stag. The common stag or red deer is of a uniform blackish brown, whilst the Elk has all its upper parts and lower jaw yellowish brown. It has also a black mark on the angle of the mouth which is wanting in the other. In the European species the circle around the eye is white, in the American it is brown. There are other marks of difference which it is unnecessary to point out, as the species are now regarded by all naturalists as distinct.

Our esteemed friend Dr. RICHARDSON has applied to this species the name of *Cervus strongyloceros* of SCHREBER, because the figure of PERRAULT (Mem. sur les an. vol. 2, p. 45) did not exhibit the pale mark on the rump, and he thought it not improbable that PERRAULT's figure was that of the black-tailed deer (*Cervus macrotis*). We do not believe that the latter species ever reaches the latitude where PERRAULT's specimen was procured; but as we have already stated in this article, younger specimens of our Elk exhibit only faint traces of this pale mark on the rump, and in some they are entirely wanting. We have scarcely a doubt that RAY's description was intended to apply to our American Elk, and we have therefore adopted his specific name.

Plate LXIII.

N° 13.

On Stone by W. E. Hitchcock

Lith. Printed & Col^d by J.T. Bowen, Philad.

Black-tailed Hare.

Drawn from Nature by J.J. Audubon, F.R.S.F.L.S.

LEPUS CALLOTIS.—Wagler.

Black Tailed Hare.

PLATE LXIII.—Male.

L. magnitudine, L. glacialem adæquans, supra flavescente fusco canoque varius, subtus albus; auribus pedibusque prælongis, cauda longa, nigra.

CHARACTERS.

Size of the polar hare; ears and legs, very long; tail, long and black; mottled with gray and yellowish-brown above, beneath, white.

SYNONYMES.

Lepus Callotis, Wagler, 1832.
    "   Nigricaudatus, Bennett, Proceedings of the Zoological Society of London, 1833, p. 41, marked in the Catalogue of the Zoological Society, 582.
Lepus Nigricaudatus, Bachman, Journal of the Academy Nat. Sciences, Philadelphia, vol. viii., pt. 1, p. 84, an. 1839.

DESCRIPTION.

This interesting species is similar to others composing a certain group of hares found in America, characterized by being large, and having very long ears, and long and slender legs and bodies, the whole form indicating capacity for long leaps and rapid locomotion. In all these characteristics, *Lepus Callotis* approaches nearest to Townsend's hare, (*Lepus Townsendii,*) which may be considered the type of this group.

COLOUR.

The whole of the upper surface, fawn colour, tipped with black; hairs on the back, silvery gray for one-third of their length, then pale fawn, then black, then fawn, tipped with black. Back of the neck, brownish black, slightly tipped with fawn. A number of hairs of unusual length, (two and one-fourth inches,) and delicately interspersed along the sides; in the greatest abundance along the shoulders. These hairs are black from the base for two-thirds of their length, the remainder pale fawn;

sides, and under parts of the neck, dingy pale fawn, gradually becoming
white on the chest; haunches, legs and under surface white; the hairs
on the rump annulated with black, and near the root of the tail almost
entirely black; the whole of the tail on the upper surface to the extremity black; on the under surface the hairs are black from the roots, slightly
tipped with grayish brown. Hairs on the under surface of the feet, in
some specimens red, in others a soiled yellowish-brown. Ears, posteriorly
for two-thirds of their breadth black at the roots, gradually blending into
fawn, and on the inner third the longitudinal line of demarcation being
very distinct; this fawn colour is mixed with black hairs, edged at the tip
with black, the remainder of the edge fawn; the outer margin of the posterior surface to its apex pure white. Inner surface of the ears nearly
naked, except at the outer edge, where they are clothed with short grizzled brown hairs. Whiskers white and black, the former predominating;
chin and throat, white. The marginal line of demarcation between the
colour of the back and that of the under surface, is somewhat abrupt
across the upper portion of the thighs, and very distinctly marked.

DIMENSIONS.

		Inches.
Length from point of nose to root of tail, - - -		20
Tail (vertebræ), - - - - - - - -		1½
" including fur, - - - - - - -		2¼
From heel to longest nail, - - - - - -		4¾
Head over the curve, - - - - - - -		4½
From eye to nose, - - - - - - -		1¾
Ears posteriorly, - - - - - - -		4¾
Greatest breadth, - - - - - - -		2⅜

HABITS.

Our account of this species is principally derived from the journals of J.
W. Audubon, kept during his journey through part of Texas, made for the
purpose of procuring the animals of that State, and obtaining some knowledge of their habits for our present work, in 1845 and 1846, with an extract from which we now present our readers.

"One fine morning in January, 1845, at San Antonio de Bexar, as I
mounted my faithful one-eyed chesnut horse, admiring his thin neck and
bony legs, his delicate head and flowing flaxen tail and mane, I was
saluted with a friendly good morning by Mr. CALAHAN, then holding the
important office of mayor of the little village; and on his ascertaining
that my purpose was to have a morning hunt on the prairies and through

the chapparal, which I did day after day, he agreed to accompany me in search of the animals I was anxiously trying to obtain, and in quest of which I rode over miles of prairie with my bridle on the knobbed pummel of my Texan saddle, the most comfortable saddle I have ever tried, (being a sort of half Spanish, half English build,) my horse with his neck stretched out and his head about on a level with his shoulders, walking between four and five miles an hour, turning to the right or to the left agreeably to the slightest movement of my body, so well was he trained, leaving both hands and eyes free, so that I could search with the latter every twig, tussock or thicket, and part the thick branches of the chapparal of musquit, prickly holly, and other shrubs, which I am inclined to think quite equal to any East-Indian jungle in offering obstructions to the progress of either horse or man.

Mr. CALAHAN having mounted, we set out, and after about an hour's hard work, occupied in crossing one of the thickest covers near the town, gained the broad and nearly level prairie beyond, across which to the west we could see varied swelling undulations, gradually fading into the faint outline of a distant spur, perhaps of the rocky chain of mountains that in this latitude lie between the water courses flowing toward the Gulf of Mexico, and the streams that empty into the Gulf of California : so far away indeed seemed these faint blue peaks that it required but a little stretch of the imagination to fancy the plains of California but just at the other side. I was enchanted with the scene, scarcely knowing whether the brilliant fore-ground of cacti and tropical plants, the soft indefinite distance, or the clear summer blue sky, was most beautiful. My companion observing my enthusiasm, warmed into praises of his adopted country he had, he said, fought hard for it, and exclaimed, it is a country worth fighting for ; when my reply, of whatever nature it might have been, was prevented, and all ideas of blue mountains, vast rolling prairies, &c., were cut short by a jackass rabbit bounding from under our horses' feet ; he was instantly followed by my worthy friend the mayor at full speed on his white pony, to my great annoyance, for otherwise he would have stopped in a hundred yards or so. Away they went, and as my friend's horse was a running nag, he doubtless expected to overtake the Hare, which had only gained about fifty yards start during our momentary surprise. The Hare, as I quickly observed, did not make much shorter leaps than the horse. I could see it at each bound appear like a jack-o'-lantern floating with the breeze over a swamp, but in less time than I have taken to write this, they had ran a mile, the Hare doubled and was a hundred yards in advance, but could not stop and look behind, for he had such a race that he knew well no time was to be lost in gaining some bed of cactus or chapparal. Now on came both Hare and hunter, and the race was of the swiftest when another double caused the rider to pull up with such force that his stirrup leather broke, and the space between the mayor and the object of his pursuit was widened to a quarter of a mile, and the chase ended ; our friend dismounting to refit. We had not the good fortune to start another of these hares that day.

Some time afterwards while at Castroville, a little place of about a dozen huts and one house, this Hare was procured by a party of Indians and brought to J. W. AUDUBON, who writes : " I chanced to be visited by some of the Shawnee Indians who were in the neighbourhood on a hunting expedition. They were highly astonished and pleased with my drawings, which I exhibited to them while trying to explain what animals I wanted. I made a hasty sketch of a hare with immensely long ears, at which I pointed with an approving nod of the head, and then made another sketch smaller and with shorter ears, at which last I shook my head and made wry faces ; the Indians laughed, and by their gutteral eugh, haugh, li, gave me to understand that they comprehended me ; and in a day or two, I had a beautiful specimen of the Black-tailed Hare brought to me, but with the head shot off by a rifle ball. The Indians were quite disappointed that it did not answer my purpose, and smoothed down the fur on the body, which is the only part of the skin they generally preserve, and what they thought I wanted.

The specimen I drew from was shot by POWEL, one of Colonel HAYS' rangers, from whom I received many attentions and who acted most kindly while with me on one of my excursions from San Antonio. This Hare is so rare in those parts of Texas that I visited, that I can say little of its habits. It appears to be solitary, or nearly so, fond of high open prairie with clumps of trees, or rather bushes and thickets about them, trusting to its speed for safety and only taking cover from hawks and eagles. Near San Petruchio, as I was informed, this Hare is more abundant than in this vicinity, and two or three of them can occasionally be started in a morning's ride."

The specimen from which Mr. BENNETT described and named this Hare (*Lepus nigricaudatus*, Bennett, Zoological Proceedings, 1833, p. 41), has a more definitely marked line of white along the sides and legs than the one I drew from ; but this species varies so much in its markings, that one figure with the characters given is probably as like the majority as another.

The line of white and black near the tip of the ears extended longitudinally, is by many considered a good specific character, but it does not, I think, hold out in respect to this animal.

It is singular that this fine species of Hare should be so rare in the collections of Europe ; I saw only two, and did not hear of the existence of any in the museums which I had not an opportunity of examining.

Since the Mexican war broke out, several have been sent home by our officers.   We have the pleasure of acknowledging the receipt of a fine skin from Lieutenant ABERT, who also favoured us with some skins of quadrupeds from the vicinity of Santa Fe, which we shall have occasion to notice elsewhere, and for which we return him our best thanks.

This species is called the *Jackass* Rabbit in Texas, owing to the length of its ears.

GEOGRAPHICAL DISTRIBUTION.

This Hare is found as far north as Santa Fe, in the great prairies ; it does not, however, occur near the shores of the lower Red River, nor near the Gulf of Mexico indeed, until we get as far south as about latitude 30°, from which parallel to the southward it becomes more abundant, and may be said to be the common Hare of Mexico.   Whether it is found beyond the limits of North America we are unable to say, but suppose not, as the museums of Europe have been better supplied with South American species than with those of our northern portion of the Western hemisphere, and as already observed, do not contain more than the two specimens mentioned above, one of which is stated to have been received from Mexico and the other from California.

GENERAL REMARKS.

There is a specimen in the Berlin Museum, labelled Lepus Callotis, WAG-LER, described by him in 1832.   This specimen corresponds in all essential particulars with that which exists in the Zoological Museum of London, described by BENNETT.   Hence we are obliged to adopt WAGLER's name, he having the priority as the first scientific describer.

PUTORIUS PUSILLUS. — Dekay.

The Small Weasel.

PLATE LXIV.

P. erminia tertia parti minore ; caudâ breviuscula.   Supra rufo-fuscus subtus albus.

CHARACTERS.

A third smaller than the Ermine ; tail rather short ; Colour, brown above white beneath.

SYNONYMES.

Mustela (putorius) Vulgaris, Bach., Fauna Bor. Am., vol. i., p. 45.
P. Vulgaris, Emmons, Mass. Report, 1840, p. 44.
Mustela Pusilla, Dekay, Nat. Hist. N. Y., p. 34.

DESCRIPTION.

This is much the smallest of all our species of Weasel, if we are to judge from two specimens that are in our possession, which appear to be full grown.   The tail is about one-fourth the length of the body, and is a little longer than that of the common Weasel (*M. Vulgaris*) of Europe.   It is, however, a still smaller animal, and differs from it in several other particulars : its ears are less broad, its feet smaller, the colour on the back is a shade darker, the white on the under surface extends much farther along the sides, towards the back, and the dividing line between the colours on the upper and lower surface is more distinct.   The head is small, neck slender, and the body vermiform.   Whiskers the length of the head, ears very small, toes and nails slender, covered with hairs.

COLOUR.

We are inclined to believe that this species does not become white in winter.   We kept a small weasel alive throughout a winter in our boyhood, but cannot now decide whether it was this species or another, (*P. Fuscus,*) which we will describe in our next volume.   That species underwent no change in winter.   It is more glossy than the ermine in

Plate LXIV

On Stone by W. H. Hitchcock

Drawn from Nature by J.W. Audubon

Lith. Printed & Col.d by J. T. Bowen, Phil.

Little American Brown Weasel.

summer pelage and a shade paler in colour. It is light yellowish brown
on the head, neck, and the whole of the upper surface ; this colour prevails on the outer portions of the fore-legs to near the feet, the outer surface of the hind-legs, the rump, and the whole of the tail, which is not
tipped with black as in the ermine. The white on the under surface commences on the upper lips and extends along the neck, inner surface of
the legs, rises high up along the sides, including the outer and inner
surfaces of the feet. The moustaches are white and black, the former
colour predominating.

DIMENSIONS.

								Inches.
Length from point of nose to root of tail,	-	-	-					7
Head and neck,	-	-	-	-	-	-	-	3
Tail (vertebræ),	-	-	-	-	-	-	-	2
" including fur,	-	-	-	-	-	-	-	$2\frac{1}{8}$

HABITS

From the form and structure of this species, we might naturally presume that it possesses all the habits of the ermine. It feeds on insects,
eggs of birds, and mice, but from its diminutive size we are led to suppose that it is not mischievous in the poultry house, and would scarcely
venture to attack a full-grown Norway rat.

GEOGRAPHICAL DISTRIBUTION.

The specimens from which our descriptions were made, were obtained
in the State of New-York, one at the Catskills, and the other at Long
Island. If it should prove to be the species we once had in captivity, it
exists also in the northern part of New-York, where we captured it.
RICHARDSON asserts that it exists as far to the North as the Saskatchewan
river, and Captain Bayfield obtained specimens at Lake Superior.

GENERAL REMARKS.

Sir John RICHARDSON states that this species, like the ermine, becomes
white in winter in the fur countries. We are disposed to believe that
this is not the case in the latitude of New-York. This fact, however,
is no evidence that the species in those widely separated localities
are different. The ermine in the northern part of Virginia seldom undergoes a perfect change, and in Carolina remains brown throughout the
whole year. Sir John RICHARDSON states (p. 45) that the specimens presented to the Zoological Society by Capt. BAYFIELD, agreed in all respects

with the common weasel of Europe.   We, however, examined these specimens and compared them with the European weasel, and found no difficulty in discovering characters by which the species are separated. We have an indistinct recollection that the prince of Musignano named the specimen in the Zoological Society; but as he did not, as far as we know, describe it, we have, according to our views on these subjects, assigned to Dr. DEKAY the credit of the specific name.

Plate LXV

Drawn from Nature by J. Audubon, FRS FLS

On Stone by W E Hitchcock

Lith. Printed & Col.d by J.T. Bowen, Phil.

Little Harvest Mouse

MUS HUMILIS.—Bachman.

Little Harvest Mouse.

PLATE LXV.—Males and Females.

M. corpore supra rutilo-cinereo, et quoad baccas et lineam in utrisque lateribus ferrugineo; subtus flavo-albente. M. musculus minor.

CHARACTERS.

Smaller than the house mouse; colour, reddish-gray above; cheeks and line along the side, light ferruginous; beneath, white with a yellowish tinge.

SYNONYMES.

Mus humilis, Bach. Read before the Academy of Nat. Sciences, 1837. Journal Acad., vol. vii.
Mus humilis, Bach., Acad. Nat. Sciences, Oct. 5th, 1841.

DESCRIPTION.

Incisors, small and short; head, much more rounded, nose, less pointed, and skull, proportionably broader than the corresponding portions in the common house-mouse; legs, rather short, and slender; there are four toes on the fore-feet, with a minute and almost imperceptible nail in the place of a thumb; on the hind-foot there are five toes; claws short, weak, sharp, and slightly hooked; nose, short and pointed; the moustaches are composed of a few hairs, not rigid, of the length of the head; the eyes are smaller and less prominent than those of the white-footed mouse, resembling those of the common house-mouse; the ears are of moderate size, broad at base, erect, ovate, clothed on both surfaces and around the edges with short adpressed hairs, extending a little beyond the fur; palms naked; upper surface of feet covered with hairs to the end of nails; the tail is round when the animal is in a living state, but after the specimens are dried, becomes square; it is thinly clothed with short hairs; the fur on the whole body is short, glossy, and very fine.

Teeth, yellow; nails, white; eyes, black; moustaches, mostly white; a few near the nostrils black; nose, cheeks, ears on both surfaces, and a line extending from the sides of the neck running along the shoulder and separating the colours of the back and under surface, dark buff; on the back, the hairs are plumbeous at the roots, then yellowish fawn colour; upper lips, chin, and throat, white; neck and under surface of body white shaded with buff.

DIMENSIONS.

					Inches
From point of nose to root of tail,`	-	-	-	-	$2\frac{3}{4}$
Tail,	-	-	-	-	2
Height of ear,	-	-	-	-	$\frac{3}{8}$

HABITS.

By the casual observer, this diminutive little species, on being started from its retreat in the long grass, or under some fence or pile of brushwood, might be mistaken for the young of the white-footed mouse (*Mus leucopus*), or that of the jumping mouse (*Meriones Americanus*). It however differs widely from either, and bears but a general resemblance to any of our American species.

About twenty years ago, whilst we were endeavouring to make ourselves acquainted with the species of smaller rodentia existing in the Southern States, we discovered this little Mouse in the grass fields and along the fences of the plantations a few miles from Charleston, S. C. We procured it in the way in which field mice and other small quadrupeds in all countries can be most easily obtained, by having what are denominated figure of 4 traps, set along fences and ditches in the evening, baited with meat and seeds of various kinds. On the following morning we usually were rewarded with a number of several interesting species. We on two occasions preserved this Mouse in a domestic state, once for a year, during which time it produced two broods of young: the first consisting of four were born in May, the second of three in July. They reared all their young. We fed them at first on pea or ground nuts, (*Hypogea arachis*,) cornmeal, (maize,) the latter they preferred boiled, but after having tempted their appetites with the seeds of the Egyptian Millet, (*Pennisitum tiphoideum*,) we discovered that they relished it so well, we allowed it finally to become their exclusive food. They refused meat on all occasions. They were very gentle, allowed themselves to be taken into the hand, and made no attempt to bite, or scarcely any to escape. The young, when born, were naked and blind, but in a very few days became covered with hair, and at a week old were seen peeping out of their nests. We did not discover that the female dragged the young, attached to the teats, in the manner of the white-footed mouse. We placed a female in a cage with a male of the white-footed mouse: they lived on tolerably good terms for six months, but produced no young. We then placed the same female with the male of the common mouse. The latter immediately commenced fighting with our little pet, and in the morning she was found dead in the cage, bitten and mutilated in various places.

This to us is a rare species; after a search of twenty years we have obtained only a dozen specimens from the fields. The nests, which we have oftener seen than their occupants, were placed on the surface of the ground among the long grass, composed of soft withered grasses, and covered over in the manner of the nest of WILSON's meadow mouse. We have also seen the nests of this species under brush-heaps and beneath the rails of fences, similarly constructed.

We doubt whether this species is of much injury to the farmer. It consumes but little grain, is more fond of residing near grass fields, on the seeds of which it subsists, than among the wheat fields. We have observed in its nest small stores of grass seeds—the outer husks and other remains of the Broom grass (*Andropogon dissitiflorum*)—also that of the Crab grass (*Digitaria sanguinalis*,) and small heaps of the seeds of several species of *paspalum*, *poa* and *panicum*, especially those of *panicum Italicum.*

The specimen from which this description was taken was a little the largest of any we have seen. It was a female captured on the 10th December, and containing four young in its matrix; we presume therefore that this species, like the field mice in general, produce young several times during the summer.

GEOGRAPHICAL DISTRIBUTION.

We have met with this species sparingly in South Carolina along the seaboard, and received it from Dr. BARRATT, of Abbeville, S. C. We procured a specimen in Ebenezer, (Georgia,) where the inhabitants stated they had never before observed it. A specimen was sent to us by our friend Mr. RUFFIN, who obtained it in Virginia. If we have not inadvertently blended two species, this animal can be traced as far to the north-east as the State of New-York, several having been procured in traps on the farms in the vicinity of the city.

GENERAL REMARKS.

We sent a minute description of this species to the Academy of Natural Sciences in 1837, which was read by our friend Dr. Morton ; although informed that it was published in the transactions of the Society, we have not seen it in print. A second description was published in the transactions of the same Society, October, 1841. We have not ascertained that the species has been noticed by any other naturalist.

In examining the teeth of this species, we have found that the tuberculous summits on the molars were less distinct than in those which legitimately belong to the genus *Mus*, and that there are angular ridges on the enamel by which it approaches the genus *Arvicola ;* it is in fact an intermediate species, but in the aggregate of its characteristics perhaps approaches nearest to *Mus*, where we for the present have concluded to leave it.

On Stone by Wm E Hitchcock

Virginian Opossum.

Drawn from Nature   Lith Printed & Col'd by   Bowen Phil

GENUS DIDELPHIS.—Linnæus.

DENTAL FORMULA.

$$Incisive \frac{10}{8}; \quad Canine \frac{1-1}{1-1}; \quad Molar \frac{7-7}{7-7} \text{ or } \frac{6-6}{7-7} = 48 \text{ or } 50.$$

Head, long and conical ; muzzle, pointed ; ears, large, membraneous, rounded, and almost naked ; tongue, acculeated ; internal toe of the hind foot, opposable to the fingers, and destitute of a nail, pendactylous; nails, curved ; tail, long, scaly, and slightly covered with rigid hair ; stomach, simple.   Female, with a pouch.

The generic name is derived from the Greek, *dis*, twice or double, and *delphis*, a womb.

The interesting group of the Marsupialia has recently been arranged by Owen into five tribes and families, and sixteen genera ; these include about seventy known species, to which additions are continually making ; the Virginian Opossum being, however, the only species known in America north of Mexico.   Most of the other species of this genus (as at present restricted,) inhabit tropical America.   It is composed of fifteen species, some of which are still doubtful.

DIDELPHIS VIRGINIANA.—Shaw.

Virginian Opossum.

PLATE LXVI.—Female, and Young Male seven months old.

D. pilis laneis basi albis, apice fuscis ; sericeis longis albis ; facie, rostro colloque pure albis ; auriculis nigris apice flavicantibus ; caudâ corpore breviore basi pilosâ tota albicante.

CHARACTERS.

Hair soft and woolly, white near the roots, tipped with brown; the long hairs white and silky ; face near the snout, pure white ; ears, black ; base and margin, whitish ; tail, shorter than the body ; base, covered with whitish hair.

SYNONYMES.

VIRGINIAN OPOSSUM, Pennant, Hist. Quad., vol. ii., p. 18, pl. 63.
    "   "   " Arctic Zoology, vol. i., p. 73.
SARIGUE DES ILLINOIS, Buff., sup. 6.
OPOSSUM AMERICANUS, D'Azara, Quad. du Paraguay.
DIDELPHIS VIRGINIANA, Shaw's Zool., vol. i., p. 73.
MARSUPIALL AMERICANUM, Tyson, in Phil. Trans., No. 239, p. 105.
COWPER, bid., No. 290, p. 1565.
OPOSSUM, Catesby's Carolina, p. 120, fig. e.
    " Barton's Facts, Observations and Conjectures relative to the gene
        ration of the Opossum of N. Am., London, 1809 and 1813.
POSSUM, Lawson's Carolina, p. 120, fig. e.
D. VIRGINIANUS, Harlan, Fauna, p. 119.
    "   " Godman, vol. ii., p. 7, fig.
VIRG. OPOSSUM, Griffith, vol. iii., p. 24.
    "   " Dekay, Nat. Hist. N. Y., p. 3, fig. 2, pl. 15.
OPOSSUM, Notes on the generation of the Virginian Opossum, (Didelphis Virginiana,)
        J. Bachman, D. D., Transactions of the Acad. of Nat. Sciences, April,
        1848, p. 40.
        Letter from M. Michel, M. D., on the same subject, Trans. Acad. Nat.
        Sciences, April, 1848, p. 46.

DESCRIPTION.

Body, stout and clumsy ; head, long and conical ; snout, pointed : the nostrils at the extremity of the long muzzle open on the sides of a protruberant naked and glandulous surface. Ears, large, thin, and membraneous ; mouth, wide, and borders rounded ; jaws, weak ; eyes, placed high on the forehead, small, and without external lids, oblique ; moustaches, on the sides of the face, and a few over the eye, strong and rigid. The tongue is covered with rough papillæ. Nails, of moderate length, curved ; inner toe on the posterior extremities destitute of a nail and opposable to the other toes, thus forming a kind of hand. Tail, (which may be considered a useful appendage to the legs in aiding the motions of the animal), prehensile and very strong, but capable of involution only on the under side, long, round, and scaly, covered with a few coarse hairs for a few inches from the base, the remainder with here and there a hair scattered between. Soles of the hind feet, covered with large tubercles. The female is furnished with a pouch containing thirteen mammæ arranged in a circle, with one in the centre.

The fur is of two kinds, a soft woolly hair beneath, covered by much longer hairs, which are, however, not sufficiently dense to conceal the under coat. The woolly hair is of considerable length and fineness, especially in winter.

COLOUR.

The woolly hair on the upper surface of the body, when blown aside, is white at the base and black at the tips; the long interspersed hairs are mostly white; a few towards the points exhibit shades of dark brown and black; moustaches, white, and black; eyes, black; ears, black, at base, the borders edged with white to near the extremities, where they are broadly patched with white; snout and toes, flesh coloured; face, neck, and nails, yellowish white; a line of dark brown commences on the forehead, widens on the head, and extends to the shoulders—there is also a line of dark brown under the chest; the feet in most specimens are brownish black; we have seen an occasional one where they were reddish brown; tail, brown.

The young differ somewhat in colour from the old: they are uniformly lighter in colour, the head being quite white, with a very distinct black dorsal line commencing faintly on the hind head, and running down the back to near the rump.

DIMENSIONS.

	Inches.
A well grown female:	
From point of nose to root of tail,	15½
Length of tail,	12
Height of ear,	1⅞
Breadth of ear,	1¼
Orifice of the distended pouch in diameter,	15¼
Teats measured immediately after the young had been withdrawn,	1

Weight, 12lbs.

Young, ten days old, nostrils open, ears pretty well developed:

	Inches.
Length of head and body,	1½
Tail,	⅓

Weight, 22 grains.

HABITS.

In our first volume (pp. 111, 112) we have spoken of the curiosity eagerly indulged, and the sensations excited, in the minds of the discoverers of our country, on seeing the strange animals that they met with. Travellers in unexplored regions are likely to find many unheard-of objects in nature that awaken in their minds feelings of wonder and admiration. We can imagine to ourselves the surprise with which the Opossum was regarded by Europeans when they first saw it. Scarcely any thing was known of the marsupial animals, as New Holland had not as yet opened its unrivalled stores of singularities to astonish the world. Here was a strange animal, with the head and ears of the pig, sometimes hanging on the limb of a tree, and occasionally swinging like the monkey by the tail! Around that prehensile appendage a dozen sharp-nosed, sleek-headed young, had entwined their own tails, and were sitting on the mother's back! The astonished traveller approaches this extraordinary compound of an animal and touches it cautiously with a stick. Instantly it seems to be struck with some mortal disease : its eyes close, it falls to the ground, ceases to move, and appears to be dead! He turns it on its back, and perceives on its stomach a strange apparently artificial opening. He puts his fingers into the extraordinary pocket, and lo! another brood of a dozen or more young, scarcely larger than a pea, are hanging in clusters on the teats. In pulling the creature about, in great amazement, he suddenly receives a gripe on the hand—the twinkling of the half-closed eye and the breathing of the creature, evince that it is not dead, and he adds a new term to the vocabulary of his language, that of " playing 'possum."

Like the great majority of predacious animals, the Opossum is nocturnal in its habits. It suits its nightly wanderings to the particular state of the weather. On a bright starlight or moonlight night, in autumn or winter, when the weather is warm and the air calm, the Opossum may every where be found in the Southern States, prowling around the outskirts of the plantation, in old deserted rice fields, along water courses, and on the edges of low grounds and swamps ; but if the night should prove windy or very cold, the best nosed dog can scarcely strike a trail, and in such cases the hunt for that night is soon abandoned.

The gait of the Opossum is slow, rather heavy, and awkward ; it is not a trot like that of the fox, but an amble or pace, moving the two legs on one side at a time. Its walk on the ground is plantigrade, resting the whole heel on the earth. When pursued, it by no means stops at once and feigns death, as has often been supposed, but goes forward at a rather slow speed, it is true, but as fast as it is able, never, that we are aware of, increasing it to a leap or canter, but striving to avoid its pursuers by sneaking off to some thicket or briar patch ; when, however, it discovers that the dog is in close pursuit, it flies for safety to the nearest tree, usually a sapling, and unless molested does not ascend to the top, but seeks an easy resting place in some crotch not twenty feet from the ground, where it waits silently and immoveably, till the dog, finding that his master will not come to his aid, and becoming weary of barking at the foot of the tree, leaves the Opossum to follow the bent of his inclinations, and conclude his nightly round in search of food. Although a slow traveller, the Opossum, by keeping perseveringly on foot during the greater part of the night, hunts over much ground, and has been known to make a circle of a mile or two in one night. Its ranges, however, appear to be restricted or extended according to its necessities, as when it has taken up its residence near a corn field, or a clump of ripe persimmon trees, (*Diosperos Virginiana*,) the wants of nature are soon satisfied, and it early and slowly carries its fat and heavy body to its quiet home, to spend the remainder of the night and the succeeding day in the enjoyment of a quiet rest and sleep.

The whole structure of the Opossum is admirably adapted to the wants of a sluggish animal. It possesses strong powers of smell, which aid it in its search after food ; its mouth is capacious, and its jaws possessing a greater number and variety of teeth than any other of our animals, evidencing its omnivorous habits ; its fore-paws, although not armed with retractile claws, aid in seizing its prey and conveying it to the mouth. The construction of the hind-foot with its soft yielding tubercles on the palms and its long nailless opposing thumb, enable it to use these feet as hands, and the prehensile tail aids it in holding on to the limbs of trees whilst its body is swinging in the air ; in this manner we have observed it gathering persimmons with its mouth and fore-paws, and devouring them whilst its head was downwards and its body suspended in the air, holding on sometimes with its hind-feet and tail, but often by the tail alone.

We have observed in this species a habit which is not uncommon among a few other species of quadrupeds, as we have seen it in the raccoon and occasionally in the common house dog—that of lying on its back for hours in the sun, being apparently dozing, and seeming to enjoy this position as a change. Its usual posture, however, when asleep, is either lying at full length on the side, or sitting doubled up with its head under its fore-legs, and its nose touching the stomach, in the manner of the raccoon.

The Opossum cannot be called a gregarious animal. During summer, a brood composing a large family may be found together, but when the young are well grown, they usually separate, and each individual shifts for himself ; we have seldom found two together in the same retreat in autumn or winter.

Although not often seen abroad in very cold weather in winter, this animal is far from falling into that state of torpidity to which the marmots, jumping mice, and several other species of quadrupeds are subject. In the Southern States, there are not many clear nights of starlight or moonshine

in which they may not be found roaming about ; and although in their farthest northern range they are seldom seen when the ground is covered with
snow, yet we recollect having come upon the track of one in snow a foot
deep, in the month of March, in Pennsylvania ; we pursued it, and captured
the Opossum in its retreat—a hollow tree.   It may be remarked, that animals like the Opossum, raccoon, skunk, &c., that become very fat in autumn
require but little food to support them through the winter, particularly
when the weather is cold.

The Opossum, although nocturnal in its general habits, is not unfrequently, particularly in spring and summer, found moving about by day.   We
have on several occasions met with it in the woods at mid-day, in places
where it was seldom molested.

Nature has wisely provided this species with teeth and organs indicating its omnivorous character and its possessing an appetite for nearly
all kinds of food ; and in this particular it exhibits many of the propensities and tastes of the raccoon.   It enters the corn fields (maize), crawls
up the stalks, and sometimes breaks them down in the manner of the raccoon, to feed on the young and tender grains ; it picks up chesnuts, acorns,
chinquapins and beach nuts, and munches them in the manner of the bear.
We have, on dissection, ascertained that it had devoured blackberries,
whortleberries, and wild cherries, and its resort to the persimmon tree is pro-
verbial.   It is also insectivorous, and is seen scratching up the leaves in
search of worms, and the larvæ of insects, of which it is very fond.   In
early spring it lays the vegetable kingdom under contribution for its
support, and we have observed it digging up the roots of the small atama-
masco lily, (*Zepherina atamasco*,) and the young and tender shoots of the
China brier, (*Smilax rotundifolia*,) as they shoot out of the ground like as-
paragus.   It is moreover decidedly carnivorous, eating young birds that
it may detect on the ground, sucking the eggs in all the partridge, towhee-
bunting and other nests, it can find in its persevering search.   It destroys
mice and other rodentia, and devours whole broods of young rabbits,
scratching about the nest and scattering the hair and other materials of
which it was composed.   We have observed it squatting in the grass and
brier thickets in Carolina, which are the common resort of the very abundant cotton rat, (*Sigmodon hispidum*,) and from patches of skin and other
mutilated remains, we satisfied ourselves that the Opossum was one among
many other species designed by Providence to keep in check the too rapid increase of these troublesome rats.   We must admit that it sometimes
makes a sly visit to the poultry house, killing a few of the hens and
playing havoc among the eggs.   The annoyances of the farmer, however,
from this mischievous propensity, are not as great as those sustained from

some of the other species, and cannot for a moment be compared with the destruction caused by the weasel, the mink, or the skunk.

The domicile of the Opossum in which it is concealed during the day, and where it brings forth its young, which we have often examined, is found in various localities. This animal is a tolerable digger, although far less expert in this quality than the Maryland marmot, its den is usually under the roots of trees or stumps, when the ground is so elevated as to secure it from rains and inundations. The hollow of a large fallen tree, or an opening at the roots of a standing one, also serve as a convenient place for its nest. The material which we have usually found composing this nest along the seaboard of Carolina is the long moss (*Tillandsia usnoides*); although we have sometimes found it composed of a bushel or more of oak and other leaves.

On firing into a squirrel's nest which was situated in the fork of a tree some forty feet from the ground, we brought down an Opossum, which had evidently expelled its legitimate occupant. The Florida rat is known to collect heaps of sticks and leaves, and construct nests sometimes a yard in diameter and two feet high : these are usually placed on the ground, but very frequently on the entangled vines of the grape, smilax, and supple jack, (*Ziziphus volubilis*.) In these nests an Opossum may occasionally be found, dozing as cozily as if he had a better right than that of mere possession.

Hunting the Opossum is a very favourite amusement among domestics and field labourers on our Southern plantations, of lads broke loose from school in the holidays, and even of gentlemen, who are sometimes more fond of this sport than of the less profitable and more dangerous and fatiguing one of hunting the gray fox by moonlight. Although we have never participated in an Opossum hunt, yet we have observed that it afforded much amusement to the sable group that in the majority of instances make up the hunting party, and we have on two or three occasions been the silent and gratified observers of the preparations that were going on, the anticipations indulged in, and the excitement apparent around us.

On a bright autumnal day, when the abundant rice crop has yielded to the sickle, and the maize has just been gathered in, when one or two slight white frosts have tinged the fields and woods with a yellowish hue, ripened the persimmon, and caused the acorns, chesnuts and chinquepins (*Castanea pumilla*) to rattle down from the trees and strewed them over the ground, we hear arrangements entered into for the hunt. The Opossums have been living on the delicacies of the season, and are now in fine order, and some are found excessively fat ; a double enjoyment is anticipated, the fun of catching and the pleasure of eating this excellent substitute for roast pig.

"Come, men," says one, " be lively, let us finish our tasks by four o'clock, and after sundown we will have a 'possum hunt." " Done," says another, " and if an old coon comes in the way of my smart dog, Pincher, I be bound for it, he will shake de life out of him." The labourers work with increased alacrity, their faces are brightened with anticipated enjoyment, and ever and anon the old familiar song of " 'Possum up the gum tree " is hummed, whilst the black driver can scarcely restrain the whole gang from breaking out into a loud chorus.

The paraphernalia belonging to this hunt are neither showy nor expensive. There are no horses caparisoned with elegant trappings—no costly guns imported to order—no pack of hounds answering to the echoing horn ; two or three curs, half hound or terriers, each having his appropriate name, and each regarded by his owner as the best dog on the plantation, are whistled up. They obey the call with alacrity, and their looks and intelligent actions give evidence that they too are well aware of the pleasure that awaits them. One of these humble rustic sportsmen shoulders an axe and another a torch, and the whole arrangement for the hunt is completed. The glaring torch-light is soon seen dispersing the shadows of the forest, and like a jack o'lantern, gleaming along the skirts of the distant meadows and copses. Here are no old trails on which the cold-nosed hound tries his nose for half an hour to catch the scent. The tongues of the curs are by no means silent—ever and anon there is a sudden start and an uproarious outbreak : " A rabbit in a hollow, wait, boys, till I twist him out with a hickory." The rabbit is secured and tied with a string around the neck : another start, and the pack runs off for a quarter of a mile, at a rapid rate, then double around the cotton fields and among the ponds in the pine lands—" Call off your worthless dog, Jim, my Pincher has too much sense to bother after a fox." A loud scream and a whistle brings the pack to a halt, and presently they come panting to the call of the black huntsman. After some scolding and threatening, and resting a quarter of an hour to recover their breath and scent, they are once more hied forwards. Soon a trusty old dog, by an occasional shrill yelp, gives evidence that he has struck some trail in the swamp. The pack gradually make out the scent on the edges of the pond, and marshes of the rice fields, grown up with willows and myrtle bushes (*Myrica cerifera*). At length the mingled notes of shrill and discordant tongues give evidence that the game is up. The race, though rapid, is a long one, through the deep swamp, crossing the muddy branch into the pine lands, where the dogs come to a halt, unite in conclave, and set up an incessant barking at the foot of a pine. "A coon, a coon ! din't I tell you," says Monday, " that if Pincher come across a coon, he would do he work ?" An additional piece of split light-wood is added to the torch, and the coon is seen doubled up in the form of a hornet's nest in the very top of the long-leaved pine. (*P. palustris*). The tree is without a branch for forty feet or upwards, and it is at once decided that it must be cut down : the axe is soon at work, and the tree felled. The glorious battle that ensues, the prowess of the dogs, and the capture of the coon, follow as a matter of course. See our article on the raccoon, pp. 80, 81, where we have briefly described such a scene.

Another trail is soon struck, and the dogs all open upon it at once : in an instant they rush, pell mell, with a loud burst of mingled tongues, upon some animal along the edge of an old field destitute of trees. It proves to be an Opossum, detected in its nightly prowling expedition. At first, it feigns death, and, rolling itself into a ball, lies still on the ground ; but the dogs are up to this "'possum playing," and seize upon it at once. It now feels that they are in earnest, and are not to be deceived. It utters a low growl or two, shows no fight, opens wide its large mouth, and, with few struggles, surrenders itself to its fate. But our hunters are not yet satisfied, either with the sport or the meat : they have large families and a host of friends on the plantation, the game is abundant, and the labour in procuring it not fatiguing, so they once more hie on the dogs. The Opossum, by its slow gait and heavy tread, leaves its foot-prints and scent behind it on the soft mud and damp grass. Another is soon started, and hastens up the first small gum, oak, or persimmon tree, within its reach ; it has clambered up to the highest limb, and sits crouching up with eyes closed to avoid the light. "Off jacket, Jim, and shake him down ; show that you know more about 'possum than your good-for-nutten fox-dog." As the fellow ascends, the animal continues mounting higher to get beyond his reach ; still he continues in pursuit, until the affrighted Opossum has reached the farthest twig on the extreme branches of the tree. The negro now commences shaking the tall pliant tree top ; while with its hind hands rendered convenient and flexible by its opposing thumb, and with its prehensile tail, the Opossum holds on with great tenacity. But it cannot long resist the rapidly accumulating jerks and shocks : suddenly the feet slip from the smooth tiny limb, and it hangs suspended for a few moments only by its tail, in the meantime trying to regain its hold with its hind hands ; but another sudden jerk breaks the twig, and down comes the poor animal, doubled up like a ball, into the opened jaws of eager and relentless canine foes ; the poor creature drops, and yields to fate without a struggle.

In this manner half a dozen or more Opossums are sometimes captured before midnight. The subsequent boasts about the superior noses, speed and courage of the several dogs that composed this small motley pack—

the fat feast that succeeded on the following evening, prolonged beyond the hour of midnight, the boisterous laugh and the merry song, we leave to be detailed by others, although we confess we have not been uninterested spectators of such scenes.

> " Let not ambition mock their useful toil,
> " Their homely joys and destiny obscure,
> " Nor grandeur hear with a disdainful smile,
> " The simple pleasures of the humble poor."

The habit of feigning death to deceive an enemy is common to several species of quadrupeds, and we on several occasions witnessed it in our common red fox (*V. Fulvus*). But it is more strikingly exhibited in the Opossum than in any other animal with which we are acquainted. When it is shaken from a tree and falls among grass and shubbery, or when detected in such situations, it doubles itself into a heap and feigns death so artfully, that we have known some schoolboys carrying home for a quarter of a mile an individual of this species, stating that when they first saw it, it was running on the ground, and they could not tell what had killed it. We would not, however, advise that the hand should on such occasions be suffered to come too familiarly in contact with the mouth, lest the too curious meddler should on a sudden be startled with an unexpected and unwelcome gripe.

This species has scarcely any note of recognition, and is remarkably silent ; when molested, it utters a low growl ; at other times its voice resembles the hissing of a cat. The Opossum displays no cunning in avoiding traps set to capture it, entering almost any kind of trap, very commonly being taken in a log trap called a dead fall.

From its very prolific nature it can afford to have many enemies. In addition to the incessant war waged against it by men and dogs, we have ascertained that its chief enemy among rapacious birds is the Virginian owl, (*Strix Virginiana*,) which flying abroad at the same hour in which the Opossum is on foot, pounces on it, and kills it with great ease. We have heard of an instance in which it was seen in the talons of the white-headed eagle, (*Halietus leucocephalus*,) and of two or three in which the great hen-hawk (*F. Borealis*) was observed feeding upon it. We recollect no instance of its having been killed by the wild cat or the fox. The wolf, it is said, seizes on every Opossum it can find, and we have heard of two instances where half-grown animals of this species were found to have been swallowed by the rattlesnake.

Although the dog hunts it so eagerly, yet we have never been able to ascertain that it ever feeds upon its flesh ; indeed, we have witnessed the dog passing by the body of a fresh killed Opossum, and going off half a mile farther to feed on some offensive carcase.

The Opossum is easily domesticated when captured young. We have, in endeavouring to investigate one of the very extraordinary characteristics of this species, preserved a considerable number in confinement, and our experiments were continued through a succession of years. Their nocturnal habits were in a considerable degree relinquished, and they followed the servants about the premises, becoming troublesome by their familiarity and their mischievous habits. They associated familiarly with a dog on the premises, which seemed to regard them as necessary appendages of the motley group that constituted the family of brutes in the yard. They devoured all kinds of food : vegetables, boiled rice, hominy, meat both raw and boiled, and the scraps thrown from the kitchen ; giving the preference to those that contained any fatty substance.

On one occasion a brood of young with their mother made their escape, concealed themselves under a stable, and became partially wild ; they were in the habit of coming out at night, and eating scraps of food, but we never discovered that they committed any depredations on the poultry or pigeons. They appeared however to have effectually driven off the rats, as during the whole time they were occupants of the stable, we did not observe a single rat on the premises. It was ascertained that they were in the habit of clambering over fences and visiting the neighbouring lots and gardens, and we occasionally found that we had repurchased one of our own vagrant animals. They usually, however, returned towards daylight to their snug retreat, and we believe would have continued in the neighbourhood and multiplied the species had they not in their nightly prowlings been detected and destroyed by the neighbouring dogs.

A most interesting part of the history of this animal, which has led to the adoption of many vulgar errors, remains to be considered, viz., the generation of the Opossum.

Our investigations on this subject were commenced in early life, and resumed as time and opportunity were afforded, at irregular, and sometimes after long intervals, and were not satisfactorily concluded until within a month of the period of our writing this article, (June, 1849). The process by which we were enabled to obtain the facts and arrive at our conclusions is detailed in an article published in the Transactions of the Academy of Natural Sciences, April, 1848, p. 40. Subsequent investigations have enabled us to verify some of these facts, to remove some obscurities in which the subject was yet involved, and finally to be prepared to give a correct and detailed history of a peculiarity in the natural history of this quadruped, around which there has hitherto been thrown a cloud of mystery and doubt.

Our early authors—Marcgrave, Pison, Valentine, Beverly. the Marquis of Chastellux, Pennant, and others, contended that "the pouch was the matrix of the young Opossum, and that the mammæ are. with regard to the young, what stalks are to their fruits." De Blainville and Dr. Barton speak of two sorts of gestation, one uterine and the other mammary. Blumenbach calls the young when they are first seen on the mammæ, abortions; and Dr. Barton's views (we quote from Griffith) are surprisingly inaccurate: "The Didelphes," he says, "put forth, not foetuses but gelatinous bodies; they weigh at their first appearance generally about a grain, some a little more, and seven of them together weighed ten grains." In 1819, Geoffroy St. Hillaire propounded to naturalists the following question : "Are the pouched animals born attached to the teats of the mother?" Godman, in his American Natural History, published in 1826, gave to the world a very interesting article on the Opossum, full of information in respect to the habits, &c., comprising all the knowledge that existed at that day in regard to this species. He was obliged, however, to admit, vol. 2, p. 7, "the peculiarities of its sexual intercourse, gestation, and parturition, are to this day involved in profound obscurity. Volumes of facts and conjectures have been written on the subject, in which the proportion of conjecture to fact has been as a thousand to one, and the difficulties still remain to be surmounted." And De-kay, in the work on the Quadrupeds of the State of N. York, (Nat. Hist. of N.York, 1842, p. 4,) states : "The young are found in the external abdominal sac, firmly attached to the teat in the form of a small gelatinous body, not weighing more than a grain. It was along time believed that there existed a direct passage from the uterus to the teat, but this has been disproved by dissection. Another opinion is, that the embryo is excluded from the uterus in the usual manner and placed by the mother to the teat ; and a third, that the embryo is formed where it is first found. Whether this transfer actually takes place, and if so, the physiological considerations connected with it, still remain involved in great obscurity."

The approaches to truth in these investigations have been very gradual, and the whole unusually slow. Cowper, Tyson, De Blainville, Home and others, by their examinations and descriptions of the organs of the Marsupialiæ, prepared the way for farther developments. A more judicious examination and scientific description by Owen and others, of the corresponding organs in the kangaroo, the largest of all the species composing these genera, and the discovery of the foetus in utero, enabled naturalists to conclude, that the similar structure in the Opossum would indicate a corresponding result. No one, however, was entitled to speak with positive certainty until the young were actually detected in the uterus, nor could an explanation of the peculiarity in the growth of the fœtus be made until it was examined in its original bed.

We have been so fortunate in five instances as to have procured specimens in which the young were observed in this position, and therefore feel prepared to speak with certainty. We are not aware that the young of the Virginian Opossum had been previously detected in the uterus.

All our investigations were made in South Carolina, where this is a very abundant species. For some years we attempted to arrive at the object of our researches by preserving these animals in a state of confinement. But they were subject to many accidents : they frequently made their escape from their cages, and some of them became overburdened with fat and proved sterile, so that we did not succeed in a single instance in obtaining young from females in a state of confinement. From this cause the naturalists of Europe, and especially those of France, who were desirous of making investigations in regard to our Opossum, have been so long unsuccessful. Their usual complaint has been, "Your Opossums do not breed in confinement." In this, Dr. Barton and our young friend Dr. Michel were more fortunate, but in both cases the young were produced before they were enabled to detect them in their previous existing position. We varied our experiments by endeavouring to discern the precise period when young were usually produced. We ascertained, by having a number of females procured with young in their pouches, that about the close of the first week in March, a little earlier or later, according to the age of the individual, or warmth, or coldness, of the previous winter, was the time when in this latitude this event usually occurs. Here, however, another difficulty presented itself, which for several successive seasons, thwarted us in our investigations. In the third week of February 1847, by offering premiums to the servants on several neighbouring plantations we obtained in three nights thirty-five Opossums, but of that number there was not a single female. A week afterwards, however, when the young were contained in the pouch, we received more females than males. From this circumstance we came to the conclusion that during the short period of gestation, the females, like those of some other species of quadrupeds, particularly the American black bear, conceal themselves in their burrows and can seldom be found. We then changed our instructions for capturing them, by recommending that they should be searched for in the day time, in hollow logs and trees and places where they had been previously known to burrow. By this means we were enabled at different times to obtain a small number in the state in which we were desirous of examining them. We feel under great obligations to several gentlemen of Carolina for aiding us in our investigations by procuring specimens, especially our relative Colonel HASKELL, Mr. JOHNSON, and JAMES FISHER, Esq., a close observer and intelligent naturalist. The latter, by his persevering efforts, pursued for some years at Jordan's Mills, on the upper waters of the Edisto, obtained two females in May, 1849, in the particular state in which he knew we were anxious to procure them, and brought them to us without having been previously aware that we had published the facts a year before.

The Opossums we were enabled to examine were dissected on the 11th, 14th and 18th February, 1848, and on the 12th and 22d May, 1849. Some of these had advanced to near the time of parturition. The young of those brought us by Mr. FISHER each weighed $2\frac{1}{2}$ grains. Those of one, sent us by Col. HASKELL, weighed 3 grains; and the young of another which we obtained by a Cæsarian operation, at a moment when all the rest had been excluded, and this individual alone remained, weighed 4 grains.

We remarked, that this however was a little the largest of six that composed the family, five of which were already in the pouch and attached to the teats. The largest one weighed $3\frac{3}{4}$, and another $3\frac{1}{4}$ grains. The weight, then, of the young Opossum at the moment of birth, is between 3 and 4 grains, varying a little in different specimens as is the case in the young of all animals.

The degree of life and animation in young Opossums at the moment of birth has been greatly underrated. They are neither abortions, as BLUMENBACH represented them, nor as Dr. BARTON has described them— " not foetuses, but gelatinous bodies, weighing about a grain more or less, seven of them together weighing 10 grains "—but little creatures that are nearly as well developed at birth as the young of the white-footed mouse and several other species of rodentia. They are covered by an integument, nourished by the mammæ, breathe through nostrils, perform the operations of nature, are capable of a progressive movement at the moment of their birth, and are remarkably tenacious of life. The individual which was dissected from the parent in the manner above detailed, moved several inches on the table by crawling and rolling, and survived two hours; the thermometer in the room was at the time standing at 66° Fahrenheit. The period of gestation is from fifteen to sixteen days. We received a female from a servant who informed us, that he had that morning seen it in intercourse with the male. We first saw the young on the morning of the 17th day. Our friend Dr. MIDDLETON MICHEL, a gentleman of high scientific attainments, and who had long been engaged in investigating the characters and habits of this species, in a communication made to us, (Trans. of the Acad. Nat. Sciences, April, 1848, p. 46,) assured us from his personal observation in which he was careful to note the hour of the day, the exact period is

15 days. As he possessed better opportunities of deciding in regard to the time, the animals being in a state of domestication, we are rather more disposed to yield to his observations than to our own; there is, however, only the difference of a day between us.

The young, when first born, are naked and flesh-coloured; the eyes, together with the ears, are covered by a thin integument through which these organs and the protuberances of the ears are distinctly visible. The mouth is closed, with the exception of a small orifice, sufficiently large to receive the teat, which is so thin and attenuated that it seems no larger than the body of a pin. Length of body, 7-12ths of an inch; of tail, 2-10ths. The nails, which can be seen with the naked eye, are very distinct when viewed with a microscope, and are of a dark brown colour, small and much hooked. The nostrils are open; the lungs filled with air, and when placed in water, the young float on the surface.

The number of young usually found in the pouch appear to be less than those that are born. The highest number we have found in the pouch was thirteen, the smallest six; whereas the preserved uterus brought to us by Mr. FISHER, contained fifteen. In all such cases, where a greater number of young are produced than there are teats, the last of the brood must inevitably perish, as those that are attached appear incapable of relinquishing their hold.

The manner in which the young at birth reach the pouch, and become attached to the teats, has been the subject of much speculation and inquiry. We had an opportunity of examining this process in part, without, however, having been aware at the time that it was going on. We intended to dissect a small female Opossum, which had been a few days in our possession, but ascertained in the morning at seven o'clock on the day our examination was to have been made, that she had three young in her pouch; supposing from her small size, that she would produce no additional number, we concluded to spare her life. She was confined in a box in our study; when we occasionally looked at her, we found her lying on one side, her shoulders elevated, her body drawn up in the shape of a ball; the pouch was occasionally distended with her paws—in this position the parts reached the edge of the pouch; she was busily employed with her nose and mouth licking, as we thought, her pouch, but in which we afterwards ascertained, were her young.

At six o'clock in the afternoon we were induced to examine her again, in consequence of having observed that she had for several hours appeared very restless, when we discovered that she had added four more to her previous number, making her young family now to consist of seven. With no inconsiderable labour and the exercise of much patience, we removed three of the young from the teats, one of which perished under the process, we replaced the two living ones in the pouch ; at nine o'clock examined her again and found both the young once more attached. We came to the conclusion, that she shoved them into the pouch, and with her nose or tongue moved them to the vicinity of the teats, where by an instinct of nature, the teat was drawn into the small orifice of the mouth by suction. We observed subsequently, that a young one that had been extracted from its parent a few moments before the time when it would have been born, and which had been rolled up in warm cotton, was instinctively engaged in sucking at the fibres of the cotton, and had succeeded in drawing into its mouth a considerable length of thread. A nearly similar process was observed by our friend Dr. MICHEL. He . states : "The female stood on her hind legs, and the body being much bent, the young appeared and were licked into the pouch."

There is a great difficulty in deciding the question, whether the mother aids the young in finding the teats, in consequence of the impossibility of the spectators being able to know what she is actually doing, whilst her nose is in the pouch. We believe the majority of naturalists who had an opportunity of witnessing our experiments came to the conclusion, that the mother, after shoving them into the pouch, left them to their own instinct, and they became attached without her assistance. We tried another experiment that suggested itself to us. Believing that the mother would not readily adopt the young of another, or afford them any assistance, we removed six out of ten that composed her brood, returned two of her own to the pouch, together with three others fully double the size, that had been obtained from another female. She was soon observed doubled up with her nose in the pouch, and continued so for an hour, when she was examined and one of her own small ones was found attached to the teat. Seven hours afterwards she was examined again, and both the small ones were attached, but the three larger ones still remained crawling about the pouch. On the following morning, it was ascertained that the mother had adopted the strangers, as the whole family of different sizes were deriving sustenance from her.

On another occasion, a female Opossum had been sent to us caught by a dog and much wounded, in consequence of which she died a few days afterwards, but first producing seven young which to every appearance had been still born. Yet they were in the pouch, and it appeared to us that the mother's uncontrollable attachment to her young, induced her to place her offspring in the pouch, even after they were dead.

An interesting inquiry remains to be answered: Is the Opossum a placental or non-placental animal ? Until we were favoured with a recent opportunity of carefully examining a uterus, containing nine young on one side, and six on the other, kindly brought to us by our friend JAMES FISHER, we were unable fully to answer this question. Our dissections and examinations were witnessed by Professors MOUL-TRIE, HUME, Drs. HORLBECK MICHEL, PORCHER and others.

The Opossum is, as far as we are able to judge from the specimens examined, a non-placental animal, inasmuch as there could not be detected the slightest adhesion between the exterior membrane of the fœtus and the internal surface of the mother. The membranes consisted of a vitelline sac, filled with ramifications of omphalo-mesenteric vessels, there was a slight appearance of an umbilical cord and umbilical vessels, constituting a true allantois, but no portions of them were attached to the uterus. There was no appearance of a placenta.

The growth of the young Opossum is suprisingly rapid. We weighed the largest young one at a week old and found it had increased from $3\frac{3}{4}$ grains to 30 grains. Length of head and body exclusive of tail, $1\frac{1}{4}$ inch; tail, $\frac{1}{2}$ inch. The young at this age were very tenacious of life, as on removing two, they remained alive on the floor without any covering through a cool night, in a room containing no fire, and still exhibited a slight motion at twelve o'clock on the following day. The teats of the mother after the young had been gently drawn off measured an inch in length, having been much distended, and appeared to have been drawn into the stomach of the young. The pouches of the young females were quite apparent; they used their prehensile tails, which could now be frequently seen entwined around the necks of others. At twelve days old the eyes were still closed, a few hairs had made their appearance on the moustache; the orifice of the ears were beginning to be developed, and the nails were quite visible and sharp.

When the young are four weeks old, they begin from time to time to relax their hold on the teats, and may now be seen with their heads occasionally out of the pouch. A week later, and they venture to steal occasionally from their snug retreat in the pouch, and are often seen on the mother's back securing themselves by entwining their tails around hers. In this situation she moves from place to place in search of food, carrying her whole family along with her, to which she is much attached, and in whose defence she exhibits a considerable degree of courage, growling at any intruder, and ready to use her teeth with great severity on man or dog. In travelling, it is amusing to see this large family moving about. Some of the young, nearly the size of rats, have their tails entwined around the legs of the mother, and some around her neck, thus they are dragged along. They have a mild and innocent look, and are sleek, and in fine condition, and this is the only age in which the word pretty can be applied to the Opossum. At this period, the mother, in giving sustenance to so large a family, becomes thin, and is reduced to one half of her previous weight. The whole family of young remain with her about two months, and continue in the vicinity till autumn. In the meantime, a second and often a third brood is produced, and thus two or more broods of different ages may be seen, sometimes with the mother, and at other times not far off.

The Opossum, with the exception of our gray rabbit, is one of the most prolific of our quadrupeds. We consider the early parts of the three months of March, May and July, as the periods in South Carolina when they successively bring forth ; it is even probable that they breed still more frequently, as we have observed the young during all the spring and summer months. In the month of May, 1830, whilst searching for a rare species of coleoptera, in removing with our foot some sticks composing the nest of the Florida rat, we were startled on finding our boot unceremoniously and rudely seized by an animal which we soon ascertained was a female Opossum. She had in her pouch five very small young whilst, seven others, about the size of full grown rats were detected peeping from under the rubbish. The females produce young at a year old. The young born in July do not bring forth as early as those born in March, but have their young as soon as the middle of the succeeding May. There is, of course, in this as well as in other species, some degree of irregularity in the time of their producing, as well as in the number of their young. We have reason to believe, also, that this species is more prolific in the southern than in the Middle States.

GEOGRAPHICAL DISTRIBUTION.

The Hudson River may be regarded as the farthest eastern limit of the Opossum. We have no doubt but that it will in time be found existing to the east of the Hudson, in the southern counties of New-York as well as on Long-Island and the warmer parts of the Eastern States, as the living animals are constantly carried there, and we have little doubt that if it was considered important it could be encouraged to multiply there. It has been stated to us that in New-Jersey, within five or ten miles of New-York, as many as ten or fourteen of these animals have within a few years past been taken in an autumn by means of traps, but that their number is gradually diminishing. It is common in New-Jersey and Pennsylvania, becoming more abundant as we proceed southwardly through North Carolina, South Carolina, Georgia, Louisiana and Texas, to Mexico ; inhabiting in great numbers the inter-tropical regions. To the west we have traced it in all the south-western states. It exists in Indiana, Mississippi, Missouri, and Arkansas, and extends to the Pacific ; it is said to exist in California. It is somewhat singular, that in every part of America, as far as we have been able to observe, the geographical range of the Opossum is very nearly the same as that of the persimon tree, of whose fruit it is so fond. This we regard, however, as merely accidental, as this food is not essential to its support. The Opossum neither ceases to multiply or to thrive in seasons in which the persimon has failed.

GENERAL REMARKS.

In our plate, we gave PENNANT as the originator of the scientific name of this species. We find, however, that he only calls it the Virginia Opossum, with a reference to the *Didelphys marsupialis,* LINNEUS. GMELIN subsequently arranged it under *Didelphys marsupialis.* As SHAW, in 1800, as far as we have been able to ascertain, seems to have been the first who applied the Latin specific name, *D. Virginiana,* we have, in accordance with the rules laid down by naturalists, given him the credit of the specific name.

GENUS CANIS.—Linnæus.

DENTAL FORMULA.

$$Incisive \; \frac{6}{6}; \quad Canine \; \frac{1-1}{1-1}; \quad Molar \; \frac{6-6}{6-6} = 40.$$

The three first in the upper jaw, and the four in the lower, trenchant but small, and called also false molars. The great carnivorous tooth above bi-cuspid, with a small tubercle on the inner side, that below with the posterior lobe altogether tubercular, and two tuberculous teeth behind each of the great carnivorous teeth. Muzzle, elongate; tongue, soft; ears, erect, (sometimes pendant in the domestic varieties.) Fore feet, pendacty-lous; hind feet, tetradactylous. Teats, both inguinal and vental.

CANIS LUPUS.—Linn.—(Var. Ater.)

Black American Wolf.

PLATE LXVII. Male.

C. niger, magnitudine, formaque C. lupi.

CHARACTERS.

Size and shape of the Common American Wolf; Canis, lupus occidentalis; colour black.

SYNONYMES.

Loup Noir de Canada, Buffon, vol. ix., p. 364-41.
Black Wolf, Long's Expd., vol. i., p, 95.
    "   Say, Frankl. Jour., vol. i., p. 172.
    "   Griffith, Anim. King., vol. 2., p. 348.
    "   Godman, Nat. Hist., vol. i., p. 267.
Canis Lyacon, Harlan's Fauna, p. 82.
Var. E. Lupus ater, Black Amer. Wolf, Richardson, Fauna Boreali Amer., p. 70.

Plate LXVII

Black American Wolf

On Stone by W.E. Hitchcock.

Lith. Printed & Col.d by J.T. Bowen Philad.a

Drawn from Nature by J.W. Audubon.

DESCRIPTION.

We regard this animal as a mere variety of the Common American Wolf, to be hereafter described, and need only here observe, that all the Wolves we have examined, such as the *Canis nubilis* of SAY, the White Wolf, the Red Texan Wolf and the Black Wolf, are of the same form, although in size the White Wolf is considerably the largest.

COLOUR.

Face, legs, point of tail and under jaw, black; body, irregularly and transversely barred with blackish brown and greyish; sides of the neck, greyish brown; behind the shoulders, under the belly and on the forehead, greyish brown. Some specimens are darker than others—we have examined several that were perfectly black on the whole surface of the body.

DIMENSIONS.

	Feet.	Inches.
Length of head and body - - - -	3	2
Do. of tail vertebræ - - - -		11
Do. including fur - - - -	1	1
Height of ear - - - - -		3

HABITS.

Not an individual of the party saw a Black Wolf during our trip up the Missouri, on the prairies near Fort Union, or along the shores of that portion of the Yellow Stone River that we visited. Mr. SAY speaks of its being the most common variety on the banks of the Missouri, but, unfortunately, does not state precisely where.

Wolves of this colour were abundant near Henderson, Kentucky, when we removed to that place, and we saw them frequently during our rambles through the woods after birds.

We found a Black Wolf in one of our wild turkey pens, early one morning. He observed us, as we approached, but instead of making his escape, squatted close down, like a dog which does not wish to be seen. We came up within a few yards of the pen, and shot him dead, through an opening between the logs. This Wolf had killed several fine turkeys, and was in the act of devouring one, which was, doubtless, the reason he did not attempt to make his escape when we approached him.

There is a strong feeling of hostility entertained by the settlers of the

wild portions of the country, toward the Wolf, as his strength, agility, and cunning, (in which last qualification, he is scarcely inferior to his relative, the fox,) tend to render him the most destructive enemy of their pigs, sheep, or young calves, which range in the forest ; therefore, in our country, he is not more mercifully dealt with than in any other part of the world.   Traps and snares of various sorts are set for catching him in those districts in which he still abounds.   Being more fleet and perhaps better winded than the fox, the Wolf is seldom pursued with hounds or any other dogs in open chase, unless wounded.   Although Wolves are bold and savage, few instances occur in our temperate regions of their making an attack on man ; and we have only had one such case come under our own notice.   Two young negroes, who resided near the banks of the Ohio, in the lower part of the State of Kentucky, about thirty years ago, had sweethearts living on another plantation, four miles distant.   After the labours of the day were over, they frequently visited the fair ladies of their choice, the nearest way to whose dwelling lay directly across a large cane brake.   As to the lover every moment is precious, they usually took this route to save time.   Winter had set in cold, dark and gloomy, and after sunset scarcely a glimpse of light or glow of warmth were to be found in that dreary swamp, except in the eyes and bosoms of the ardent youths who traversed these gloomy solitudes.   One night, they set forth over a thin crust of snow.   Prudent, to a certain degree, the lovers carried their axes on their shoulders, and walked as briskly as the narrow path would allow.   Some transient glimpses of light now and then met their eyes in the more open spaces between the trees, or when the heavy drifting clouds parting at times allowed a star to peep forth on the desolate scene.   Fearfully, a long and frightful howl burst upon them, and they were instantly aware that it proceeded from a troop of hungry and perhaps desperate wolves. They paused for a moment and a dismal silence succeeded.   All was dark, save a few feet of the snow-covered ground immediately in front of them.   They resumed their pace hastily, with their axes in their hands prepared for an attack.   Suddenly, the foremost man was assailed by several wolves which seized on him, and inflicted terrible wounds with their fangs on his legs and arms, and as they were followed by many others as ravenous as themselves, several sprung at the breast of his companion, and dragged him to the ground.   Both struggled manfully against their foes, but in a short time one of the negroes had ceased to move ; and the other, reduced in strength and perhaps despairing of aiding his unfortunate comrade or even saving his own life, threw down his axe, sprang on to the branch of a tree, and speedily gained a place of safety amid the boughs.   Here he passed a miserable night, and the next morning the bones of his friend lay scattered around on the snow, which was stained with his blood. Three dead wolves lay near, but the rest of the pack had disappeared ; and Scipio sliding to the ground, recovered the axes and returned home to relate the terrible catastrophe.

· About two years after this occurrence, as we were travelling between Henderson and Vincennes, we chanced to stop for the night at the house of a farmer, (for in those days hotels were scarce in that part of the good State of Indiana.) After putting up our horses and refreshing ourself, we entered into conversation with our worthy host, and were invited by him to visit the wolf pits which he had constructed about half a mile from the house. Glad of the opportunity, we accompanied him across the fields to the skirts of the adjoining forest, where he had three pits within a few hundred yards of each other. They were about eight feet deep, broadest at the bottom, so as to render it impossible for the most active animal to escape from them. The mouth of each pit was covered with a revolving platform of boughs and twigs, interlaced together and attached to a cross piece of timber, which served for an axle. On this light sort of platform, which was balanced by a heavy stick of wood fastened to the under side, a large piece of putrid venison was tied for bait. After examining all the pits, we returned to the house, our companion remarking that he was in the habit of visiting his pits daily, in order to see that all was right ; that the wolves had been very bad that season ; had destroyed nearly all his sheep, and had killed one of his colts. " But," added he, " I am now paying them off in full, and if I have any luck, you will see some fun in the morning." With this expectation we retired to rest, and were up at day-light. " I think," said our host, " that all is right ; for I see the dogs are anxious to get away to the pits, and although they are nothing but curs, their noses are pretty keen for wolves." As he took up his gun and axe and a large knife, the dogs began to howl and bark, and whisked around us as if full of delight. When we reached the first pit, we found the bait had been disturbed and the platform was somewhat injured, but the animal was not in the pit. On examining the second pit, we discovered three famous fellows safe enough in it, two black and one brindled, all of good size. They were lying flat on the earth, with their ears close down to their heads, their eyes indicating fear more than anger. To our astonishment, the farmer proposed descending into the pit to hamstring them, in order to haul them up, and then allow them to be killed by the dogs, which, he said, would sharpen his curs for an encounter with the wolves, should any come near his house in future. Being novices in this kind of business, we begged to be lookers on. " With all my heart," cried the farmer, " stand here, and look at me," whereupon he glided down, on a knobbed pole, taking his axe and knife with him, and leaving his rifle to our care. We were not a little surprised at the cowardice of the wolves. The woodman stretched out their hind legs, in succession, and with a stroke of the knife cut the principal tendon above the joint, exhibiting as little fear, as if he had been marking lambs. As soon as he had thus disabled the wolves, he got out, but had to return to the house for a rope, which he had not thought of. He returned quickly, and, whilst I secured the platform in a perpendicular position on its axis, he made a slip knot at one end of the rope, and threw it over the head of one of the wolves. We now hauled the terrified animal up; and motionless with fright, half choked, and disabled in its hind legs, the farmer slipped the rope from its neck, and left it to the mercy of the dogs, who set upon it with great fury and worried it to death. The second was dealt with in the same manner; but the third, which was probably oldest, showed some spirit the moment the dogs were set upon it, and scuffled along on its forelegs, at a surprising rate, snapping all the while furiously at the dogs, several of which it bit severely; and so well did the desperate animal defend itself, that the farmer, apprehensive of its killing some of his pack, ran up and knocked it on the head with his axe. This wolf was a female, and was blacker than the other dark-coloured one.

Once, when we were travelling on foot not far from the southern boundary of Kentucky, we fell in with a Black Wolf, following a man with a rifle on his shoulders. On speaking with him about this animal, he assured us that it was as tame and as gentle as any dog, and that he had never met with a dog that could trail a deer better. We were so much struck with this account and the noble appearance of the wolf, that we offered him one hundred dollars for it; but the owner would not part with it for any price.

Our plate was drawn from a fine specimen, although not so black a one as we have seen. We consider the Dusky Wolf and the Black Wolf as identically the same.

As we shall have occasion to refer to the characteristics of Wolves generally again, we shall not prolong this article; the Black, as already stated, being, in fact, only a variety. In our account of the Common Gray Wolf of the North, and the White Wolf of the Prairies, which last is very common, we shall give farther and more specific details of their breeding and other matters.

GEOGRAPHICAL DISTRIBUTION

All packs of American Wolves usually consist of various shades of colour and varieties, nearly black, have occasionally been found in every part of the United States. The varieties, with more or less of black, continue to increase as we proceed farther to the south, and in Florida the prevailing colour of the wolves is black. We have seen two or three skins procured in N. Carolina. There is a specimen in the Museum of the Philosophical Society of Charleston, obtained at Goose Creek, a few years ago, that is several shades darker than the specimen from which our drawing was made ; and in a gang of seventeen wolves, which existed in Colleton District, S. C., a few years ago, (sixteen of which were killed by the hunters in eighteen months), we were informed that about one fifth were black and the others of every shade of colour—from black to dusky grey and yellowish white. We have heard of this variety in the southern part of Missouri, Louisiana, and the northern parts of Texas.

SCIURUS CAPISTRATUS.—Bosc.

Fox Squirrel.

PLATE LXVIII

S. magnus, colorem variens; naso auriculisque albis; pilis crassis; cauda corpore longiore.

CHARACTERS.

Size, large; tail, longer than the body; hair, coarse; ears and nose, white; subject to great variety in colour.

SYNONYMES.

Sciurus Capistratus; Bosc, Ann. du Mus., vol. i., p. 281.
     "   Vulpinus? Linn. Ed. Gmel., 1788.
     "   Niger; Catesby.
Black Squirrel; Bartram's Travels in North America.
Sciurus Capistratus; Desm. Mammalogie, p. 332.
     "   Variegatus; Desm. Mammalogie, p. 333.
     "   Capistratus; Cuv., Regne Animal, vol. i., p. 139.
Fox Squirrel, Lawson's Carolina, p. 124.
Sciurus Capistratus; Harlan.
Sciurus Vulpinus; Godman.

DESCRIPTION.

This is the largest and most interesting species of the genus, found in the United States. Although it is subject to great varieties of colour, occasioning no little confusion by the creation of several nominal species, yet it possesses several striking and uniform markings by which it may, through all its varieties, be distinguished at a glance from any other.

The Fox Squirrel is furnished with the following teeth, viz :—

$$Incisive \frac{2}{2}; \quad Canine \frac{00}{00}; \quad Molar \frac{4-4}{4-4} = 20.$$

But although we have thus given to this species but four grinders in the upper jaw, which peculiarity applies to nearly all the specimens that may

Fox Squirrel.

be examined,—yet, in a very young animal, obtained on the 5th of April, in South Carolina, and which had apparently left the nest but a day or two, we observed a very minute, round, deciduous, anterior grinder on each side. These teeth, however, must be shed at a very early period ; as in two other specimens, obtained on the 20th of the same month, they were entirely wanting. The teeth of all our squirrels present so great a similarity, that it will be found impossible to designate the species from these alone, without referring to other peculiarities which the eye of the practical naturalist may detect. In young animals of this species, the tuberculous crowns on the molars are prominent and acute ; these sharp points, however, are soon worn off, and the tubercles in the adult are round and blunt. The first molar in the upper jaw is the smallest, and is triangular in shape ; the second and third one a little larger and square ; and the posterior one, which is about the size of the third, is rounded on its posterior surface. The upper incisors, which are of a deep orange colour anteriorly, are strong and compressed, deep at their roots, flat on their sides ; in some specimens there is a groove anteriorly running longitudinally through the middle, presenting the appearance of a double tooth ; in others, this tooth is wanting. In the lower jaw, the anterior grinder is the smallest ; the rest increase in size to the last, which is the largest.

Nose, obtuse ; forehead, slightly arched ; whiskers, a little longer than the head ; ears, rounded, covered with short hairs on both surfaces ; there is scarcely any projection of fur beyond the outer surface, as is the case in nearly all the other species ; the hair is very coarse, appearing in some specimens geniculate ; tail, broad and distichous ; legs and feet, stout ; and the whole body has more the appearance of strength than of agility.

COLOUR.

In the grey variety of this species, which is—as far as we have observed—the most common, the nose, extending to within four or five lines of the eyes, the ears, feet, and belly, are white ; forehead and cheeks, brownish black ; the hairs on the back are dark plumbeous near the roots, then a broad line of cinereous, then black, and broadly tipped with white, with an occasional black hair interspersed, especially on the neck and fore shoulder, giving the animal a light grey appearance ; the hairs of the tail are, for three-fourths of their length, white from the roots, then a ring of black, with the tips white. This is the variety given by Bosc and other authors as *Sciurus capistratus*.

Second variety : the Black Fox Squirrel. Nose and ears, white ; a few light-coloured hairs on the feet ; the rest of the body and tail, black ; there are, occasionally, a few white hairs in the tail. This is the original Black Squirrel of CATESBY and BARTRAM, (*Sci. Niger.*)

Third variety. Nose, mouth, under jaw and ears, white; head, thighs and belly, black; back and tail, dark grey. This is the variety alluded to by DESMAREST, (Ency. Method, Mammalogie, 333.)

There is a fourth variety, which is very common in Alabama, and also occasionally seen in the upper districts of South Carolina and Georgia, which has on several occasions been sent to us as a distinct species. The ears and nose, as in all the other varieties, are white. This, indeed, is a permanent mark, running through all the varieties, by which this species may be easily distinguished. Head and neck, black; back, a rusty blackish brown; neck, thighs, and belly, bright rusty colour; tail, annulated with black and red. This is the variety erroneously considered by the author of the notes on McMURTRIE's "Translation of Cuvier," (see vol. i., Appendix, p. 433,) as *Sciurus rufiventer.*

The three first noted above are common in the lower and middle districts of South Carolina; and, although they are known to breed together, yet it is very rare to find any specimens indicating an intermediate variety. Where the parents are both black, the young are invariably of the same colour—the same may be said of the other varieties; where, on the other hand, there is one parent of each colour, an almost equal number are of the colour of the male, the other of the female. On three occasions, we had an opportunity of examining the young produced by progenitors of both colours. The first nest contained two black and two grey; and the third, three black and two grey. The colour of the young did not, in a majority of instances, correspond with that of the parent of the same sex: although the male parent was black, the young males were frequently grey, and *vice versa.*

DIMENSIONS.

						Inches.	Lines.
Length of	head and body	-	-	-	-	14	5
"	tail vertebræ	-	-	·	-	12	4
"	tail to tip	-	-	-	-	15	2
"	palm and middle fore claw -		-		-	1	9
"	sole and middle hind claw -		-		-	2	11
"	fur on the back	-	-	-	-		8
Height of ear, posteriorly		-	-	-	-		7

HABITS.

Although there is a general similarity of habit in all the species of *Sciurus,* yet the present has some peculiarities which we have never noticed in any other. The Fox Squirrel, instead of preferring rich low lands, thickly clothed with timber, as is the case with the Carolina Grey Squirrel, is seldom seen in such situations; but prefers elevated pine ridges, where the trees are not crowded near each other, and where there is an occasional oak and hickory interspersed. It is also frequently found in the vicinity of rich valleys, to which it resorts for nuts, acorns and chinque-pins, (*castanea pumila*,) which such soils produce. In some aged and par-tially decayed oak, this Squirrel finds a safe retreat for itself and mate ; a hollow tree of any kind is sufficient for its purpose if Nature has prepared a hole, it is occupied, if otherwise, the animal finds no difficulty in gnaw-ing one or several, for its accommodation. The tree selected is in all cases hollow, and the Squirrel only gnaws through the outer shell in order to find a residence, which requires but little labour and skill to render it secure and comfortable. At other times, it takes possession of the deserted hole of the ivory-billed woodpecker, (*Picus principalis*).) The summer duck (*Anas sponsa*) too, is frequently a competitor for the same residence ; contests for possession occasionally take place between these three species, and we have generally observed, that the tenant that has already deposited its eggs or young in such situations is seldom ejected. The male and female summer duck unite in chasing and beating with their wings any Squirrel that may approach their nests, nor are they idle with their bills and tongues, but continue biting, hissing and clapping their wings until the intruder is expelled. On the other hand, when the Squirrel has its young in the hole of a tree, and is intruded on, either by a woodpecker or a summer duck, it immediately rushes to its hole, and after having entered remains at the mouth of it, occasionally protruding its head, and with a low angry bark keeps possession, until the intruder, weary of the contest, leaves it unmolested. Thus Nature imparts to each species additional spirit and vigour in defence of its young; whilst at the same time, the in-truder on the possessions of others, as if conscious of the injustice of his acts, evinces a degree of pusillanimity and cowardice.

In the vicinity of the permanent residence of the Fox Squirrel, several nests, composed of sticks, leaves and mosses, are usually seen on the pine trees. These are seldom placed on the summits, but in the forks, and more frequently where several branches unite and afford a secure basis for them. These nests may be called their summer home, for they seem to be occupied only in fine weather, and are deserted during wintry and stormy seasons.

In December and January, the season of sexual intercourse, the male chases the female for hours together on the same tree, running up one side and descending on the other, making at the same time a low guttural noise, that scarcely bears any resemblance to the barking which they utter on other occasions. The young are produced from the beginning of March, and sometimes earlier, to April. The nests containing them, which we have had opportunities of examining, were always in hollow trees. They receive the nourishment of the mother for four or five weeks, when they are left to shift for themselves, but continue to reside in the vicinity of, and even to occupy the same nests with, their parents till autumn. It has been asserted by several planters of Carolina, that this species has two broods during the season.

The food of the Fox Squirrel is various; besides acorns, and different kinds of nuts, its principal subsistence for many weeks in autumn is the fruit extracted from the cones of the pine, especially the long-leaved pitch pine, (*Pinus palustris.*) Whilst the green corn is yet in its milky state, this Squirrel makes long journeys to visit the fields, and for the sake of convenience frequently builds a temporary summer-house in the vicinity, in order to share with the little Carolina squirrel and the crow a portion of the delicacies and treasures of the husbandman; where he is also exposed to the risks incurred by the thief and plunderer: for these fields are usually guarded by a gunner, and in this way thousands of squirrels are destroyed during the green corn season. The Fox Squirrel does not appear to lay up any winter stores—there appears to be no food in any of his nests, nor does he, like the red squirrel, (*Sciurus hudsonius*), resort to any hoards which in the season of abundance were buried in the earth, or concealed under logs and leaves. During the winter season he leaves his retreat but seldom, and then only for a little while and in fine weather in the middle of the day. He has evidently the power, like the marmot and racoon, of being sustained for a considerable length of time without much suffering in the absence of food. When this animal makes his appearance in winter, he is seen searching among the leaves where the wild turkey has been busy at work, and gleaning the refuse acorns which have escaped its search; at such times, also, this squirrel does not reject worms and insects which he may detect beneath the bark of fallen or decayed trees. Towards spring, he feeds on the buds of hickory, oak, and various other trees, as well as on several kinds of roots, especially the wild potato, (*Apios tuberosa.*) As the spring advances farther, he is a constant visitor to the black mulberry tree, (*Morus rubra,*) where he finds a supply for several weeks. From this time till winter, the fruits of the field and forest enable him to revel in abundance.

Most other species of this genus when alarmed in the woods immediately betake themselves to the first convenient tree that presents itself,—not so with the Fox Squirrel. When he is aware of being discovered whilst on the ground, he pushes directly for a hollow tree, which is often a quarter of a mile distant, and it requires a good dog, a man on horseback, or a very swift runner, to induce him to alter his course, or compel him to ascend any other tree. When he is silently seated on a tree and imagines himself unperceived by the person approaching him, he suddenly spreads himself flatly on the limb, and gently moving to the opposite side, often by this stratagem escapes detection. When, however, he is on a small tree, and is made aware of being observed, he utters a few querulous barking notes, and immediately leaps to the ground, and hastens to a more secure retreat. If overtaken by a dog, he defends himself with great spirit, and is often an overmatch for the small terriers which are used for the purpose of treeing him.

He is very tenacious of life, and an ordinary shot gun, although it may wound him repeatedly, will seldom bring him down from the tops of the high pines to which he retreats when pursued, and in such situations the rifle is the only certain enemy he has to dread.

This Squirrel is seldom seen out of its retreat early in the morning and evening, as is the habit of other species. He seems to be a late riser, and usually makes his appearance at 10 or 11 o'clock, and retires to his domicile long before evening. He does not appear to indulge so frequently in the barking propensities of the genus as the other and smaller species. This note, when heard, is not very loud, but hoarse and gutteral. He is easily domesticated, and is occasionally seen in cages, but is less active and sprightly than the smaller species.

As an article of food, the Fox Squirrel is apparently equally good with any other species, although we have observed that the little Carolina squirrel is usually preferred, as being more tender and delicate. Where, however, squirrels are very abundant, men soon become surfeited with this kind of game, and in Carolina, even among the poorer class, it is not generally considered a great delicacy.

This species, like all the rest of the squirrels, is infested during the summer months with a troublesome larva (*Oestrus*), which fastening itself on the neck or shoulders, must be very annoying, as those most affected in this manner are usually poor and their fur appears thin and disordered. It is, however, less exposed to destruction from birds of prey and wild beasts than the other species. It leaves its retreat so late in the morning, and retires so early in the afternoon, that it is wholly exempt from the rapacity of owls, so destructive to the Carolina squirrel. We have seen it bid defiance to the attacks of the red-shouldered hawk (*Falco lineatus*), the only abundant species in the south; and it frequents high grounds and open woods, to which the fox and wild cat seldom resort, during the middle of the day, so that man is almost the only enemy it has to dread.

GEOGRAPHICAL DISTRIBUTION.

This species is said to exist sparingly in New Jersey. We have not observed it farther north than Virginia, nor could we find it in the mountainous districts of that state. In the pine forests of North Carolina, it becomes more common. In the middle and maritime districts of South Carolina it is almost daily met with, although it cannot be said to be a very abundant species anywhere. It exists in Georgia, Alabama, Mississippi, Florida and Louisiana.

GENERAL REMARKS.

This Squirrel has been frequently described under different names. Bosc appears to be entitled to the credit of having bestowed on it the earliest specific name. GMELLIN, in 1788, named it *S. vulpinus*. The black squirrel of CATESBY is the black variety of the present species.

Plate LXIX

Drawn from Nature by J. J. Audubon, F.R.S. F.L.S.

On Stone by W. E. Hitchcock.

Lith. Printed & Col'd by J. T. Bowen, Phil.

Common Star = Nose Mole.

GENUS CONDYLURA.—Illiger.

DENTAL FORMULA.

$$Incisive\ \tfrac{2}{4};\quad Canine\ \tfrac{1-1}{1-1};\quad Molar\ \tfrac{8-8}{7-7} = 40.$$

Muzzle, long, extremity ciliated ; ears, none ; external eyes, small ; feet, pendactylous ; nails before, formed for digging—those behind, weak and small.

The generic name *Condylura* was given by Illiger, founded on an accidental character. A figure of Delafaille erroneously represents the tail as knobbed : hence the genus was formed from two Greek words—Χορδαδας (nodus) and ωθη (cauda) " knobbed tail."

There is but one well determined species of this genus at present known.

CONDYLURA CRISTATA.—Linn.

Common Star-Nosed Mole.

PLATE LXIX.

C. naribus carunculatus ; caudâ corpore breviore ; vellus obscure cinereo, nigricans, subtus dilutior.

CHARACTERS.

Nostrils, surrounded by a circle of membraneous processes ; tail, shorter than the body ; colour, brownish black above, a shade lighter beneath.

SYNONYMES.

Sorex cristatus, Linn., Ed. 12, p. 73.
Long-tailed Mole, Pennant's Hist. Quad., vol. ii., p. 232 to 90, f. 2.
    "   " Pennant's Arct. Zool., vol. i., p. 140.
Talpa longicaudata erx. Syst., tom. i., p. 188.
Long-tailed Mole, Condylura a lonquequeue, Desm. Mamm., f. i., p. 158.
    "   " Condylura cristata, Harlan, p. 36.
    "   " Godm. vol. i., p, 100.
    "   " C. macroura, Harlan, p. 39.
    "   " C. longicaudata, Richardson Fauna, p. 13 ; C. macroura, p. 234.
    "   " C. cristata, De Kay, N. Hist. N. Y., p. 12.

DESCRIPTION.

In the upper jaw there are two large incisive teeth hollowed in front in the shape of a spoon. The next tooth on each side is long, pointed, conical, with two tubercles, one before and the other behind at the base, resembling in all its characters a canine tooth : these are succeeded by five small molars on each side, the posterior one being the largest. There are three true molars on each side, with two acute tubercles on the inner side —the first or anterior of these molars is the largest, the second a little smaller, and the third or posterior one the smallest. In the lower jaw there are four large incisors, spoon shaped, and bearing a strong resemblance to those in the upper jaw. The next on each side are tolerably long sharp, conical teeth, corresponding with those above which we have set down as canine. The four succeeding teeth on each side, which may be regarded as false molars, are lobed and increase in size as they approach the true molars ; the three molars on each side resemble those above, having two folds of enamel forming a point.

In the shape of its body this species bears a considerable resemblance to the Common Mole of Europe (*Talpa Europea*) and to BREWER's Shrew Mole (*Scalops Brewerii*); in the indications on the nose, however, it differs widely from both. The body is cylindrical, about as stout as that of our Common Shrew Mole, and has the appearance of being attached to the head without any distinct neck. Muzzle, slender and elongated, termi nated with a cartilaginous fringe which originated its English name—the Star-nosed Mole. This circular disk is composed of twenty cartilaginous fibres, two of which situated beneath the nostrils are shortest. The eyes are very small. Moustaches, few and short. There is an orifice in place of an external ear, which does not project beyond the skin. Fore feet, longer and narrower than those of the Common Shrew, feet longer and narrower than those of the Common Mole ; palms, naked, covered with scales ; claws, flattened, acute, channelled beneath ; hind extremities longer than the fore ones, placed far back ; feet nearly naked, scaly; tail, subcylindrical, sparingly covered with coarser hair. It is clothed with dense soft fur.

COLOUR.

Eyes, black ; nose and feet, flesh colour ; point of nails and end of cartilaginous fringe, roseate. The fur on the whole body, dark plumbeous at the roots, and without any annulations, deepening towards the apex into a brownish black. In some shades of light the Star Nose appears perfectly black throughout. On the under surface it is a shade lighter. In the colour of the feet we have seen some variations: a specimen before us, has dark brown feet, another pale ashy brown, and a third yellowish white ; the majority of specimens, however, have their feet brownish white. One specimen is marked under the chin, throat and neck with light yellowish brown, the others are darker in those parts.

DIMENSIONS.

							Inches.
From point of nose to root of tail	-	-	-	-			5
Tail	-	-	-	-	-	-	-	3
From heel to end of claw	-	-	-	-		-	$\frac{7}{8}$
Breadth of palm	-	-	-	-	-	-	-	$\frac{3}{8}$

HABITS.

As far as we have been able to ascertain, the habits of this species do not differ very widely from those of our Common Shrew Mole. We doubt, however, whether its galleries ever run to so great a distance as those of the latter animal, nor does it appear to be in the habit of visiting high grounds. It burrows and forms galleries under ground, and appears to be able to make rapid progress in soft earth. Its food is of the same nature as that of the Common Mole, and it appears to prefer the vicinity of brooks or swampy places, doubtless because in such localities earth worms and the larvæ of various insects are generally abundant.

The proper use of the radiating process at the end of the nose has not been fully ascertained, but as the animal has the power of moving these tendrils in various directions, they may be useful in its search after worms or other prey, as is the moveable snout of the Shrew Mole. When confined in a box, or on the floor of a room, this Mole feeds on meat of almost any kind. It is not as strong as the Common Mole, nor as injurious to the farmer, since it avoids cultivated fields, and confines itself to meadows and low swampy places.

During the rutting season the tail of the Star-nosed Mole is greatly enlarged, which circumstance caused Dr. HARLAN to describe a specimen taken at that season as a new species, under the name *Condylura macroura*.

Dr. GODMAN's account of the abundance of this species does not coincide with our own experience on this subject. He says, " In many places it is scarcely possible to advance a step without breaking down their galleries, by which the surface is thrown into ridges and the surface of the green sward in no slight degree disfigured." We have sometimes supposed that he might have mistaken the galleries of the Common Shrew

Mole for those made by the Star-Nose, as to us it has always appeared a rare species in every part of our Union.

In a few localities where we were in the habit, many years ago, of obtaining the Star-nosed Mole, it was always found on the banks of rich meadows near running streams. The galleries did not run so near the surface as those of the Common Shrew Mole. We caused one of the galleries to be dug out, and obtained a nest containing three young, apparently a week old. The radiations on the nose were so slightly developed that until we carefully examined them we supposed they were the young of the Common Shrew Mole. The nest was spacious, composed of withered grasses, and situated in a large excavation under a stump. The old ones had made their escape, and we endeavoured to preserve the young ; but the want of proper nourishment caused their death in a couple of days.

The specimen of the Star-nosed Mole, from which our plate was drawn, was sent to us by our highly esteemed friend JAMES G. KING, Esq., having been captured on a moist piece of ground at his country seat in New Jersey, opposite the city of New-York.

GEOGRAPHICAL DISTRIBUTION.

This species is found sparingly in all the northern and eastern states. Dr. RICHARDSON supposes it to exist as far north as Lake Superior. We obtained a specimen five miles from the Falls of Niagara, on the Canada side, and have traced it in all the New-England States. We received specimens from Dr. BREWER, obtained near Boston, and from W. O. AYRES, Esq., from Long Island. We caught a few of these animals near New-York, and obtained others from various parts of the state. We saw a specimen at York, Pennsylvania, and found another at Frankfort, east of Philadelphia. We captured one in the valleys of the Virginia Mountains, near the Red Sulphur Springs, and received another from the valleys in the mountains of North Carolina, near the borders of South Carolina, and presume it may follow the valleys of the Alleghany ridge as far to the south as those latitudes. We have never found it in South Carolina or Georgia, but to the west we have traced it in Ohio and the northern parts of Tennessee.

GENERAL REMARKS.

We have been induced to undertake a careful examination of the teeth of this species, which forms the type of the genus, in consequence of the wide differences existing among authors in regard to the characters of the teeth. DEMAREST gave six incisors above and four below in the under jaw, cheek-teeth fourteen above and sixteen beneath. In this arrangement he is followed by HARLAN, GODMAN, GRIFFITH, DE KAY and others. The description of the teeth, by DESMAREST, is very accurate, and so is the very recent one of Dr. DE KAY. F. CUVIER, on whose judgment, in regard to characters founded on dentition, we would sooner rely than on that of any other naturalist, has on the other hand, (*Des dents des Mammifères*, 1825, p. 56,) given descriptions and figures of these teeth, there being two incisive, two canine, and sixteen molar above, and two incisive, two canine, and fourteen molar below. Our recent examination of a series of skulls is in accordance with his views, and we have adopted his dental arrangement. The difference, however, between these authors is more in appearance than in reality. The incisors, canine, and false molars, in their character so nearly approach each other, that it is exceedingly difficult to assign to the several grades of teeth their true position in the dental system.

LINNÆUS described this species under the name of *Sorex cristatus*, in 1776, (12th edition, p. 73); PENNANT, in 1771, gave a description and poor figure of what he called the Long-tailed Mole; and in 1777, ERXLEBEN bestowed on the animal thus figured, the name of *S. longicaudata*. PENNANT's specimen was received from New-York, and although it was badly figured it was correctly characterized " Long tailed Mole, with a radiated nose," and in his " Arctic Zoology " he describes it as " the nose long, the end radiated with short tendrils." The whole mistake we conceive was made by DESMAREST, whose work we have found exceedingly inaccurate, misled, probably, by PENNANT's figure, without looking at his description. He gives one of the characters "*point des crêtes nasales,*" when PENNANT had stated quite the reverse. Hence the error of HARLAN, whose article on *Condylura longicaudata* is a translation of DESMAREST. We feel confident that this supposed species must be struck from the list of true species in our *Fauna*.

The *Condylura macroura* of HARLAN, (*Fauna Americana*, p. 30,) was regarded as a new species, in consequence of a specimen with the tail greatly enlarged. It was a second time published by RICHARDSON, who adopted HARLAN's name; GODMAN first suggested the idea that this might be traced to a peculiarity in the animal at a particular season. It is known that a similar enlargement takes place annually in the neck of the male deer during the rutting season. We have examined several specimens where the tail was only slightly enlarged, and the swelling was just commencing, and we possess one where one half of the tail from the root is of the usual large size of *C. macroura*, and the other half towards the end is abruptly diminished so as to leave one half of the tail to designate a new species and the other half forcing it back to its legitimate place in the system of nature.

The singular character (knobbed tail) on which this Genus was erroneously founded should suggest to the naturalist the necessity of caution. The tails of quadrupeds in drying often assume a very different shape from that which they originally possessed. This is especially the case among the Shrews and mice, that are described from dried specimens, as square-tailed, angular or knobbed, whereas in nature their tails were round.

Plate LXX

On Stone by W. E. Hitchcock

Drawn from Nature by J. J. Audubon F.R.S. &c.

Say's Least Shrew.

Lith. Printed & Col'd by J. T. Bowen, Philad.

GENUS SOREX.—Linn.

DENTAL FORMULA.

Incisive $\frac{2}{2}$; *Lateral incisive or false Canine from* $\frac{3\ to\ 5}{2-2}$; *Molar from* $\frac{4\ to\ 5}{3-3}$; *from* 26 *to* 34 *teeth.*

Incisive teeth in the upper jaw indented at their base ; in the lower, proceeding horizontally from their aveoli and turned upwards towards their points where they are usually of a brown colour ; lateral incisive or false canine, conical, small, shorter than the cheek-teeth.

Muzzle and nose, much elongated ; snout, moveable. Ears and eyes, small ; pendactylous ; nails, hooked. A series of glands along the flanks, exuding a scented unctuous matter.

The generic name is derived from the Latin word *Sorex*, a Shrew, field rat.

Authors have described about twenty-three species of Shrews, twenty existing on the Eastern continent and thirteen in N. America. Many of these species are not as yet determined, we can scarcely doubt from past discoveries that this number will in time be greatly increased. They are, no doubt, susceptible of being arranged into different groups and genera.

We know no genus in which the American naturalist has a greater prospect of success in adding new species than that of Sorex.

SOREX PARVUS.—Say.

Say's Least Shrew.

PLATE LXX.

S. supra fuscenti-cinereus, infra cinereus ; dentibus nigricantibus ; cauda brevi, sub-cylindrica.

CHARACTERS.

Body above brownish ash, cinereous beneath. **Teeth black, tail short,** *sub-cylindrical.*

SYNONYMES.

Sorex Parvus, Say, Long's Exped., vol. i., p. 163.
"   "   Linsby, Am. Journal, vol. xxxix., p. 388.
"   "   Harlan, p. 28. Godman, vol. i., p. 78, pl., fig. 2.
"   "   Dekay, Nat. Hist. N. Y., p. 19.

DESCRIPTION.

DENTAL SYSTEM.

$$Incisive\ \frac{2}{2};\quad Lateral\ incisive\ \frac{4-4}{2-2};\quad Molar\ \frac{4-4}{4-4} = 32.$$

In the upper jaws the incisors are small, much hooked, and have a posterior lobe; the succeeding lateral incisors, are minute, conical, not lobed, the two anterior ones much the largest. The first grinder is smaller than the second and third, the fourth is the smallest. In the lower jaw the incisors are a little smaller than those in the upper. They are much more hooked and have each a large posterior lobe. The two lateral incisors are small not lobed—the grinders have each two sharp points rising above the enamel. The second tooth is largest and the third smallest. Nose slender and long, but less so than that of many other species, especially that of *S. longirostris* and *S. Richardsonii.* Muzzle, bi-lobate, naked; moustaches, numerous, long, reaching to the shoulders; body, slender; eyes, very small, ears, none; the auditory opening being covered by a round lobe, without any folds above; feet sparsely clothed with minute hairs, palms naked; tail thickly clothed with minute hairs, fur, short, close, soft, and silky.

COLOUR.

All the teeth are at their points intensely black; whiskers, white and black; point of nose, feet, and nails, whitish; the hair is, on the upper surface plumbeous from the roots, and of an ashy-brown at the tips; a shade lighter on the under surface: under the chin it is of an ashy grey gradually blending with the colours on the back.

DIMENSIONS.

Inches.

From point of nose to root of tail,   .   .   .   2⅞
Tail, ...   ¾

HABITS.

This little creature, to which the above name was attached by SAY, was first captured by Mr. TITIAN R. PEALE, during LONG's Expedition to the Rocky Mountains, at Engineer Cantonment on the Missouri, where it was found in a pit-fall excavated for catching wolves.

Look at the plate, reader, and imagine the astonishment of the hunter on examining the pit intended for the destruction of the savage prowlers of the prairies, when, instead of the game that he intended to entrap, he perceived this, the Least Shrew, timidly running across the bottom.

The family to which this Shrew belongs, is somewhat allied in form and habits to the mole, but many species are now probably extinct.

We have seen a fragment of a fossil remainder of the tooth of a Sorex, found by our young friend Dr. LECONTE, of New-York, in the mining region adjoining Lake Superior, from the size of which, the animal must have been at least a yard long, and no doubt was, with its carnivorous teeth, a formidable beast of prey; whether it had insects and worms of a corresponding size to feed upon, in its day and generation, is a matter of mere conjecture, as even the wonderful discoveries of geologists have thrown but little light on the modes of life of the inhabitants of the ancient world, although some whole skeletons are found from time to time by their researches.

The Least Shrew feeds upon insects and larvæ, worms and the flesh of any dead bird or beast that it may chance to discover.

It also eats seeds and grains of different kinds. It burrows in the earth, but seeks its food more upon the surface of the ground than the mole, and runs with ease around its burrow about fences and logs. Some birds of prey pounce upon the Shrew, whilst it is playing or seeking its food on the grass, but as it has a musky, disagreeable smell, it is commonly left after being killed, to rot on the ground, as we have picked up a good many of these little quadrupeds, which to all appearance had been killed by either cats, owls or hawks. This smell arises from a secretion exuded from glands which are placed on the sides of the animal (Geoffroy, Mem. Mus. Hist. Nat., Vol. i., 1815), This secretion, like that of most animals, varies according to the age, the season, &c., and prevails more in males than females.

Of the mode in which the Least Shrew passes the winter we have no very positive information. It is capable of sustaining a great degree of cold. We have never found one of these animals in a torpid state, when examining burrows, holes, or cavities in and under rocks or stones, &c., for the purpose of ascertaining, if possible, the manner in which they passed the winter. We have seen minute tracks on the surface of the snow where it was four feet in depth in the Northern parts of New-York, which we ascertained were the foot-prints of a Shrew which was afterwards captured, although we cannot be certain that it was this species. It had sought the dried stalks of the pig weed (*chenopodium album*) on which the ripened seeds were still hanging and upon which it had evidently been feeding.

We are unacquainted with any other habits of this minute species.

GEOGRAPHICAL DISTRIBUTION.

If authors have made no mistake in the designation of this species, as we strongly suspect, it has a wide geographical range : according to RICHARDSON, it is found as far to the north as Behring's Straits. The specimens from which our figures were taken, were obtained in the immediate vicinity of New-York. Dr. DEKAY, in his Nat. Hist. of New-York, p. 20, mentions that although he had been unsuccessful in obtaining it in New-York, a specimen was found in Connecticut, by Mr. LINSLEY. We have not ascertained its southern range, all we know of its existence in the west, is from SAY's short description of the only specimen obtained west of the Missouri.

GENERAL REMARKS.

All our authors seem anxious to obtain SAY's Least Shrew, and we have seen dozens of specimens of young Shrews of several species, labeled in the cabinets "*Sorex Parvus.*"

Although there were few more accurate describers than SAY, yet his description of *S. parvus*, is too imperfect, to enable us to feel confident of the species. There was no examination of its dental system, and his description would easily apply to half a dozen other species. The characters by which we may separate the different Shrews are not easily detected, they very much resemble each other in form, colour and habits; they are minute nocturnal animals and not easily procured.

There exist but few specimens in our cabinets to enable us to institute comparisons, and a century will pass away before all our species are discovered. We have very little doubt, that when the species which

was obtained in the far West and described by SAY, and that of RICHARD-
SON from the far north, and ours from the vicinity of New-York, are
obtained and compared and their dental system carefully examined, it
will be ascertained that they are three distinct species, and our successors will be surprised that the old authors gave to the Shrews so
wide a geographical range.

SAY's description is subjoined for convenient comparison. " Body
above brownish cinereous, beneath cinereous: head elongated, eyes and
ears concealed; whiskers long, the longest nearly attaining the back
of the head; nose naked emarginate; front teeth black, lateral ones
piceous; feet whitish, five-toed; nails prominent, acute, white; tail
short, sub-cylindrical, of moderate thickness, slightly thicker in the middle—whitish beneath. Length of head and body, two inches four
lines, of tail, 0.75." RICHARDSON's animal was according to his description, dark brownish grey above, and grey beneath. Length of head
and body two inches three lines, tail one inch.

CANIS LATRANS.—Say.

PRAIRIE WOLF,—BARKING WOLF.

PLATE LXXI.—MALE.

C. cano cinereus nigris et opace pulvo-cinnameo-variegatus; lateribus pallidioribus; fasciâ taise lâta brevinigrâ; cauda rectâ fusiformi cineraceo-cinnameoque variegata apice nigra.

CHARACTERS.

Hair cinereous grey, varied with black above and dull fulvous cinnamon; sides paler than the back, obsoletely fasciate, with black above the legs; tail straight, bushy, fusiform, varied with grey and cinnamon, tip black.

SYNONYMES.

SMALL WOLVES, Dr Praly, Louisiana, vol. ii., p. 54.
PRAIRIE WOLF, Gass. Journal, p. 56.
PRAIRIE WOLF and BURROWING DOG, Lewis and Clark, vol. i., p. 102, 13, 203.
     vol. iii., pp. 102, 136, 203.
   "   "   Schoolcraft's Travels, 285.
CANIS LATRANS, Say, Long's Exped. i., p. 168.
  "   "   Harlan, p. 33.
  "   "   God., 1 vol., 26.
  "   "   Richardson, F. B. Ar. 75.
LYCISCUS CAJOTTIS, Hamilton Smith, Nat. Lib., vol. iv., p. 164, p. 6.

DESCRIPTION.

The Barking or Prairie Wolf is intermediate in size, between the large American Wolf and the grey Fox (*V. virginianus.*) It is a more lively animal than the former, and possesses a cunning fox-like countenance. In seeing it on the prairies, and also in menageries, in a state of domestication, we have often been struck with its quick, restless manner, and with many traits of character that reminded us of sly reynard.

The nose is sharp and pointed; nostrils moderately dilated and naked —the upper surface to the forehead covered with compact short hairs; eyelids placed obliquely on the sides of the head. Eyes rather small—

Plate LXXI

On Stone by Wᵐ E Hitchcock

Prairie Wolf

Drawn from Nature by J W Audubon

Lith. Printed & Colᵈ by J.T. Bowen, Philadᵃ

moustaches few, very rigid, extending to the eyes, four or five stiff hairs rising on the sides of the neck below the ears. Head rather broad; Ears, erect, broad at base, running to an obtuse point, clothed with compact soft fur in which but few of the longer hairs exist; body, tolerably stout; legs, of moderate length, shorter in proportion than those of the common Wolf; Tail, large and bushy, composed like the covering of the body of two kinds of hair, the inner soft and woolly, the outer longer and coarser and from two to three and a half inches in length. Soles of the feet naked, nails rather stout, shaped like those of the dog. The whole structure of the animal is indicative of speed, but from its compact shape and rather short legs we would be led to suppose that it was rather intended for a short race than a long heat.

COLOUR.

Nostrils, around the edges of the mouth, and moustaches, black; upper surface of nose, and around the eyes, reddish brown; upper lip, around the edges of the mouth, and throat, white; eye-lids, yellowish white; hairs on the forehead, at the roots reddish brown, then a line of yellowish white tipped with black, giving it a reddish grey appearance. Inner surface of the ears (which are thinly clothed with hair) white; outer surface, yellowish brown; the fore legs reddish brown, with a stripe of blackish extending from the fore shoulder in an irregular black line over the knee to near the pans. Outer surface of the ʻhind legs, reddish brown, inner surface a little lighter.

On the back the soft under fur is dingy yellow; the longer hair from the roots to two-thirds of its length black, then a broad line of yellowish brown, broadly tipped with black. Neck, reddish brown; throat and all beneath, yellowish white, with bars under the throat and on the chest and belly of a reddish tinge. On the tail the softer hair is plumbeous, the longer hairs are like those on the back, except on the tip of the tail where they are black for nearly their whole length. The description here given is from a very fine specimen obtained at San Antonio in Texas. There is not however a uniformity of colour in these animals, although they vary less than the large wolves. The specimen which RICHARDSON described was obtained on the Saskatchewan. We examined it in the Zoological Museum of London: it differs in some shades of colours from ours — its ears are a little shorter, its nose less pointed, and the skull less in breadth—but it was evidently the same species, and could not even be regarded as a distinct variety. The many specimens we examined and compared, in various tints of colour differed considerably, some wanting the brown tints, being nearly grey, while many had black markings on the shin and forelegs which were absent in others. In all descriptions of wolves, colour is a very uncertain guide in the designation of species.

DIMENSIONS.

	Ft.	Inches.
From point of nose to root of tail	2	10
Tail vertebræ, ...	.	11
Do. to end of hair, ...	1	3
Height of ear, ...	.	3
Breadth of do. at the base, ...	.	3
From heel to end of longest nail, ...	.	6
Point of nose to corner of eye, ...	.	$3\frac{1}{2}$
Breadth of skull, ...	.	4
Fore shoulder to end of longest nail, . . .	1	1
Breadth across the forehead, ...	.	$2\frac{1}{8}$

HABITS.

We saw a good number of these small wolves on our trip up the Missouri river, as well as during our excursions through those portions of the country which we visited bordering on the Yellow Stone.

This species is well known throughout the western parts of the States of Arkansas and Missouri, and is a familiar acquaintance of the " voyageurs" on the upper Missouri and Mississippi rivers. It is also found on the Saskatchewan. It has much the appearance of the common grey Wolf in colour, but differs from it in size and manners.

The Prairie Wolf hunts in packs, but is also often seen prowling singly over the plains in search of food. During one of our morning rambles near Fort Union, we happened to start one of these wolves suddenly. It made off at a very swift pace and we fired at it without any effect, our guns being loaded with small shot at the time; after running about one hundred yards it suddenly stopped and shook itself violently, by which we perceived that it had been touched; in a few moments it again started and soon disappeared beyond a high range of hills, galloping along like a hare or an antelope.

The bark or howl of this wolf greatly resembles that of the dog, and on one occasion the party travelling with us were impressed by the idea that Indians were in our vicinity, as a great many of these wolves were about us and barked during the night like Indian dogs. We were all on the alert, and our guns were loaded with ball in readiness for an attack.

In Texas the Prairie Wolves are perhaps more abundant than the other species; they hunt in packs of six or eight, which are seen to most advantage in the evening, in pursuit of deer. It is amusing to see them cut across the curves made by the latter when trying to escape, the hindmost Wolves thus saving some distance, and finally striking in ahead of the poor deer and surrounding it, when a single Wolf would fail in the attempt to capture it. By its predatory and destructive habits, this Wolf is a great annoyance to the settlers in the new territories of the west. Travellers and hunters on the prairies, dislike it for killing the deer, which supply these wanderers with their best meals, and furnish them with part of their clothing, the buck-skin breeches, the most durable garment, for the woods or plains. The bark or call-note of this Wolf, although a wild sound to the inhabitant of any settled and cultivated part of the country, is sometimes welcomed, as it often announces the near approach of daylight; and if the wanderer, aroused from his slumbers by the howling of this animal, raises his blanket and turns his head toward the east, from his camping-ground underneath the branches of some broad spreading live-oak, he can see the red glow, perchance, that fringes the misty morning vapours, giving the promise of a clear and calm sunrise in the mild climate of Texas, even in the depth of winter. Should day-light thus be at hand, the true hunter is at once a-foot, short space of time does he require for the duties of the toilet, and soon he has made a fire, boiled his coffee, and broiled a bit of venison or wild turkey.

This Wolf feeds on birds, small and large quadrupeds, and when hard pressed by hunger, even upon carrion or carcasses of buffaloes, &c. It is easily tamed when caught young, and makes a tolerable companion, though not gifted with the good qualities of the dog. We had one once, which was kept in a friend's store in the west, and we discovered it to be something of a rat catcher. This individual was very desirous of being on friendly terms with all the dogs about the premises, especially with a large French poodle that belonged to our friend, but the poodle would not permit our half-savage barking Wolf to play with him, and generally returned its attempted caresses with an angry snap, which put all further friendly demonstrations out of the question. One day we missed our pet from his accustomed place near the back part of the ware-house, and while we were wondering what had become of him, were attracted by an unusual uproar in the street. In a moment we perceived the noise was occasioned by a whole pack of curs of high and low degree, which were in full cry, and in pursuit of our Prairie Wolf. The creature thus hard beset, before we could interfere, had reached a point opposite a raised window, and to our surprise, made a sudden spring at it and jumped into the warehouse without touching the edges of the sills, in the most admirable manner, while his foes were completely baffled.

After this adventure the Wolf would no longer go out in the town and seemed to give up his wish to extend the circle of his acquaintance.

The Barking or Prairie Wolf digs its burrows upon the prairies on some slight elevation, to prevent them from being filled with water. These dens have several entrances, like those of the red fox. The young, from five to seven and occasionally more in number, are brought forth in March and April. They associate in greater numbers than the larger Wolves, hunt in packs, and are said by RICHARDSON to be fleeter than the common Wolf. A gentleman, an experienced hunter on the Saskatchewan, informed him that the only animal on the plains which he could not overtake when mounted on a good horse, was the prong-horned antelope, and that the Prairie Wolf was next in speed.

All our travellers have informed us, that on the report of a gun on the prairies, numbers of these Wolves start from the earth, and warily approach the hunter, under an expectation of obtaining the offal of the animal he has killed.

The skins of the Prairie Wolves are of some value, the fur being soft and warm ; they form a part of the Hudson Bay Company's exportations, to what extent we are not informed. RICHARDSON says they go under the name of cased-wolves skins, not split open like those of the large Wolf, but stripped off and inverted or cased, like the skin of a fox or rabbit.

GEOGRAPHICAL DISTRIBUTION.

According to RICHARDSON, the northern range of this species is about the fifty-fifth degree of latitude. It is found abundantly on the plains of the western prairies and sparingly on the plains adjoining the woody shores of the Columbia river. It exists in California, and is found in Texas and on the eastern side of the mountains in New Mexico. We have traced it to within the tropics, but are not aware that it reaches as far south as Panama. The eastern branches of the Missouri river appear to be its farthest eastern range.

GENERAL REMARKS.

There has been but little difficulty in the nomenclature of this species. Hamilton Smith, we perceive, has given it a new name, from a specimen obtained in Mexico. The description of its habits, by Lewis and Clarke, is full and accurate and in accordance with our own observations.

CANIS LUPUS.—Linn.—(Var. Albus.)

WHITE AMERICAN WOLF.

PLATE LXXII.—MALE.

C. magnitudine formaque C. lupi; vellere flavido-albo; naso canescente.

CHARACTERS.

Size and shape of the grey wolf, fur over the whole body of a yellowish-white colour, with a slight tinge of grey on the nose.

SYNONYMES.

WHITE WOLF, Lewis and Clark, vol. i., p. 107, vol. iii., p. 263.
CANIS LUPUS, Albus, Sabine, Frank. Journ., p. 652.
WHITE WOLF, Frank. Journal, p. 312.
"   "   Lyon's Private Journal, p. 279.
LUPUS ALBUS VAR. B. WHITE WOLF, Richardson, F. B. A., p. 68.

DESCRIPTION.

In shape, this Wolf resembles all the other varieties of large North American Wolves. (The prairie or barking Wolf, a distinct and different species, excepted.) It is large, stout, and compactly built; the canine teeth are long; others stout, large, rather short. Eyes, small. Ears, short and triangular. Feet, stout. Nails, strong and trenchant. Tail, long and bushy. Hairs on the body, of two kinds; the under coat composed of short, soft and woolly hair, interspersed with longer coarse hair five inches in length. The hairs on the head and legs are short and smooth, having none of the woolly appearance of those on other portions of the body.

COLOUR.

The short fur beneath the long white coat, yellowish white, the whole outer surface white, there is a slight tinge of greyish on the nose. Nails black; teeth white.

Another Specimen.—Snow-white on every part of the body except the tail, which is slightly tipped with black.

Plate LXXII

White American Wolf.

Drawn from Nature by J. W. Audubon.

On Stone by W.E. Hitchcock.

Lith. Printed & Col.d by J.T. Bowen Phila.

Another.—Light grey on the sides legs and tail; a dark brown stripe on the back, through which many white hairs protrude, giving it the appearance of being spotted with brown and white. This variety resembles the young Wolf noticed by RICHARDSON, (p. 68) which he denominates the pied Wolf.

DIMENSIONS.

	Feet.	Inches.
From point of nose to root of tail, - - - -	4	6
Do. tail, vertebræ, - - - - - - -	1	2
Do. do. end of hair, - - - - - -	1	8
Height of ear, - - - - - - - -		3½

HABITS.

The White Wolf is far the most common variety of the Wolf tribe to be met with around Fort Union, on the prairies, and on the plains bordering the Yellow Stone river. When we first reached Fort Union we found Wolves in great abundance, of several different colours, white, grey, and brindled. A good many were shot from the walls during our residence there, by EDWARD HARRIS, Esq., and Mr. J. G. BELL. We arrived at this post on the 12th of June, and although it might be supposed at that season the Wolves could procure food with ease, they seemed to be enticed to the vicinity of the Fort by the cravings of hunger. One day soon after our arrival, Mr. CULBERTSON told us that if a Wolf made its appearance on the prairie, near the Fort, he would give chase to it on horseback, and bring it to us alive or dead. Shortly after, a Wolf coming in view, he had his horse saddled and brought up, but in the meantime the Wolf became frightened and began to make off, and we thought Mr. CULBERTSON would never succeed in capturing him. We waited, however, with our companions on the platform inside the walls, with our heads only projecting above the pickets, to observe the result. In a few moments we saw Mr. CULBERTSON on his prancing steed as he rode out of the gate of the Fort with gun in hand, attired only in his shirt, breeches and boots. He put spurs to his horse and went off with the swiftness of a jockey bent upon winning a race. The Wolf trotted on and every now and then stopped to gaze at the horse and his rider, but soon finding that he could no longer indulge his curiosity with safety, he suddenly gallopped off with all his speed, but he was too late in taking the alarm, and the gallant steed soon began to gain on the poor cur, as we saw the horse rapidly shorten the distance between the Wolf and his enemy. Mr. CULBERTSON fired off his gun as a signal to us that he felt sure of bringing in the beast, and although the hills were gained by the fugitive, he had not time to make for the broken ground and deep ravines, which he would have reached in few minutes, when we heard the crack of the gun again, and Mr. CULBERTSON galloping along dexterously picked up the slain Wolf without dismounting from his horse, threw him across the pummel of his saddle, wheeled round and rode back to the Fort, as fast as he had gone forth, a hard shower of rain being an additional motive for quickening his pace, and triumphantly placed the trophy of his chase at our disposal. The time occupied, from the start of the hunter, until his return with his prize did not exceed twenty minutes. The jaws of the animal had become fixed, and it was quite dead. Its teeth had scarified one of Mr. CULBERTSON's fingers considerably, but we were assured that this was of no importance, and that such feats as the capture of this wolf were so very common, that no one considered it worthy of being called an exploit.

Immediately after this real wolf hunt, a sham Buffalo chase took place, a prize of a suit of clothes being provided for the rider who should load and shoot the greatest number of times in a given distance. The horses were mounted, and the riders started with their guns empty— loaded in a trice, while at speed, and fired first on one side and then on the other, as if after Buffaloes. Mr. CULBERTSON fired eleven times in less than half a mile's run, the others fired less rapidly, and one of them snapped several times, but as a snap never brings down a Buffalo, these mishaps did not count. We were all well pleased to see these feats performed with much ease and grace. None of the riders were thrown, although they suffered their bridles to drop on their horses necks, and plied the whip all the time. Mr. CULBERTSON's mare, which was of the full, black foot Indian breed, about five years old, was highly valued by that gentleman, and could not have been purchased of him for less than four hundred dollars.

To return to the wolves. — These animals were in the habit of coming at almost every hour of the night, to feed in the troughs where the offal from the Fort was deposited for the hogs. On one occasion, a wolf killed by our party was devoured during the night, probably by other prowlers of the same species.

The white wolves are generally fond of sitting on the tops of the eminences, or small hills in the prairies, from which points of vantage they can easily discover any passing object on the plain at a considerable distance.

We subjoin a few notes on wolves generally, taken from our journals, made during our voyage up the Missouri in 1843.

These animals are extremely abundant on the Missouri river, and in the adjacent country. On our way up that extraordinary stream, we first heard of wolves being troublesome to the farmers who own sheep, calves, young colts, or any other stock on which these ravenous beasts feed, at Jefferson city, the seat of goverment of the State of Missouri; but to our great surprise, while there not a black wolf was seen.

Wolves are said to feed at times, when very hard pressed by hunger, on certain roots which they dig out of the earth with their forepaws, scratching like a common dog in the ground. When they have killed a Buffalo or other large animal, they drag the remains of the carcass to a concealed spot if at hand, then scrape out the loose soil and bury it, and often lie down on the top of the grave they have thus made for their victim, until urged again by hunger, they exume the body and feast upon it. Along the banks of the river, where occasionally many Buffaloes perish, their weight and bulk preventing them from ascending where the shore is precipitous, wolves are to be seen in considerable numbers feeding upon the drowned Bisons.

Although extremely cunning in hiding themselves, at the report of a gun wolves soon come forth from different quarters, and when the alarm is over, you have only to conceal yourself, and you will soon see them advancing towards you, giving you a fair chance of shooting them, sometimes at not more than thirty yards distance. It is said that although they frequently pursue Buffalo, &c., to the river, they seldom if ever follow them after they take to the water. Their gait and movements are precisely the same as those of the common dog, and their mode of copulating, and the number of young brought forth at a litter is about the same. The diversity of their size and colour is quite remarkable, no two being quite alike.

Some days while ascending the river, we saw from twelve to twenty-five wolves; on one occasion we observed one apparently bent on crossing the river, it swam toward our boat and was fired at, upon which it wheeled round and soon made to the shore from which it had started.

At another time we saw a wolf attempting to climb a very steep and high bank of clay, when, after falling back thrice, it at last reached the top and disappeared at once. On the opposite shore another was seen lying down on a sand bar like a dog, and any one might have supposed it to be one of those attendants on man. Mr. BELL shot at it, but too low, and the fellow scampered off to the margin of the woods, there stopped to take a last lingering look, and then vanished.

In hot weather when wolves go to the river, they usually walk in up to their sides, and cool themselves while lapping the water, precisely in the manner of a dog.   They do not cry out or howl when wounded or when suddenly surprised, but snarl, and snap their jaws together furiously.   It is said when suffering for want of food, the strongest will fall upon the young or weak ones, and kill and eat them.   Whilst prowling over the prairies (and we had many opportunities of seeing them at such times) they travel slowly, look around them cautiously, and will not disdain even a chance bone that may fall in their way; they bite so voraciously at the bones thus left by the hunter that in many cases their teeth are broken off short, and we have seen a number of specimens in which the jaws showed several teeth to have been fractured in this way.

After a hearty meal, the wolf always lies down when he supposes himself in a place of safety.   We were told that occasionally when they had gorged themselves, they slept so soundly that they could be approached and knocked on the head.

The common wolf is not unfrequently met with in company with the Prairie wolf (*Canis latrans.*) On the afternoon of the 13th of July, as Mr. BELL and ourselves were returning to Fort Union, we counted eighteen wolves in one gang, which had been satiating themselves on the carcass of a Buffalo on the river's bank, and were returning to the hills to spend the night. Some of them had their stomachs distended with food and appeared rather lazy.

We were assured at Fort Union that wolves had not been known to attack men or horses in that vicinity, but they will pursue and kill mules and colts even near a trading post, always selecting the fattest. The number of tracks or rather paths made by the wolves from among and around the hills to that station are almost beyond credibility, and it is curious to observe their sagacity in choosing the shortest course and the most favourable ground in travelling.

We saw hybrids, the offspring of the wolf and the cur dog, and also their mixed broods: some of which resemble the wolf, and others the dog.   Many of the Assiniboin Indians who visited Fort Union during our stay there, had both wolves and their crosses with the common dog in their trains, and their dog carts (if they may be so called) were drawn alike by both.

The natural gait of the American wolf resembles that of the Newfoundland dog, as it ambles, moving two of its legs on the same side at a time.   When there is any appearance of danger, the wolf trots off, and generally makes for unfrequented hilly grounds, and if pursued, gallops at a quick pace, almost equal to that of a good horse, as the reader will perceive from the following account. On the 16th of July 1843, whilst we were on a Buffalo hunt near the banks of the Yellow Stone river, and all eyes were bent upon the hills and the prairie, which is very broad, we saw a wolf about a quarter of a mile from our encampment, and Mr. Owen McKenzie was sent after it. The wolf however ran very swiftly and was not overtaken and shot until it had ran several miles. It dodged about in various directions, and at one time got out of sight behind the hills. This wolf was captured, and a piece of its flesh was boiled for supper; but as we had in the mean time caught about eighteen or twenty Cat-fish, we had an abundant meal and did not judge for ourselves whether the wolf was good eating or not, or if its flesh was like that of the Indian dogs, which we have had several opportunities of tasting.

Wolves are frequently deterred from feeding on animals shot by the hunters on the prairies, who, aware of the cautious and timid character of these rapacious beasts, attach to the game they are obliged to leave behind them a part of their clothing, a handkerchief, &c., or scatter gun powder around the carcass, which the cowardly animals dare not approach although they will watch it for hours at a time, and as soon as the hunter returns and takes out the entrails of the game he had left thus protected, and carries off the pieces he wishes, leaving the coarser parts for the benefit of these hungry animals, they come forward and enjoy the feast. The hunters who occasionally assisted us when we were at Fort Union, related numerous stratagems of this kind to which they had resorted to keep off the wolves when on a hunt.

The wolves of the prairies form burrows, wherein they bring forth their young, and which have more than one entrance ; they produce from six to eleven at a birth, of which there are very seldom two alike in colour. The wolf lives to a great age and does not change its colour with increase of years.

GEOGRAPHICAL DISTRIBUTION.

This variety of wolf is found as far north in the Arctic regions of America as they have been traversed by man. The journals of Hearne, Franklin, Sabine Richardson, and others, abound with accounts of their presence amid the snows of the polar regions. They exist in the colder parts of Canada, in the Russian possessions on the western coast of America, in Oregon, and along both sides of the Rocky Mountains, to California on the west side and Arkansas on the east. We examined a specimen of the White Wolf killed in Erie county, N. Y., about forty years ago; on the Atlantic coast they do not appear; although we have seen some specimens of a light grey colour they could not when compared with those of Missouri, be called white wolves.

GENERAL REMARKS

Cold seems necessary to produce the Wolves of white variety. Alpine regions from their altitudes effect the same change. REGNARD informs us that in Lapland, Wolves are almost all of a whitish grey colour—there are some of them white. In Siberia, wolves assume the same colour. The Alps, on the other hand, by their elevation, may be compared to the regions around the Rocky Mountains of America. In both countries wolves become white. We devoted some hours to comparing the large American, European, and Asiatic Wolves, assisted by eminent British Naturalists, in the British Museum and the Museum of the Zoological Society. We found specimens from the Northern and Alpine regions of both continents bore a strong resemblance to each other in form and size, their shades of colour differed only in different specimens from either country, and we finally came to the conclusion that the naturalist who should be able to find distinctive characters to separate the wolves into different species, should have credit for more penetration than we possess.

Plate LXXIII

Drawn from Nature by J. W. Audubon.

On Stone by Wm E. Hitchcock.

Lith Printd & Cold by J.T.Bowen, Philad.

Rocky Mountain Sheep.

GENUS OVIS.—Linn., Briss., Erxleben, Cuv., Bodd., Geoff.

DENTAL FORMULA.

$$Incisive \ \frac{0}{8}; \quad Canine \ \frac{0-0}{0-0}; \quad Molar \ \frac{6-6}{6-6} = 32.$$

Horns common to both sexes, sometimes wanting in the females, they are voluminous, more or less angular, transversely wrinkled, turned laterally in spiral directions, and enveloping an osseous arch, cellular in structure.

They have no lachrymal sinus, no true beard to the chin, the females have two mammæ; tail, rather short; ears, small, erect; legs, rather slender; hair, of two kinds, one hard and close, the other woolly; gregareous. Habit analogous to the goats. Inhabit the highest mountains of the four quarters of the globe.

The generic name is derived from the latin *Ovis*—a sheep.

There are four well determined species, one the Mouflon of Buffon, Musmon (*Ovis Musmon*) is received as the parent of the domesticated races. It is found in Corsica, Sardinia, and the highest mountain chains of Europe. One inhabiting the mountains and steppes of northern Asia, Tartary, Siberia and the Kurile Islands, one the mountains of Egypt, and one America.

OVIS MONTANA.—Desm.

Rocky Mountain Sheep.

PLATE LXXIII. Male and Female.

O. cornibus crassissimis spiralibus; corpore gracile; artubus elevatis; pilo brevi rigido rudi badio; clunibus albis o ariete major; rufo cinereus.

CHARACTERS.

Longer than the domestic sheep, horns of the male long, strong and triangular, those of the female compressed; colour deep rufous grey, a large white disk on the rump.

SYNONYMES.

ARGALI, COOK'S third voyage in 1778.
WILD SHEEP OF CALIFORNIA. Venegus.
    "   "   "   Clavigero.
WHITE BUFFALO, McKenzie voy. p. 76. An. 1789.
MOUNTAIN GOAT, Umfreville, Hudson's Bay. p. 164.
MOUNTAIN RAM, McGillivary, N. York. Med. Reposit. vol. 6. p. 238.
BIG HORN, Lewis and Clark. vol. 1. p. 144.
BELIER SAUVAGE d'AMERIQUE. Geoff, An. du. mus. t. 2. pl. 60.
ROCKY MOUNTAIN SHEEP. Warden. U. S. vol. 1. p. 217.
MOUFFLON d'AMERIQUE. Desm. Mamm. p. 487.
BIG HORNED SHEEP. (Ord.)
    "   "   Blainv. in Jour. de Physic. 1817.
OVIS AMMON. Harlan. Fauna. p. 259.
THE ARGALI, Godm. Nat. Hist. vol. 2. p. 329.
OVIS MONTANA. Richardson. F. B. Amer. p. 271.
OVIS PYGARJAS VAR OVIS AMMON. Griffith An. King. Spec. 873.

DESCRIPTION.

Male. This is a much larger animal than any variety of our largest sized sheep. It is also considerably larger than the Argali on the eastern continent.

The horns of the male are of immense size. They arise immediately above the eyes, and occupy nearly the whole head, they being only separated from each other by a space of three-fourths of an inch at the base. They form a regular curve, first backwards, then downwards and outward—the extremities being eighteen inches apart. They are flattened on the sides and deeply corrugated, the horns rising immediately behind.

The ears, are short and oval, clothed with hair on both surfaces. The general form of the animal is rather elegant, resembling the stag more than the Sheep. The tail is short.

The hair bears no resemblance to wool, but is similar to that of the American Elk and Reindeer. It is coarse, but soft to the touch, and slightly crimped throughout its whole length ; the hairs on the back are about two inches in length, those on the sides one and a half inches. At the roots of these hairs, especially about the shoulders and sides of the neck, a small quantity of short soft fur is perceptible. The legs are covered with short compact hairs.

The female Rocky Mountain Sheep resembles some of the finest specimens of the common Ram. Its neck is a little longer, as are also the head and legs, and in consequence it stands much higher. Its horns resemble more those of the goat than of the Sheep, in fact, whilst the fine erect body of the male reminds us of a large deer with the head of a ram, the female looks like a fine specimen of the antelope. The horns bend backwards and a little outwards, and are corrugated from the roots to near the points. Tail very short and pointed, covered with short hairs. Mammæ two ventral.

COLOUR.

The whole upper surface of the body, outer surface of the thighs, legs, sides and under the throat, light greyish brown, forehead and ears a little lighter. Rump, under the belly and inner surface of hind legs, greyish white; the front legs, instead of being darker on the outside and lighter on the inside, are darker in front, the dark extending round to the inside of the legs, and covering nearly a third of the inner surface. Tail and hoofs black. A narrow dorsal line from the neck to near the rump, conspicuous in the male, but comparatively quite obscure in the female. Richardson states that the old males are almost totally white in spring.

DIMENSIONS.

	Ft.	Inches.
Male figure in our plate.		
Length	6	
Height at shoulder	3	5
Length of tail	0	5
Girth of body behind the shoulders . .	3	11
Height to rump	3	$10\frac{3}{4}$
Length of horn around the curve . . .	2	$10\frac{1}{2}$
Do. of eye	.	$1\frac{3}{4}$
Weight 344 lbs. including horns.		
Female figure in our plate.		
Nose to root of tail	4	7
Tail	0	5
Height of rump	3	$4\frac{1}{2}$
Girth back of shoulders	3	$4\frac{1}{2}$
Horns—$44\frac{1}{2}$ lbs.		
Weight 240 lbs. (Killed July 3d, 1843.)		

HABITS.

It was on the 12th of June, 1843, that we first saw this remarkable animal; we were near the confluence of the Yellow Stone river with the Missouri, when a group of them, numbering twenty-two in all, came in sight. This flock was composed of rams and ewes, with only one young one or lamb among them. They scampered up and down the hills much in the manner of common sheep, but notwithstanding all our anxious efforts to get within gun-shot, we were unable to do so, and were obliged to content ourselves with this first sight of the Rocky Mountain Ram.

The parts of the country usually chosen by these animals for their pastures, are the most extraordinary broken and precipitous clay hills or stony eminences that exist in the wild regions belonging to the Rocky Mountain chain. They never resort to the low lands or plains except when about to remove their quarters, or swim across rivers, which they do well and tolerably fast. Perhaps some idea of the country they inhabit (which is called by the French Canadians and hunters, "mauvaise terres") may be formed by imagining some hundreds of loaves of sugar of different sizes, irregularly broken and truncated at top, placed somewhat apart, and magnifying them into hills of considerable size. Over these hills and ravines the Rocky Mountain Sheep bound up and down among the sugar loaf shaped peaks, and you may estimate the difficulty of approaching them, and conceive the great activity and sure-footedness of this species, which, together with their extreme wildness and keen sense of smell, enable them to baffle the most vigorous and agile hunter.

They form paths around these irregular clay cones that are at times from six to eight hundred feet high, and in some situations are even fifteen hundred feet or more above the adjacent prairies, and along these they run at full speed, while to the eye of the spectator below, these tracks do not appear to be more than a few inches wide, although they are generally from a foot to eighteen inches in breadth. In many places columns or piles of clay, or hardened earth, are to be seen eight or ten feet above the adjacent surface, covered or coped with a slaty flat rock, thus resembling gigantic toad stools, and upon these singular places the big horns are frequently seen, gazing at the hunter who is winding about far below, looking like so many statues on their elevated pedestals. One cannot imagine how these animals reach these curious places, especially with their young along with them, which are sometimes brought forth on these inaccessible points, beyond the reach of their greatest enemies, the wolves, which prey upon them whenever they stray into the plains below.

The "mauvaise terres" are mostly formed of greyish white clay, very sparsely covered with small patches of thin grass, on which the Rocky Mountain Sheep feed. In wet weather it is almost impossible for any man to climb up one of these extraordinary conical hills, as they are slippery, greasy and treacherous. Often when a big horn is seen on the top of a hill, the hunter has to ramble round three or four miles before he can reach a position within gun-shot of the game, and if perceived by the animal, it is useless for him to pursue him any further that day.

The tops of some of the hills in the "mauvaise terres" are composed of a conglomerated mass of stones, sand, clay and various coloured earths, frequently of the appearance and colour of bricks. We also observed in these masses a quantity of pumice stone, and these hills, we are inclined to think are the result of volcanic action. Their bases often cover an area of twenty acres; there are regular horizontal strata running across the whole chain of these hills, composed of different coloured clay, coal and earth, more or less impregnated with salt and other minerals, and occasionally intermixed with lava, sulphur, oxide and sulphate of iron; and in the sandy parts at the top of the highest hills, we found shells, but so soft and crumbling as to fall to pieces when we attempted to pick them out. We found in the "mauvaise terres," also, globular shaped masses of heavy stone and pieces of petrified wood, from fragments two or three inches wide, to stumps of three or four feet thick, apparently cotton wood and cedar. On the sides of some of the hills at various heights, are shelf-like ledges or rock projecting from the surface in a level direction, from two to six and even ten feet, generally square or flat. These ledges are much resorted to by the big horns during the heat of the day. Between these hills there is sometimes a growth of stunted cedar trees, underneath which there is a fine sweet grass, and on the summits in some cases a short dry wiry grass is found, and quantities of that pest of the Upper Missouri country, the flat-broad-leaved Cactus, the spines of which often lame the hunter. Occasionally the hills in the "mauvaise terres" are separated by numerous ravines, often not more than ten or fifteen feet wide, but sometimes from ten to fifty feet deep, and now and then the hunter comes to the brink of one so deep and wide as to make his head giddy as he looks down into the abyss below. The edges of the cañons (as these sort of channels are called in Mexico) are overgrown with bushes, wild cherries, &c., and here and there the Bison will manage to cut paths to cross them, descending in an oblique and zig-zag direction; these paths however are rarely found except where the ravine is of great length, and in general the only mode of crossing the ravine is to go along the margin of it until you come to the head, which is generally at the base of some hill, and thus get round.

These ravines exist between nearly every two neighbouring hills, although there are occasionally places where three or more hills form only one.

All of them however run to meet each other and connect with the largest, the size of which bears its proportion to that of its tributaries and their number.

Where these ravines have no outlet into a spring or water course they have subterranean drains, and in some of the valleys and even on the tops of the hills, there are cavities called "sink holes;" the earth near these holes is occasionally undermined by the water running round in circles underneath, leaving a crust insufficient to bear the weight of a man, and when an unfortunate hunter treads on the deceitful surface it gives way, and he finds himself in an unpleasant and at times dangerous predicament. These holes sometimes gradually enlarge and run into ravines below them. It is almost impossible to traverse the " mauvaise terres" with a horse, unless with great care, and with a thorough knowledge of the country. The chase or hunt after the big horn, owing to the character of the country, (as we have described it,) is attended with much danger, as the least slip might precipitate one headlong into the ravine below, the sides of the hills being destitute of every thing to hold on by excepting a projecting stone or tuft of worm wood, scattered here and there, without which even the most daring hunter could not ascend them.

In some cases the water has washed out caves of different shapes and sizes, some of which present the most fantastic forms and are naked and barren to a great degree. The water that is found in the springs in these broken lands is mostly impregnated with salts, sulphur, magnesia, &c.; but unpleasant as it tastes, it is frequently the only beverage for the hunter, and luckily is often almost as cold as ice, which renders it less disagreeable. In general this water has the effect very soon of a cathartic and emetic. Venomous snakes of various kinds inhabit the "mauvaise terres," but we saw only one copper-head.

Conceiving that a more particular account of these countries may be interesting, we will here insert a notice of them given to us by Mr. DEWEY, the principal clerk at Fort Union. He begins as follows:

" This curious country is situated, or rather begins half way up White river, and runs from south east to north west for about sixty miles in length, and varying from fifteen to forty miles in width. It touches the head of the Teton river and branches of Chicune, and joins the Black Hills at the south fork of the latter river. The hills are in some places five or six hundred yards high and upwards. They are composed of clay of various colours, arranged in layers or strata running nearly horizontally, each layer being of a different colour, white, red, blue, green, black, yellow, and almost every other colour, appearing at exactly the same height on every hill.

" From the quantity of pumice stone and melted ores found throughout them, one might suppose that they had been reduced to this state by volcanic action. From the head of the Teton river, to cross these hills to White river is about fifteen miles; there is but one place to descend, and the road is not known; the only way to proceed is to go round the end of them on the banks of the White river, and following that stream ascend to the desired point. In four day's march a man will make about fifteen miles in crossing through the "mauvaise terres." At first sight these hills look like some ancient city in ruins, and but little imagination is necessary to give them the appearance of castles, walls, towers, steeples, &c. The descent is by a road about five feet broad, winding around and among the hills, made at first probably by the bisons and the big horn sheep, and now rendered practicable by the Indians and others who have occasion to use it. It is however too steep to travel down with a loaded horse or mule, say about one foot in three, for a mile or so, after which the bases of the hills are about level with each other, but the valleys between them are cut up by great ravines in almost every direction from five to twenty and even fifty feet deep."

"In going over this part of the country great precaution is necessary, for a slip of the foot would precipitate either man or horse into the gulf below. When I descended, the interpreter, B. Daumine, a half breed, (having his eyes bandaged) was led by the hand of an Indian." Something like copperas in taste and appearance is found in large quantities, as well as pumice stone, every where. This country is the principal residence of the big horn sheep, the panther and grizzly bear; big horns especially are numerous, being in bands of from twenty to thirty, and are frequently seen at the tops of the highest peaks, completely inaccessible to any other animal. There is but one step from the prairie to the barren clay, and this step marks the difference for nearly its whole length. These "mauvaise terres" have no connexion or affinity to the surrounding country, but are, as it were, set apart for the habitation of the big horns and bears. The sight of this barren country causes one to think that thousands of square miles of earth have been carried off, and nothing left behind but the ruins of what was once a beautiful range of mountains. The principal part of these hills is white clay, which when wet is soft and adhesive, but the coloured strata are quite hard and are never discoloured by the rain, at least not to any extent, for after a hard rain the streams of water are of a pure milk white colour, untinged by any other, and so thick that ten gallons when settled will only yield about two gallons of pure limpid water, which, however, although clear when allowed to stand awhile, is scarcely drinkable, being salt and sulphurous in taste. The sediment has all the appearance of the clay already mentioned, which is nearly as white as chalk. There is only one place where wood and pure sweet water can be found in the whole range, which is at a spring nearly in the centre of the tract, and one day's journey from the White river, towards the Chicune. This appears a little singular, for if it were not for this the voyageur would be obliged to take a circuitous route of from four to five days. This spring is surrounded by a grove of ash trees, about two hundred yards in circumference. It immediately loses itself in the clay at the edge of the timber, and near the spring the road descends about sixty feet and runs through a sort of avenue at least half a mile wide, on each side of which are walls of clay extending horizontally about fifteen miles, and eighty feet high, for nearly the whole distance. Between these walls are small sugar-loaf shaped hills, and deep ravines, such as I have already described. The colours of the strata are preserved throughout. The principal volcano is the "Côte de tonnerre," from the mouth of which smoke and fire are seen to issue nearly at all times. In the neighbourhood and all around, an immense quantity of pumice stone is deposited, and from the noises to be heard, no doubt whatever exists that eruptions may from time to time be expected. There is another smaller hill which I saw giving forth heated vapours and smoke, but in general if the weather is clear the summits of the Black hills are obscured by a mist, from which circumstance many superstitions of the Indians have arisen. The highest of the Black hills are fully as high as the Alleghany mountains, and their remarkable shapes and singular characters deserve the attention of our geologists, especially as it is chiefly among these hills that fossil petrefactions are abundantly met with.

The Rocky Mountain Sheep are gregarious, and the males fight fiercely with each other in the manner of common rams. Their horns are exceedingly heavy and strong, and some that we have seen have a battered appearance, showing that the animal to which they belonged must have butted against rocks or trees, or probably had fallen from some elevation on to the stony surface below. We have heard it said that the Rocky Mountain Sheep descend the steepest hills head foremost, and they may thus come in contact with projecting rocks, or fall from a height on their enormous horns.

As is the case with some animals of the deer tribe, the young rams of this species and the females herd together during the winter and spring, while the old rams form separate flocks, except during the rutting season in December.

In the months of June and July the ewes bring forth, usually one, and occasionally, but rarely, two.

Dr. RICHARDSON, on the authority of DRUMMOND, states that in the retired parts of the mountains where the hunters had seldom penetrated, he

(DRUMMOND) found no difficulty in approaching the Rocky Mountain Sheep, which there exhibited the simplicity of character so remarkable in the domestic species; but that where they had been often fired at, they were exceedingly wild, alarmed their companions on the approach of danger by a hissing noise, and scaled the rocks with a speed and agility that baffled pursuit. He lost several that he had mortally wounded, by their retiring to die among the secluded precipices." They are, we are farther informed on the authority of DRUMMOND, in the habit of paying daily visits to certain caves in the mountains that are encrusted with saline efflorescence. The same gentleman mentions that the horns of the old rams attain a size so enormous, and curve so much forwards and downwards, that they effectually prevent the animal from feeding on the level ground.

All our travellers who have tasted the flesh of the Rocky Mountain Sheep, represent it as very delicious when in season, superior to that of any species of deer in the west, and even exceeding in flavour the finest mutton.

We have often been surprised that no living specimen of this very interesting animal has ever been carried to Europe, or any of our Atlantic cities, where it would be an object of great interest.

GEOGRAPHICAL DISTRIBUTION.

This animal is found, according to travellers, as far to the North as lat. 68, and inhabits the whole chain of the Rocky Mountains on their highest peaks down to California. It does not exist at Hudson's Bay, nor has it been found to the eastward of the Rocky Mountain chain

GENERAL REMARKS.

The history of the early discovery of this species, of specimens transmitted to Europe from time to time, obtained in latitudes widely removed from each other, of its designation under various names, and of the figures, some of which were very unnatural, that have been given of it, are not only interesting but full of perplexity. It appears to have been known to Father PICOLO, the first Catholic missionary to California, as early as 1697, who represents it as large as a calf of one or two years old; its head much like that of a stag, and its horns, which are very large, are like those of a ram: its tail and hair are speckled and shorter than a stag's, but its hoof is large, round, and cleft as an ox's. I have eaten of these beasts; their flesh is very tender and delicious." The Californian Sheep is also mentioned by HERNANDEZ, CLAVIGERO, and other writers on California. VANEGAS has given an imperfect figure of it, which was for a long time regarded as the

Siberian Argali. Mr. David Douglass, in the Zoological Journal, in April, 1829, describes a species under the name of *Ovis Californica,* which he supposed to be the sheep mentioned by Picolo. Cook, in his third voyage evidently obtained the skin of the Rocky Mountain Sheep on the north west coast of America. Mr. McGillivery, in 1823, presented to the New-York Museum a specimen of this animal, and published an account of it in the Medical Repository of New-York. This specimen being afterwards sent to France, a description and figure of it were published. Lewis and Clark, some years afterwards, brought male and female specimens to Philadelphia, which were figured by Griffith and Godman.

Several eminent naturalists, and among the rest Baron Cuvier, considered it the same as *Ovis Ammon,* supposing it to have crossed Behring's Straits on the ice. We have never had an opportunity of comparing the two species, but have examined them separately. Our animal is considerably the largest, and differs widely in the curvature of its horns from those of the eastern continent. We have no doubt of its being a distinct species from *Ovis Ammon.*

We doubt moreover, whether *Ovis Californica* will be found distinct from *Ovis Montana;* the climate in those elevated regions is every where cold. There are no intermediate spaces where the northern species ceases to exist, and the southern to commence, and when we take into consideration the variations of colour in different individuals, as also in the same individual in summer and winter, we should pause before we admit *Ovis Californica* as a true species. We have therefore added this name as a synonyme of *Ovis Montana.*

Plate LXXIII

On Stone by W. E. Hitchcock.

Brewer's Shrew Mole.

Printed & col. by Nagel & Weingärtner, 100 B'way N.Y.

Lith. Printed & Col. by J. T. Bowen, 1844.

SCALOPS BREWERI.— Bach.

Brewer's Shrew Mole.

PLATE LXXIV.

S. lanugine sericea, vellus obscure cinereo nigricans subtus fuscescens, palmæ anguste, cauda depressa, latus pilis hirsuta.

CHARACTERS.

Glossy cinereous black above, brownish beneath, palms narrow, tail flat, broad and hairy.

DESCRIPTION.

$$\text{Teeth, } \textit{Incisive } \frac{2}{4}\text{; } \textit{false molars } \frac{12}{12}\text{; } \textit{true molars } \frac{8}{6} = 44.$$

The head of *Scalops Breweri* is narrower and more elongated than that of *Sc. Aquaticus.* The cerebral portion of the skull is less voluminous, the inter-orbital portion is narrower, each of the intermaxillary bones in *Sc. Aquaticus* throws out a process, which projects upwards and forms the upper boundary of the nasal cavity, and very slightly separated by the nasal bones, whilst in *Sc. Breweri* these processes are shorter and scarcely project upwards above the plane of the nasal bone. Thus when we view the snout of *Sc. Aquaticus,* laterally, it is distinctly recurved at the tip, whereas in *Sc. Breweri* the upper surface is almost plain. But the most striking difference between these skulls is exhibited in the dentition, inasmuch as, in our present species, there are altogether forty-four teeth, in *Sc. Aquaticus* there are but thirty-six. Thus in the number of teeth *Sc. Breweri* resembles *Sc. Townsendi.*

The body of Brewer's Shrew Mole is perhaps a little larger than that of *Sc. Aquaticus.* Its snout is less flattened and narrower; its nostrils, instead of being inserted in a kind of boutir, as in the European *Talpa,* and the swine, or on the upper surface of the muzzle, as in the common shrew mole, are placed on each side, near the extremities of the nose. This species is pentadactylous, like all the rest of the genus, claws longer, thinner and sharper than the common shrew mole. Palm much narrower Its most striking peculiarity, however, is its tail, which, instead of being round and nearly naked, like that of *Sc. Aquaticus*, is flat and broad, resembling in some respects that of the Beaver, and is very thickly clothed, above and beneath, with long stiff hairs, which extend five lines beyond the vertebræ.

COLOUR.

The colour, above and beneath, is a glossy cinereous black, like velvet ; precisely similar to that of the European mole (*Talpa Europea*) with which we compared it. Under the throat there is a slight tinge of brown, the tail is ashy brown above, light beneath. The ewe is about one-third longer than that of the common shrew mole.

DIMENSIONS.

	INCHES.	LINES.
Length of the head and body	5	11
Tail vertebræ	1	0
Do. including fur	1	5
Breadth of tail	0	4
Do. of palm	0	4
Length of do to end of middle claw . . .	0	7

In the Museum of the Zoological Society of London there is a specimen obtained from the United States, which evidently is the same species. It is marked in the printed catalogue No. 145, "*Sc. Breweri* Bachman's M. SS." It however differs in having the fur more compact, and shorter, the colour somewhat darker, and in fact almost black. The hairs of the tail, instead of being brownish ash colour, are black, and the hind feet, instead of being covered above with brownish white hairs, as in our specimens, are brownish black.

DIMENSIONS OF THE SKULL OF THE ABOVE THREE SPECIES.

	LENGTH OF SKULLS.		WIDTH.	LENGTH OF PALATE.
	INCHES.	LINES.	LINES.	LINES.
Sc. Aquaticus . . .	1	4	8	7
S. Townsendi . . .	1	$7\frac{1}{4}$	$9\frac{1}{2}$	$8\frac{1}{5}$
S. Breweri . . .	1	3	$7\frac{1}{8}$	$6\frac{1}{2}$

HABITS.

In a collection of the smaller rodentia procured for us in New England by our friend THOMAS M. BREWER, Esq. an intelligent naturalist, we were surprised and gratified at finding this new species of shrew mole ; the specimen having been obtained by Dr. L. M. YALE, at Martha's Vineyard, an island on the coast of New England. In its habits it approaches much nearer the star-nosed mole (*Condylura cristata*) than any species of shrew mole. Its burrows are neither as extensive or so near the surface of the earth as those of the common shrew mole. We observed that the meadows in the valleys of Virginia, where this species is found, seldom exhibited any traces of their galleries, which are so conspicuous where the common species exists. We only possessed one opportunity of seeing this species alive. It ran across the public road near the red sulphur springs in Virginia; in its mode of progression it reminded us of the hurried, irregular and awkward manners of the common shrew mole. It had, as we ascertained, pursued its course under ground, at about five inches from the surface, until it reached the trodden and firm gravelly road, which it attempted to cross and was captured. It evidenced no disposition to bite. From the fact of our having seen three specimens, which were accidentally procured in a week, we were led to suppose that it was quite common in that vicinity. We have not found its nest, and regret that we have nothing farther to add in regard to its habits.

GEOGRAPHICAL DISTRIBUTION.

Our first specimen, as we have stated, was received from Martha's Vineyard. Our friend, the late Dr. WRIGHT, procured four specimens in the vicinity of Troy, N. Y. We obtained specimens in Western Virginia. It no doubt exists in all the intermediate country.

GENERAL REMARKS.

We suspect that this species has hitherto been overlooked in consequence of its having been blended with the common shrew mole. We observed two specimens in the museum of the Zoological Society, London, originally marked "*Talpa Europea* from America." On examining them, however, we found them of this species.

SOREX CAROLINENSIS.— Bach.

Carolina Shrew. Males and Females.

PLATE LXXV.

S. carolinensis, corpore griseo — cinerascente; cauda brevis, depressa.

CHARACTERS.

Carolina Shrew, with a short flat tail; ears not visible; body of a nearly uniform iron grey colour.

DESCRIPTION.

$$\textit{Intermediary incisors} \frac{2}{2}$$

$$\textit{Lateral incisors} \frac{5-5}{2-2}; \quad \textit{Molars}, \frac{5-5}{3-3} = 34.$$

The four front teeth are yellowish white, with their points deeply tinged with chesnut brown; all the rest are brown, a little lighter near the sockets. The upper intermediary incisors have each, as is the case in most other species of this genus, an obtuse lobe, which gives them the appearance of having a small tooth growing out from near the roots. The three lateral incisors are largest; the posterior ones very small; the first and fifth grinders are the smallest; the other three nearly equal. In the lower jaw the two first teeth are lobed; the lateral incisors are comparatively large, and crowded near the grinders. The molars are bristled with sharp points except the last, which is a tuberculous tooth.

The muzzle is moderately long and slender, and pointed with a naked deep lobed lip. The whiskers are composed of hairs apparently all white, a few of those situated in front of the eyes extending to the occiput, the rest rather short. There are no visible ears, even where the fur is removed; the auditory opening is an orifice situated far back on the sides of the head running obliquely. The orifice of the eye is so small that it can only be discovered by the aid of a good magnifying glass. The tail is flat, thickly covered with a coat of close hair, and terminated by a small pencil of hairs. The fore feet are rather broad for this genus, measuring a line and a half in breadth, resembling in some respects those of the shrew mole, (*Scalops canadensis.*) The toes are five, the inner a little shorter than the

Plate LXXV

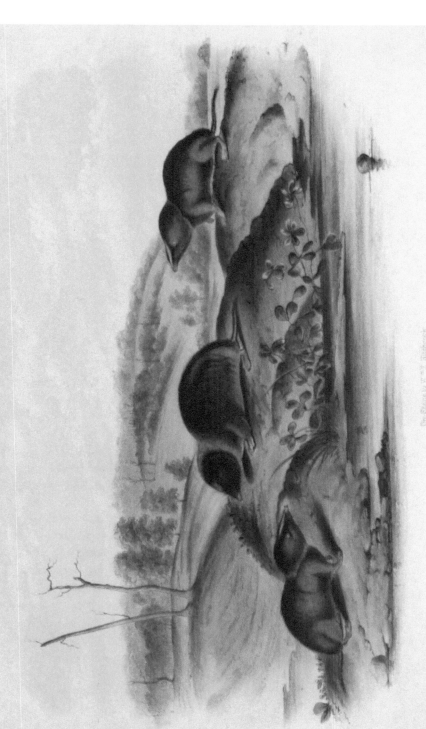

Drawn from Nature by J.J.Audubon, F.R.S.F.L.S.

On Stone by W.m E. Hitchcock.

Lith. Printed & Col.d by J.T. Bowen, Phil.

Carolina Shrew

outer one ; the third and fourth nearly equal. The nails are sharp, rather long, a little arched, but not hooked. The hind feet are more slender than the fore ones ; naked beneath, and covered above, as are also the fore feet, by a thin coat of short adpressed hairs.

COLOUR.

The fur presents the beautiful velvety appearance common to most species of this genus. The colour of the whole body is nearly uniform, considerably lustrous on the upper surface, and in most lights dark iron gray, rather darker about the head ; on the under surface the fur is of nearly the same general appearance, but is a shade lighter.

DIMENSIONS.

		INCHES.
Length of body	. . . .	3
" of tail	. . . .	$\frac{5}{8}$
" of head	. . . .	-1
" of palm to the end of nails	. . .	$\frac{5}{16}$
" of hind feet	. . . . .	$\frac{1}{2}$

HABITS.

It is difficult to know much of the habits of the little quadrupeds composing this genus. Living beneath the surface of the earth, feeding principally on worms and the larvæ of insects, shunning the light, and restricted to a little world of their own, best suited to their habits and enjoyments, they almost present a barrier to the prying curiosity of man. They are occasionally turned up by the plough on the plantations of the south, when they utter a faint, squeaking cry, like young mice, and make awkward and scrambling attempts to escape, trying to conceal themselves in any tuft of grass, or under the first clod of earth that may present itself. On two occasions, their small but compact nests were brought to us. They were composed of fibres of roots and withered blades of various kinds of grasses. They had been ploughed up from about a foot beneath the surface of the earth, and contained in one nest five, and in the other six young. In digging ditches, and ploughing in moderately high grounds, small holes are frequently seen running in all directions, in a line nearly parallel with the surface, and extending to a great distance, evidently made by this species. We observed on the sides of one of these galleries, a small cavity containing a hoard of coleopterous insects, principally composed of a rare species (*Scarabæus tityus*), fully the size of the animal itself; some of them were nearly consumed, and the rest mutilated. although still living.

GEOGRAPHICAL DISTRIBUTION.

This quadruped is found in various localities, both in the upper and maritime districts of South Carolina. We recently received specimens from our friend Dr. BARRETT, of Abbeville District ; and we have been informed by Dr. PICKERING, to whose inspection we submitted a specimen, and who pronounced it undoubtedly an undescribed species, that it had been observed as far north as Philadelphia.

Plate LXXVI

Drawn from Nature by J.W.Audubon.

On Stone by W.E.Hitchcock

Lith.ᵈ Printed & Colᵈ by J.T.Bowen.Philadᵃ

Moose Deer

CERVUS ALCES.—Linn.

Moose Deer.

PLATE LXXVI. Old Male and Young.

C. magnitudine Equi; capite permagno, labro auribusque elongatis: collo brevi, dense jubato, cornibus palmatis, cauda brevissima, vellere fusco cinereo, in nigrum vergente.

CHARACTERS.

Size of a horse. Head, very large; snout and ears, long; neck, short, with a thick mane. Horns spreading into a broad palm. Tail, short. Colour, blackish-gray.

SYNONYMES.

Elan, Stag, or Aptaptou. De Monts Nova Francia, p. 250. An. 1604.
Eslan ou Orinal. Sagard-Theodat, Canada, p. 749. An. 1636.
Orinal. La Hontan, Voy., p. 72. An. 1703.
Moose Deer. Dudley, Phil. Trans. No. 368, p. 165. An. 1721.
Orinal. Charlevoix. Nouv. France. Vol. v., p. 185. An. 1741.
    " Dupratz, Louis. Vol. i., p. 301.
Moose Deer. Pennant, Arct. Zool. Vol. i., p. 17, Fig. 1784.
Moose. Umfreville, Huds. Bay. An. 1790.
    " Herriot's Travels, 1807, Fig.
C. alces. Harlan. Fauna, p. 229.
    " Godman, Am. Nat. Hist., Vol. ii., p. 274.
The Elk. Hamilton Smith.
    " Griffith's Cuv., Vol. v., p. 303.
American Black Elk. Griffith's Cuv., Vol. iv., p. 72., plate of head.
Elk. In Nova Scotia, proceedings of the Zoological Society, 1849, p. 93.
Cervus alces. De Kay, N. Hist. N. Y., p. 115.

DESCRIPTION.

This is the largest of any known species of deer. Major Smith (Cuv. An. Kingdom, by Griffiths, Vol. iv., p. 73) says, "For us, who have the oppor-tunity of receiving the animal in all the glory of his full grown horns, amid the scenery of his own wilderness, no animal could appear more majestic or more imposing." Having ourselves on one occasion been favoured with a similar opportunity, when we had the gratification of bringing one down with a rifle and of examining him in detail as he lay before us, we confess he appeared awkward in his gait, clumsy and disproportioned in limbs, uncouth and inelegant in form, and possessing less symmetry and beauty than any other species of the deer family. His great size, enormous head, and face like a horse, and the thundering noise of the saplings bending and snapping around him as he rattled over the fallen logs, was to us the only imposing part of the spectacle. To do justice, however, to the description of the moose, by SMITH, who was a close observer and a naturalist of considerable attainments, we should quote his succeeding observations: "It is, however, the aggregate of his appearance which produces this effect; for when the proportions of its structure are considered in detail, they certainly will seem destitute of that harmony of parts which in the imagination produces the feeling of beauty."

The head forcibly reminds us of that of an enormous jackass; it is long narrow and clumsily shaped, by the swelling on the upper part of the nose and nostrils; the snout is long and almost prehensile—the muzzle extending four inches beyond the lower lip. The nostrils are narrow and long, five inches in length. The eye is deep-seated, and in proportion to the large head is small. The ears are long, 14 inches, heavy and asinine. The neck is very short, and is surmounted by a compact mane of moderate length composed of coarse rigid hairs. There is in both sexes a tuft of coarse hairs, resembling hog's bristles, beneath the throat, which is attached to a pendulous gland, more conspicuous in young than in old animals; this gland with the attached hair is ten inches long. The horns, which are found only on the males, are, when a year old, merely short knobs; they increase in size after each annual shedding, and after the fourth year become palmated, and may be termed full grown about the fifth year. The palms on the horns of the Moose are on the widest part on a moderate-sized male about 11 inches wide. The space between the roots, $6\frac{1}{2}$ inches; greatest breadth at the root, $6\frac{1}{2}$ inches; from the root to the extremity, measuring around the curve, 2 feet 10 inches. The first branch or prong on the inner side of the horn commences nine inches from the base. It here divides into two branches, one being ten and the other eleven inches in length, measuring in a curve from the root to the largest point 25 inches. These two prongs on each side incline forward, are almost round, and are pointed like those of elk horns. The palms on the main branches of the horns not only differ in different individuals, but do not often correspond on the head of the same animal. In the specimen from which we are describing, the lower and longest point on the palm is on one side 12 inches, and on the corresponding one on the opposite side only 4 inches; on the remainder of the palm there are on one side six points, on the other seven; the palm is about half-an-inch in breadth at the centre, thickening towards the base to one inch.

The horns are irregularly and slightly channelled, and are covered with whitish marks on the front surface, somewhat resembling the channels and irregular windings of grubs or sawyers between the bark and wood in old decayed trunks of trees; on the posterior surface these marks in form bear considerable resemblance to veins in the leaves of ferns. The width across the horns measuring from the outer tips rises 3 feet 4 inches; weight of the horns, 42 pounds.

The nose, including the nostrils, is thickly clothed with short hair—a triangular spot on the nose bare. The hair on the mane is coarse and compact, 10 inches in length; both surfaces of the ears are covered with dense hairs.

The outer hair is throughout coarse and angular; it is longer on the neck and shoulders than on any other part of the body: under these long hairs there is a shorter, woolly, more dense and finer coat.

COLOUR.

The teeth are white; horns brownish yellow, the extremities of the prongs becoming yellowish white. The eyes are black; nose, forehead and upper lip, yellowish fawn; inner surface of ears, yellowish white; outer surface, grayish brown. Sides of head, yellowish brown. On the neck, dark grayish brown, composed of hairs that are white, black and yellow; under the chin, yellowish brown. Hairs on the appendage under the throat, black; lower lip and chin, dark gray, formed of a mixture of white and black hairs; the softer, shorter hairs on the body are ashy gray; the long hairs when examined separately are whitish at the base, then cinereous and tipped with black, giving it a brownish black appearance.

On the under surface of the body the colour is considerably lighter than on the back, having a tinge of yellowish white; under surface of the tail, ashy white. The young animals, for the first winter, are of a reddish brown colour; individuals even of the same age often differ in colour, some being darker than others, but there is always a striking difference between the summer and winter colours, the hairs in winter becoming darker; as the moose advances in age, the colour continues to deepen until it appears black; thence it was named by HAMILTON SMITH, not inappropriately as regards colour, "the American Black Elk"

DIMENSIONS.

	Feet.	Inches.
From point of nose to root of tail, - - - -	6	11
Tail (vertebræ), - - - - - - -		8
Tail to end of hair, - - . - - - -		9½
From shoulder to point of hoof, - - -	4	6
Height of ear, - - - - - - -	1	2
From point of nose to interior canthus of eye, - -	1	10

Weight of horns, 56 pounds.

Weight of the whole animal, from 800 to 1200 pounds.

Dimensions of a Male procured in Ontario County, N. Y., in 1806.

	Feet.	Inches.
Length from point of nose to root of tail, - - -	7	2
" of tail, - - - - - - -		11
Height at shoulder, - - - - - - -	5	00
Width of horns at tip, - - - - - -	2	3
Widest part, - - - - - - . -	3	1

Weight of horns, 69 pounds.

HABITS.

We were favoured by MR. KENDALL, of the Literary Society of Quebec, with the following account of the Moose Deer, with which we will begin our article on this noble quadruped.

" The Moose are abundant to the north of Quebec and in the northern parts of the state of Maine. In the neighbourhood of Moose River and the lakes in its vicinity, they are very abundant. In the summer they are fond of frequenting lakes and rivers, not only to escape the attacks of insects which then molest them, but also to avoid injuring their antlers, which during their growth are very soft and exquisitely sensitive, and besides, such situations afford them abundance of food.

" They there feed on the water-plants, or browse upon the trees fringing the shores. In the winter they retire to the dry mountain ridges, and generally 'yard', as it is termed, on the side facing the south, where there are abundance of maple and other hard-wood trees upon which to feed, either by browsing on the tender twigs or peeling the bark from the stems of such as are only three or four inches in diameter. Their long, pendulous upper lip is admirably adapted for grasping and pulling down the branches, which are held between the fore legs until all the twigs are eaten. They peel off the bark by placing the hard pad on the roof of the mouth against the tree, and scraping upwards with their sharp, gouge-like teeth, completely denuding the tree to the height of seven or eight feet from the surface of the snow. They remain near the same spot as long as any food can be obtained, seldom breaking fresh snow, but keeping to the same tracks as long as possible.

"The antlers begin to sprout in April, and at first appear like two black knobs. They complete their growth in July, when the skin which covers them peels off and leaves them perfectly white; exposure to the sun and air, however, soon renders them brown. When we consider the immense size to which some of them grow in such a short period of time, it seems almost incredible that two such enormous excrescences could be deposited from the circulating system alone; the daily growth is distinctly marked on the velvety covering by a light shade carried around them. The first year the antlers are only about one inch long; the second year, four or five inches, with perhaps the rudiment of a point; the third year about nine inches, when each divides into a fork still round in form: the fourth year they become palmated, with a brow antler and three or four points: the fifth season they have two crown antlers and perhaps five points; the points increasing in size each year, and one or two points being added annually, until the animal arrives at its greatest vigour; after which period they decrease in size and the points are not so fully thrown out. The longest pair I ever met with had eighteen points, (others have seen them with twenty-three points,) they expanded five feet nine inches to the outside of the tips; the breadth of palm, eleven inches without the points; circumference of shaft, clear of the burr, nine inches; weight, seventy pounds! The old and vigorous animals invariably shed them in December; some of four and five years old I have known to carry them as late as March, but this is not often the case.

"The rutting season commences in September; the males then become very furious, chasing away the younger and weaker ones. They run bellowing through the forest, and when two of equal strength meet, have dreadful conflicts, and do not separate until one or both are severely injured. I bought a pair of antlers from a Penobscot Indian, with one of the brow antlers and the adjoining prong broken short off. The parts were at least 1½ inches in diameter, and nearly as hard as ivory. At that season they are constantly on the move, swimming large lakes and crossing rivers in pursuit of the female.

"The female brings forth in May. The first time she produces one fawn, but ever afterwards two. It is supposed by hunters that these twins are always one a male and the other a female.

"In summer the hair of the Moose is short and glossy—in winter long and

very coarse, attached to the skin by a very fine pelicle, and rendered warm
by a thick coat of short, fine wool. The hair on the face grows upwards
from the nose, gradually turning and ending in a thick, bushy tuft under the
jaws. The young males have generally a long, pendulous gland, growing
from the centre of this tuft, and covered with long hair, sometimes a foot
long.

"Their flesh is very coarse, though some people prefer it to any other;
it is apt to produce dysentery with persons unaccustomed to use it.
The nose or *moufle*, as it is generally called, if properly cooked is a very
delicious morsel. The tongue is also considered a delicacy; the last entrail
(called by hunters the bum-gut) is covered with round lumps of suety fat,
which they strip off and devour as it comes warm from the animal, without any cooking. Also the marrow warm from the shanks is spread upon
bread, and eaten as butter. I must confess that the disgusting luxury
was rather *too rich* to tempt me to partake of it. I have seen some
officers of the Guards enjoy it well enough !

"The seasons for hunting the Moose are March and September. In
March, when the sun melts the snow on the surface and the nights are
frosty, a *crust* is formed, which greatly impedes the animal's progress, as
it has to lift its feet perpendicularly out of the snow or cut the skin from
its shanks by coming in contact with the icy surface.

"It would be useless to follow them when the snow is soft, as their
great strength enables them to wade through it without any difficulty.
If you wish to see them previous to shooting them from their " yard," it is
necessary to make your approach to leeward, as their sense of smelling
and hearing is very acute: the crack of a breaking twig will start them,
and they are seldom seen any more, until fatigue compels them to knock
up, and thus ends the chase. Their pace is a long trot. It is necessary to have two or three small curs (the smaller the better), as they can run
upon the snow without breaking through the crust; their principal use
is to annoy the Moose by barking and snapping at their heels, without
taking hold. A large dog that would take hold would be instantly
trampled to death. The males generally stop, if pressed, and fight with
the dogs; this enables the hunter to come up unobserved and dispatch
them. Sometimes they are killed after a run of an hour, at other times
you may run them all day, and have to camp at night without a morsel
of provisions or a cloak, as everything is let go the moment the Moose
starts, and you are too much fatigued to retrace your steps to procure
them. Your only resource is to make a huge fire, and comfort yourself
upon the prospect of plenty of Moose-meat next day. As soon as the
animal finds he is no longer pursued, he lies down, and the next morning

he will be too stiff to travel far. Generally, a male, female, and two fawns
are found in a 'yard.'

"When obliged to run, the male goes first, breaking the way, the others
treading exactly in his tracks, so that you would think only one has
passed. Often they run through other 'yards,' when all join together,
still going in Indian file. Sometimes, when meeting with an obstacle they
cannot overcome, they are obliged to branch off for some distance and
again unite ; by connecting the different tracks at the place of separation
you may judge pretty correctly of their number. I have seen twelve
together, and killed seven of them.

A method of hunting this animal is as follows :

" In September, two persons in a bark canoe paddle by moonlight along
the shore of the lake imitating the call of the male, which, jealous of the
approach of a stranger, answers to the call and rushes down to the combat. The canoe is paddled by the man in the stern with the most death-
like silence, gliding along under the shade of the forest until within short
shooting distance, as it is difficult to take a sure aim by moonlight : the
man in the bow generally fires, when if the animal is only wounded,
he makes immediately for shore, dashing the water about him into foam :
he is tracked by his blood the next day to where he has lain down, and where
he is generally found unable to proceed any further. Many are killed in
this manner in the neighbourhood of Moose River every season.

Hunters sometimes find out the beaten tracks of the Moose (generally leading to the water), and bend down a sapling and attach to it a strong hempen
noose hanging across the path, while the tree is confined by another cord and
a sort of trigger. Should the animal's head pass through the dangling snare,
he generally makes a struggle which disengages the trigger, and the tree
springing upward to its perpendicular, lifts the beast off his legs, and he
is strangled !"

Mr. JOHN MARTYN, of Quebec, favoured us with the following notes on
the Moose deer: "This animal in the neighbourhood of this city (Quebec)
is mostly found in the hard woods during the winter. At this season several associate together and form groups of two, three, or four, and make
what is called 'a yard,' by beating down the snow ; and whilst in such
places they feed on all the branches they can reach, and indeed even strip
the trees of their bark, after which they are forced to extend their 'yards,'
or remove to some other place, but rather than leave the first, they will
even break branches as large as a man's thigh. In skinning off the bark,
the animal places its upper lip firmly against it, whether upward, downward or sideways, and with its teeth, which are all on its lower jaw,

takes a firm hold and tears it away in strips more or less long and broad, according to the nature of the bark of the tree.

It is ascertained by the hunter whether a Moose has been lately or not in its yard, by removing the surface of the snow from around the foot of the trees already barked above, and if they have been barked below the surface of the snow, the animal has left the spot for sometime, and it is not worth while to follow any of its tracks. The contrary, of course, takes place with different observations. At this season the female is generally accompanied by two of her calves, one two years old and generally a bull, the other the calf of the preceding spring.

These animals vary much in their colour, some being grayish brown, and others nearly black. The grayish Moose is generally the largest, often reaching the height of seven or eight feet. The females receive the males in the month of October, and at this period the latter are excessively vicious and dangerous when approached, whilst the females evince the same fierceness at the time of having calves. In some instances during the rutting season, when two males accidentally meet, they fight prodigiously hard, tearing up the earth beneath for yards around, and leaving marks of blood sufficient to prove that their encounter has been of the severest nature.

Their usual mode of defence consists in striking at their enemies with their forefeet; but in fighting with each other the males use both feet and horns, and they have sometimes been killed with marks of old wounds about their head and other parts of the body. As an instance of the force with which the Moose strikes, the following anecdote may be related: a bull-terrier in attempting to seize one by the nose, was struck by the animal with its forefoot, and knocked off to a distance of twenty feet; the dog died next day.

The Moose deer frequently turn against the hunters, even before being shot at or in the least wounded. They walk, trot, and gallop, and can leap a great distance at a single bound; like other species of deer they bend their bodies very low at times, to pass beneath branches of fallen trees, not even half their height from the earth. When pursued, they enter the most tangled thickets, and pass through them as if not feeling the impediments, the brushwood, fallen logs, &c., opposed to the hunter's progress. The calves when born are about the size of a few days old colt, but are more slender, and look very awkward on account of their apparent disproportionate long and large legs. When caught at three months old, they eat leaves, &c.; but how long they are suckled by their dam we have not been able to ascertain.

" During the summer they frequently resort to the shores of rivers, creeks or lakes, on the margins of which their tracks are seen, like those of common cattle; they enter the water and immerse their bodies to save themselves from the bites of flies, &c.

In all probability, where wolves are yet abundant, these are their most dangerous enemies besides man; but at the present time, few of these rapacious animals are to be found in the neighbourhood of Quebec. The Moose deer are frequently killed while in the water, or on the shores of some pond, lake or river; but when their young are with them, they will run and chase the hunter, and it is sometimes difficult for him to escape, unless he is so fortunate as to shoot and bring them down.

"The flesh is considered very good, especially the *moufflon*, which forms the upper lip, and is very rich, juicy and gelatinous. This is cleaned and dressed in the same manner as 'calves' head.' The hunters salt their meat for winter use. The steaks are as good as beef steaks: but the Moose are not generally fat, although their flesh is juicy and at times tender. The young at the age of twelve months are never tough, and their flesh is preferable to that of the old beasts. The inside of the mouth above, or palate, is extremely hard, and lays in folds, giving this animal the power of gripping (seizing) the bark or the branches of trees, by which means it tears them off with ease. This pad is placed immediately beneath the extremity of the *moufflon*, and is about two inches long.

"These animals feed principally on the birch, the *moose-wood*, the aspen, and various kinds of leaves and grasses; in captivity they eat hay and other dry food, even hard ship-biscuit. The females are called 'cows,' the males 'bulls,' and the young 'calves.' Their droppings resemble those of the deer kind. Although the Moose swim well they are not known to dive, they swim with the head and part of the neck above water, like cattle. When pursued in boats they frequently attempt to upset them, and at times open their mouths and make a loud snorting noise, striking at the same time with their forefeet, and occasionally sink the canoes of the Indians or hunters. Upon one occasion, a young man going fishing, and having his fowling-piece along, on turning a point of a lake, saw a large Moose in the water and fired at it with shot, tickling it severely. The Moose at once made for the canoe; and whilst the alarmed fisherman was attempting to escape, his boat became entangled in the branches of a fallen tree, when he was forced to give up the canoe and get away as he best could; the animal on reaching the boat completely demolished it. Unfortunately, the females are sometimes killed when they are with calf. They do not generally make any noise in the woods, unless when provoked, but in captivity they utter a plaintive sound, much resembling that made by the black bear,

They never are seen on the ice like the rein-deer; it would seem by the formation of their hoofs that they might walk well on the rocks, or on the ice, but they keep in the woods, and when walking over snow their feet usually sink into it until they reach the earth.

"A Mr. Bell, residing at Three Rivers, has a Moose which has been taught to draw water in a cart or in a sleigh during winter, but there is no possibility of working it during the rutting season. We have never heard of any attempt to ride on the Moose deer. Their horns, which are large, palmated, and heavy, are dropped in the months of December and January, begin to show again in the latter part of March, and in two months or thereabouts attain their full size. When covered over with 'velvet,' as it is called, they are very curious. A pair of good Moose horns sells at the high price of twenty dollars! The velvet is scraped off against trees and bushes in the manner employed by our Virginian deer. Horns have been measured when reversed and standing on the ground four feet seven inches, and ordinary pairs often measure five feet and up-wards.

"It is said that the Moose can smell at a very great distance, and that the moment they scent a man or other enemy they make off and are not easily overtaken. On the first glimpse of man, if they are lying down they rise to their feet and are off at once, and often before they are observed by the hunter. When closely pursued, they turn and make a dash at the enemy, scarcely giving him time to escape, and the hunter's best plan in such cases is to keep cool and shoot the animal as it rushes towards him, or if unpre-pared, he had best ascend a tree with all convenient dispatch. Sometimes the hunter is obliged to save himself by dodging around a tree, or by throwing down some part of his dress, upon which the Moose expends his fury, trampling on it until torn to tatters.

" Moose-hunting is followed by white or red skinned hunters in the same manner. He, however, who has been born in the woods, possesses many advantages over the 'civilized' man. The white hunters generally pro-vide themselves, previous to their starting, amply with provisions and ammunition to last them about three weeks, and sometimes go in a sleigh. The guns used are mostly single-barrelled, of ordinary size, but suited for shooting balls as well as shot,—rifles are rarely used in Canada. After leaving the settlements, the first day's journey takes them ten or twelve miles, when they select a proper place in a snowy district, as near a stream as possible.

" If the weather is fine, they cut down trees and make a camp, some of the party provide water, and others light the fires and clear off the snow for yards around, whilst evergreen trees are stripped of their branches to make up a floor and covering for them in their temporary shelter. The hunters having made all snug, cook their meat and eat it before a fire that illuminates the woods around, and causes the party to appear like a set of goblins through the darkness of night. On many such occasions the *bedding* is singed, and per chance a whisker! The feet may be partially roasted, whilst the shoulders, the hands, and probably the nose, are suffering greatly from the severity of the weather, for the thermometer may be occasionally thirty degrees below zero! The march to this spot is frequently made on snow-shoes, which are taken off, however, whilst the party are forming the encampment, clearing away the snow, and making a path to the water, which being covered with snow and ice, requires to be got at by means of shovels and axes. Before daylight, the kettles are put on the fires, tea and coffee are made, breakfast swallowed in a few moments, and the party on foot, ready to march toward the hunting-ground. On the way, every one anxiously looks out for tracks of the game, and whether hares or grouse come in the way they are shot and hung up on the trees; but if game of any kind has been thus hung up by others, whether Indians or white hunters, the party leaves it sacredly untouched—for this is the etiquette of the chase throughout this portion of country. When they at last reach the ground, the party divide, and seek for the Moose in different directions. It is agreed that no one shall shoot after separating from the rest, unless it be at the proper game, and also that in case of meeting with Moose, or with fresh signs, they are to return, and make ready to proceed to the spot together next day. Sometimes, however, this rule is broken through by some one whose anxiety (excitement) at sight of a Moose makes him forget himself and his promise. As soon as a 'yard' has been discovered, all hands sally forth, and the hunt is looked upon as fairly begun. If on approaching the 'yard,' their dogs, which are generally mongrels of all descriptions, start a Moose, the hunters, guided by their barking and the tracks of the pack and the Moose through the snow, follow with all possible celerity. The dogs frequently take hold of the Moose by the hind legs, the animal turns, and stands at bay, and the hunters thus have an opportunity to come up with the chase.

"On approaching, when at the proper distance (about sixty to eighty yards) the nearest man takes a decided aim, as nearly as possible under the forearm and through the neck, and fires, or, if fronting the beast, in the centre of the breast.

"If wounded only, the second hunter fires also, and perhaps the third, and the animal succumbs at last, though it sometimes manages to run, stumble, and scramble, for miles. After skinning the Moose, the heart and liver, and the marrow-bones, are taken out, and a good large piece of the flesh is taken to 'camp,' and is speedily well cooked and placed smoking hot before the hungry hunters. After killing all the Moose of a 'yard' or that they can find near their camp, the party pack up their material, break up the camp, and return home.

It not unfrequently happens, that a wounded Moose, or even one that has not been wounded, will turn upon the hunter, who then has to run for his life, and many instances of such incidents are related, including some hair-breadth escapes. One of these I will relate: Two Indians being on a hunt and having met with the game, one of them shot, and missed; the Moose turned upon him, and he fled as fast as he could, but when about to reach a large tree, from behind which he could defy his opponent, his snow shoes hooked in some obstacle and threw him down. The Moose set upon him furiously and began trampling on him, but the Indian drew out a knife, and succeeded in cutting the sinews of the forelegs of the animal, and finally stabbed him so repeatedly in the belly that he fell dead, but unluckily fell on the prostrate hunter, who would have been unable to extricate himself, had not his companion come to his assistance. The poor man, however, had been so much injured that he never recovered entirely, and died about two years afterwards.

During some seasons the snows are so deep, and at times so soft, that the Moose cannot go *over* the snow, but have to make their way through it, giving a great advantage to the hunters, who, on broad snow-shoes can stand or run on the surface without much difficulty. On one occasion of this nature a Moose was seen, and at once followed. The poor animal was compelled to *plough* the snow, as it were, and the hunters came up to it with ease, and actually placed their hands on its back. They then endeavoured to drive it towards their camp and secure it alive. The Moose, however, would not go in the proper direction, and they finally threw it down, and attempted to fasten its legs together; but as they had no ropes, and could not procure any better substitute for them than withes, the beast got away, and after a long chase they, being very much fatigued, shot it dead. When the snow is thus soft, the Moose deer has been known to evade the hunters by pushing ahead through tangled thickets, more especially *hackmetack* and briary places which no man can go through for any length of time without extreme labour. The Indians, however, will follow the Moose in such cases day and night, provided the moon is shining, until the animal is so fatigued that it can be overtaken and killed with ease. Instances have been known where as many as five have been killed in one day by two Indians. The Moose is not unfrequently caught in the following manner: A rope is passed over a horizontal branch of a tree, with a large noose and slip-knot at one end, whilst a heavy log is attached to the other, hanging across the limb or branch, and touching the ground. The Moose, as it walks along, passes its head through the noose, and the farther it advances, the tighter it finds itself fastened, and whilst it plunges terrified onwards, the log is raised from the ground until it reaches the branch, when it sticks, so that no matter in what manner the Moose moves, the log keeps a continued strain, rising and falling, but not giving the animal the least chance to escape, and at last the poor creature dies miserably. They are also 'pitted' at times, but their legs are so long, that this method of *securing* them seldom succeeds, as they generally manage to get out."

The Moose is well known to travellers who have crossed the Rocky Mountains, where this animal is principally called by the French name, " L'Orinal."

Whilst at Quebec, in 1842, we procured the head and neck of a very large male, (handsomely mounted) ; which was shot in the state of Maine, where the Moose is still frequently found.

Moose deer are abundant in Labrador, and even near the coast their tracks, or rather paths, may be seen, as distinctly marked as the cow-paths about a large stock-farm. In this sterile country, where the trees are so dwarfish that they only deserve the name of shrubs, and where innumerable barren hills arise, with cold clear-water ponds between, the Moose feeds luxuriously on the scanty herbage and the rank summer grasses that are found on their sides; but in winter the scene is awfully desolate, after the snows have fallen to a great depth ; the whistling winds, unimpeded by trees or forests, sweep over the country, carrying with them the light snow from the tops and windward sides of the hills in icy clouds, and soon forming tremendous drifts in the valleys. No man can face the storm-driven snows of this bleak, cold country ; the congealed particles are almost solid, and so sharp and fine that they strike upon the face or hands like small shot ; the tops of the hills are left quite bare and the straggling Moose or rein deer seek a precarious supply of mosses along their sides. At this season the Moose sometimes crosses the Gulf of St. Lawrence, on the ice to Newfoundland, or follows the coast towards the shore opposite Nova Scotia, and there passes the Gulf and wanders into more woody and favoured regions for the winter.

The following is from our friend S. W. RODMAN, Esq., of Boston, an excellent sportsman, and a lover of nature, to whom we are indebted for many kindnesses.

" Our party was returning from lake Miramichi, about the middle of July, by the marshy brook, which connects it with the Miramichi river. The canoe men were poling slowly and silently, in order not to disturb the numerous ducks which breed in those uninhabited solitudes, as we were anxious to vary our constant fish diet ; salmon either boiled or " skinned " being set before us morning, noon and night. We had not fired a gun to disturb the silence. My own and my brother's canoes were close together, when I saw an animal suddenly spring on to its feet from the long marshy grass about forty yards in advance of us. I said quickly " Cariboo," " Cariboo," "stoop low ;" which we all did and continued moving on. It was about the size of a yearling heifer, but taller, of a bright, light, red colour, with long ears pricked forward, and a large soft eye ; and stood perfectly still, looking at us. We had gone perhaps ten yards, when there appeared from the long grass by its side, first the ears, then the huge head and muffle of an old cow Moose, the first one being as I now knew her calf, of perhaps four or five months old. She gradually rose to her knees, then sat upon her haunches, and at last sprang to her feet, her eyes all the time intently fixed upon us. The calf in the meanwhile had moved slowly off. At this moment we both fired without any apparent effect, the shot being too light to penetrate the thick hide. She turned instantly, showing a large and apparently well filled udder, struck into the tremendous trot, for which the Moose is so celebrated, crossed the deep brook almost at a stride, then the narrow strip of meadow, and disappeared, crashing through the alders which intervened between the meadow and the dark evergreen forests beyond.

Our oldest woodsman, Porter, assured us that she was one of the largest of her kind, and that it was rare good fortune to approach so near to this noblest denizen of our northern forests. We were much gratified, but our regret as sportsmen was still greater, at not having been prepared to take advantage of such an opportunity as will probably never again occur to either of us. We constantly both before and afterwards saw the tracks of cariboo and Moose about our camps."

GEOGRAPHICAL DISTRIBUTION.

Capt. Franklin, in his last expedition, states that several Moose were seen at the mouth of Mackenzie River, on the shores of the Arctic Sea, in latitude 69°. Farther to the eastward towards the Copper-mine River, we are informed by Richardson, they are not found in a higher latitude than 65°. Mackenzie saw them high up on the eastern declivity of the Rocky Mountains, near the sources of the Elk River ; Lewis and Clark saw them at the mouth of the Oregon. To the east they abound in Labrador, Nova Scotia, New-Brunswick, and Lower Canada. In the United States they are found in very diminished numbers in the unsettled portions of Maine and at long intervals in New-Hampshire and Vermont. In the state of New-York, according to the observations, made by Dr. Dekay, (Nat. Hist. N. Y., p. 117), which we believe strictly correct, they yet exist in Herkimer, Hamilton, Franklin, Lewis and Warren counties, and their southern limit along the Atlantic coast is 43° 30'.

GENERAL REMARKS.

We have considerable doubts whether our Moose deer is identical with the Scandinavian elk (*Cervus alces*, of authors), and have therefore not quoted any of the synonymes of the latter, but having possessed no favourable opportunities of deciding this point, we have not ventured on the adoption of any of the specific names which have from time to time been proposed for the American Moose.

Plate LXXVII

On Stone by Wm E. Hitchcock.

Drawn from Nature by J. W. Audubon.

Lith. Printed & Col^d by J. T. Bowen, Phil.

Prong-Horned Antelope.

GENUS ANTILOCAPRA.—Ord.

DENTAL FORMULA.

$$Incisive \; \frac{0}{-}; \quad Canine \; \frac{0-0}{0-0}; \quad Molar \; \frac{6-6}{6-6} = 32.$$

Horns common to both sexes; small in the female; horns persistent, greatly compressed, rough, pearled, slightly striated, with an anterior process, and the point inclining backwards; eye large; no suborbital sinus; no inguinal pores; no muzzle; facial line, converse; no canines; no succentorial hoofs; tail very short; hair stiff, coarse, undulating, flattened; female, mammæ.

Habit, peaceable, gregarious, herbivorous, confined to North-America.

Only one well determined species belongs to this genus.

The generic name *Antilocapra*, is derived from the two genera *Antilope* and *Capra*, Goat Antelope.

ANTILOCAPRA AMERICANA.—Ord.

PRONG-HORNED ANTELOPE.

PLATE LXXVII. Male and Female.

Cornibus pedalibus compressis, intus planis, antiæ granulatis striatisque propugnaculo compresso procurvo cum cornum parte posteriore retrorsum uncinata furcam constitutiente; colore russo fuscescente, gutture, cluni-umque disco albis: statura, Cervus Virginianus.

CHARACTERS.

Horns compressea, flat on the inner side, pearled and striated, with a com-pressed snag to the front; colour, reddish dun; throat and disk on the but-tocks, white. Size of the Virginia deer.

SYNONYMES.

TEUTHLAMACAMÆ. Hernandez, Nov.-Hispan, p. 324, fig. 324. An. 1651.
LE SQUENOTON. Hist. d'Amerique, p. 175. An. 1723.
SQUINATON. Dobb's, Hudson's Bay, p. 24. An. 1744.
ANTILOPE, CABRE OR GOAT. Gass Journal, pp. 49, 111.
ANTILOPE. Lewis and Clarke Journ., Vol. i., pp. 75, 208, 396; Vol. ii., p. 169.
ANTILOPE AMERICANA. Ord, Guthrie's Geography. 1815.
CERVUS HAMATUS. Blainville, Nouv-Ball. Society. 1816.
ANTILOCAPRA AMERICANA. Ord, Jour. de Phys., p. 80. 1818.
ANTILOPE FURCIFER. C . Hamilton Smith, Lin. Trans., Vol. xiii., plate 2. An. 1823.
ANTILOPE PALMATA. Smith, Griffith, Cuv., Vol. v., p. 323.
ANTILOPE AMERICANA. Harlan Fauna, p. 250.
       "   Godman, Nat. Hist., Vol. ii., p. 321
ANTILOPE FURCIFER. Richardson, F. B. A., p. 261, plate 21.

DESCRIPTION.

The Prong-horned Antelope possesses a stately and elegant form, and resembles more the antelope than the deer family. It is shorter and more compactly built than the Virginia deer ; its head and neck are also shorter and the skull is broader at the base. The horns of the male are curved upwards and backwards with a short triangular prong about the centre, inclined inwards, not wrinkled. Immediately above the prong the horn diminishes to less than half the size, below the prong the horn is flat and very broad, extremity of the horn sharp and pointed, and of the prong blunt. There are irregular little points on the horns of the male, two or three on each side. One specimen has two on the inside of each horn and one on the outside irregularly disposed.

Nostrils large and open, placed rather far back, eyes large and prominent, ears of moderate size, acuminate in shape ; on the back of the neck in winter specimens there is a narrow ridge of coarse hairs resembling a short mane. In summer there only remains of this mane a black stripe on the upper surface of the neck; eyelashes profuse; there is no under-fur. The hairs are of a singular texture, being thick, soft, wavy and slightly crimped beneath the surface : they are brittle, and when bent do not return to their original straight form, interiorly they are white, spongy and pithy; scrotum pendulous. There is not the slightest vestige of any secondary hoofs on either of its fore or hind legs, such as are seen in deer and other animals. The hoofs are strong and compact, small and diminishing suddenly to a point.

COLOUR.

The nose is yellowish brown, eye lashes black, the orbits with a blackish brown border, outer edge and points of the ears brownish black. There is a white band about two inches wide in front of and partly encircling the throat, narrowing to a point on each side of the neck; beneath this is a brown band about the same breath, underneath which is a grayish white spot of nearly a triangular shape; this is formed by a patch on each side of the throat of yellowish brown. The chest, belly, and sides to within five or six inches of the back are grayish white. A large light-coloured patch of nine inches in breadth exists on the rump, similar to that on the Rocky Mountain sheep and the elk. This whitish patch is separated by a brown-yellowish line, running along the vertebræ of the back to the tail. Legs, pale brownish yellow, approaching to dull buff colour, all the upper surface yellowish brown; under jaw and cheek, pale or grayish white; lips, whitish.

Female.—The female is a size smaller than the male. The neck is shorter. The form is similar, except that the markings are rather fainter; the brownish yellow which surrounds the different whitish or grayish white spots and bands being much paler than in the male. The horn is destitute of a prong; it is only three inches in length, nearly straight, and running to an acute point. The female possesses no mane

DIMENSIONS.

	Feet.	Inches.
From point of nose to root of tail, ...	4	2
Height, to shoulder from end of hoof, . . .	3	1
Length of ear, ...		4
Length of prong, ...		6

HABITS.

Reader, let us carry you with us to the boundless plains over which the prong-horn speeds. Hurra for the prairies and the swift antelopes, as they fleet by the hunter like flashes or meteors, seen but for an instant, for quickly do they pass out of sight in the undulating ground, covered with tall rank grass. Observe now a flock of these beautiful animals; they are not afraid of man—they pause in their rapid course to gaze on the hunter, and stand with head erect, their ears as well as eyes directed towards him, and make a loud noise by stamping with their forefeet on the hard earth; but suddenly they become aware that he is no friend of theirs, and away they bound like a flock of frightened sheep—but far more swiftly do the graceful antelopes gallop off, even the kids running with extraordinary speed by the side of their parents—and now they turn around a steep hill and disappear, then perhaps again come in view, and once more stand and gaze at the intruder. Sometimes, eager with curiosity and anxious to examine the novel object which astonishes as well as alarms them, the antelopes on seeing a hunter, advance toward him, stopping at intervals, and then again advancing, and should the hunter partly conceal himself, and wave his handkerchief or a white or red rag on the end of his ramrod, he may draw the wondering animals quite close to him and then quickly seizing his rifle send a ball through the fattest of the group, ere the timid creatures have time to fly from the fatal spot.

The Indians, we were told, sometimes bring the antelope to within arrow-shot (bow-shot), by throwing themselves on their backs and kicking up their heels with a bit of a rag fastened to them, on seeing which moving amid the grass the antelope draws near to satisfy his curiosity.

The atmosphere on the western prairies is so pure and clear that an antelope is easily seen when fully one mile off, and you can tell whether it is feeding quietly or is alarmed ; but beautiful as the transparent thin air shews all distant objects, we have never found the great western prairies equal the *flowery* descriptions of travellers. They lack the pure streamlet wherein the hunter may assuage his thirst—the delicious copses of dark, leafy trees ; and even the thousands of fragrant flowers, which they are poetically described as possessing, are generally of the smaller varieties; and the Indian who roams over them is far from the ideal being—all grace, strength and nobleness, in his savage freedom—that we from these descriptions conceive him.   Reader, do not expect to find any of the vast prairies that border the Upper Missouri, or the Yellow-Stone rivers, and extend to the Salt Lakes amid the Californian range of the Rocky Mountains, verdant pastures ready for flocks and herds, and full of the soft perfume of the violet.   No ; you will find an immense waste of stony, gravelly, barren soil, stretched before you ; you will be tormented with thirst, half eaten up by stinging flies, and lucky will you be if at night you find wood and water enough to supply your fire and make your cup of coffee ; and should you meet a band of Indians, you will find them wrapped in old buffalo robes, their bodies filthy and covered with vermin, and by stealing or begging they will obtain from you perhaps more than you can spare from your scanty store of necessaries, and armed with bows and arrows or firearms, they are not unfrequently ready to murder, or at least rob you of all your personal property, including your ammunition, gun and butcher knife !

The Prong-horned Antelope brings forth its young about the same time as the common deer : from early in May to the middle of June ; it has generally two fawns at a birth.   We have heard of no case in which more than that number has been dropped at a time, and probably in some cases only one is fawned by the dam.   The young are not spotted like the fawn of the common deer, but are of a uniform dun colour.   The dam remains by her young for some days after they are born, feeding immediately around the spot, and afterwards gradually enlarging her range ; when the young are a fortnight old they have gained strength and speed enough to escape with their fleet-footed mother from wolves or other four-footed foes. Sometimes, however, the wolves discover and attack the young when they are too feeble to escape, and the mother then displays the most devoted courage in their defence. She rushes on them, butting and striking with her short horns, and sometimes tosses a wolf heels over head, she also uses her forefeet, with which she deals severe blows, and if the wolves are not in strong force, or desperate with hunger, puts them to flight, and then seeks with her young a safer pasturage, or some almost inaccessible rocky hill side.

The rutting season of this species commences in September, the bucks run for about six weeks, and during this period fight with great courage and even a degree of ferocity. When a male sees another approaching, or accidentally comes upon one of his rivals, both parties run at each other with their heads lowered and their eyes flashing angrily, and while they strike with their horns they wheel and bound with prodigous activity and rapidity, giving and receiving severe wounds,—sometimes like fencers, getting within each others " points," and each hooking his antagonist with the recurved branches of his horns, which bend considerably inwards and downwards.

The Prong-horned Antelope usually inhabits the low prairies adjoining the covered woody bottoms during spring and autumn, but is also found on the high or upland prairies, or amid broken hills, and is to be seen along the margins of the rivers and streams : it swims very fast and well, and occasionally a herd when startled may be seen crossing a river in straggling files, but without disorder, and apparently with ease.

Sometimes a few of these animals, or even only one or two by themselves may be seen, whilst in other instances several hundreds are congregated in a herd. They are remarkably shy, are possessed of a fine sense of smell, and have large and beautiful eyes, which enable them to scan the surface of the undulating prairie and detect the lurking Indian or wolf, creep he ever so cautiously through the grasses, unless some intervening elevation or copsewood conceal his approach. It is, therefore, necessary for the hunter to keep well to *leeward*, and to use extraordinary caution in "sneaking" after this species ; and he must also exercise a great deal of patience and move very slowly and only at intervals, when the animals with heads to the ground or averted from him, are feeding or attracted by some other object. When they discover a man thus stealthily moving near them, at first sight they fly from him with great speed, and often retire to the broken grounds of the clay hills, from which they are not often tempted to stray a great distance at any time. As we have already mentioned, there are means, however, to excite the timid antelope to draw near the hunter, by arousing his curiosity and decoying him to his ruin.   The antelopes of the Upper Missouri country are frequently shot by the Indians whilst crossing the river ; and, as we were informed, preferred the northern side of the Missouri ; which, no doubt, arises from the prevalence on that bank of the river of certain plants, trees or grasses, that they are most fond of.   Males and females are found together at all seasons of the year.   We have been told that probably a thousand or more of these animals have been seen in a single herd or flock at one time, in the spring.

It was supposed by the hunters at Fort Union, that the prong-horned antelope dropped its horns ; but as no person had ever shot or killed one without these ornamental and useful appendages, we managed to prove the contrary to the men at the fort by knocking off the bony part of the horn, and showing the hard, spongy membrane beneath, well attached to the skull and perfectly immoveable.

The Prong-horned Antelope is never found on the Missouri river below *L'eau qui court ;* but above that stream they are found along the great Missouri and its tributaries, in all the country east of the Rocky Mountains, and in many of the great valleys that are to be met with among these extraordinary " big hills."   None of these antelopes are found on the shores of the Mississippi, although on the headwaters of the Saint Peter's river they have been tolerably abundant.   Their walk is a slow and somewhat pompous gait, their trot elegant and graceful, and their gallop or " run " light and inconceivably swift ; they pass along, up or down hills, or along the level plain with the same apparent ease, while so rapidly do their legs perform their graceful movements in propelling their bodies over the ground, that like the spokes of a fast turning wheel we can hardly see them, but instead, observe a gauzy or film-like appearance where they should be visible.

In autumn, this species is fatter than at any other period.   Their liver is much prized as a delicacy, and we have heard that many of these animals are killed simply to procure this choice morsel.   This antelope feeds on the short grass of the prairies, on mosses, buds, &c. ; and suffers greatly during the hard winters experienced in the north-west ; especially when the snow is several feet in depth.   At such times they can be caught by hunters provided with snow shoes, and they are in this manner killed, even in sight of Fort Union, from time to time.

It is exceedingly difficult to rear the young of this species ; and, although many attempts have been made at Fort Union, and even an old one caught and brought within an enclosure to keep the young company, they became furious, and ran and butted alternately against the picket-wall or fence, until they were too much bruised and exhausted to recover. WILLIAM SUBLETTE, Esq., of St. Louis, Missouri, however, brought with him to that city a female antelope, caught when quite young on the prairies of the far west, which grew to maturity, and was so very gentle, that it would go all over the house, mounting or descending the stairs, and occasionally going on to the roof of the building he lived in. This female was alive when we first reached St. Louis, but not being aware of its existence, we never saw it. It was killed before we left by a buck-elk, belonging to the same gentleman.

Whilst on our journey in the far west, in 1843, on one occasion, we had the gratification of seeing an old female, in a flock of eight or ten antelopes, suckling its young. The little beauty performed this operation precisely in the manner of our common lambs, almost kneeling down, bending its head upwards, its rump elevated, it thumped the bag of its mother, from time to time, and reminded us of far distant scenes, where peaceful flocks feed and repose under the safeguard of our race, and no prowling wolf or hungry Indian defeats the hopes of the good shepherd who nightly folds his stock of the Leicester or Bakewell breed. Our wild antelopes, however, as we approached them, scampered away ; and we were delighted to see that first, and in the van of all, was the young one !

On the 21st July, 1843, whilst in company with our friend, EDWARD HARRIS, Esq., during one of our hunting excursions, we came in sight of an antelope gazing at us, and determined to stop and try if we could bring him toward us by the trick we have already mentioned, of throwing our legs up in the air and kicking them about, whilst lying on our back in the grass. We kicked away first one foot and then the other, and sure enough, the antelope walked slowly toward us, apparently with great caution and suspicion. In about twenty minutes he had advanced towards us some two or three hundred yards. He was a superb male, and we looked at him for several minutes when about sixty yards off. We could see his fine protruding eyes ; and being loaded with buck-shot, we took aim and pulled trigger. Off he went, as if pursued by a whole Black-foot Indian hunting party. Friend HARRIS sent a ball at him, but was as unsuccessful as ourselves, for he only ran the faster for several hundred yards, when he stopped for a few minutes, looked again at us, and then went off, without pausing as long as he was in sight. We have been informed by LAFLEUR, a man employed by the Company, that antelopes will escape with great ease even when they have one limb broken, as they can run fast enough upon three legs to defy any pursuit. Whilst we were encamped at the

" Three Mamelles," about sixty miles west of Fort Union, early one morning an antelope was heard *snorting*, and was seen by some of our party for a few minutes only. This snorting, as it is called, resembles a loud whistling, singing sound prolonged, and is very different from the loud and clear snorting of our common deer ; but it has always appeared to us to be almost useless to attempt to describe it ; and although at this moment we have the sound of the antelope's snort *in our ears*, we feel quite unable to give its equivalent in words or syllables.

The antelope has no lachrymal pits under the eyes, as have deer and elks, nor has it any gland on the hind leg, so curious a feature in many of those animals of the deer tribe which drop their horns annually, and only wanting (so far as our knowledge extends) in the *Cervus Richardsonii*, which we consider in consequence as approaching the genus *Antilope*, and in a small deer from Yucatan and Mexico, of which we had a living specimen for some time in our possession.

The prong-horned antelope often dies on the open prairies during severe winter weather, and the remains of shockingly poor, starved, miserable individuals of this species, in a state of the utmost emaciation, are now and then found dead in the winter, even near Fort Union and other trading posts.

The present species is caught in pens in the same manner nearly as the bison, (which we have already described at p. 97) but is generally despatched with clubs, principally by the women. In the winter of 1840, when the snow was deep in the ravines, having drifted, Mr. LAIDLAW, who was then at Fort Union, caught some of them by following them on horseback and forcing them into these drifts, which in places were as much as ten to twelve feet deep. They were brought to the fort in a sleigh, and let loose about the rooms ; they were to appearance so very gentle that the people suffered their children to handle them, although the animals were loose. They were placed in the carpenter's shop, one broke its neck by leaping over a turning-lathe, and the rest all died ; for as soon as they had appeased the cravings of hunger, they began to fret for their accustomed liberty, and regained all their original wildness. They leaped, kicked and butted themselves against every obstacle, until too much exhausted to recover.—These individuals were all captured by placing nooses, fixed on the end of long poles, round their necks, whilst they were embedded in the soft and deep snow drifts, to which they had been driven by Mr. LAIDLAW

There are some peculiarities in the gait of this species that we have not yet noticed. The moment they observe a man or other strange object producing an alarm, they bound off for some thirty or forty yards, raising all their legs at the same time. and *bouncing*, at it were, from two to three feet above the ground ; after this they stretch their bodies out and gallop at an extraordinary speed. We have seen some which, when started, would move off and run a space of several miles, in what we thought did not exceed a greater number of minutes !

From what we have already said, it will be inferred that the wolf is one of the most formidable enemies of this species. We have, however, not yet mentioned that in some very cold and backward seasons the young, when first born at such times, are destroyed by these marauders in such numbers that the hunters perceive the deficiency and call them scarce for the next season. Antelopes are remarkably fond of saline water or salt, and know well where the *salt-licks* are found. They return to them daily, if near their grazing grounds, and lay down by them, after licking the salty earth or drinking the salt water. Here they will remain for hours at a time, in fact until hunger drives them to seek in other places the juicy and nourishing grasses of the prairie. This species is fond of taking its stand, when alone, on some knoll, from which it can watch the movements of all wanderers on the plains around, and from which a fair chance to run in any direction is secured, although the object of its fear may be concealed from view occasionally by a ravine, or by another projecting ridge like its own point of sight.

We had in our employ a hunter on the Yellow-Stone River, who killed two female antelopes and broke the leg of a third at one shot from an ordinary western rifle. The ball must have passed entirely through the two first of these animals.

We have represented on our plate two males and a female in the fore ground, with a flock of these timid creatures running at full speed in the distance.

We subjoin the following account of the Antelopes seen by J. W. Audubon and his party on their overland journey through Northern Mexico and Sonora to California.

" Leaving Altar, Sonora, the country was flat and uninteresting, except that large patches of coarse grass, sometimes miles in length, took the place of the naked clay plains we had been riding through. The tall cactus, described by Fremont and Emory, in its eccentric forms was remarkable enough even by daylight, but at night, a very little superstition, with the curved and curiously distorted forms, produced in some cases by disease of the plant, or by the violent gales that periodically sweep those prairies, might make the traveller suppose this was a region in which beings supernatural stalked abroad. The shrill whistle of the Antelope, new to us all, added to the wild and unearthly character of the scene. The Maricapos Indians were said to be friendly, but we *did not know it*, and after our long watchings against Camanche, Apatche, Wako and Paramanii, who among us, as we knew how Indians sometimes personate the animals of the section they live in, but listened with intense interest to the slightest noise foreign to our previous knowledge. The short quick stampings of impatience or nervousness, continually repeated by the animals, were, however, soon distinguished in the stillness of our prairie camp at night, and feeling thus assured that only one of the deer tribe was the cause of our anxiety, blankets and tent soon covered us, and we left the beautiful and innocent creatures, now that we knew them, to their own reflections, if any they made, as to who and what we were, until morning.

At day light, RHOADES and VAN HORN, two hunters good as ever accompanied a train across the broad prairies ranged over by Buffalo, Elk. or Deer, looked out the trails, and reported Antelopes; but brought none to camp; not expecting to see any more of this herd, we started on our tramp towards the great Sonora Desert.

STEVENSON had a new horse, and as he had never been mounted without blindfolding him, after the Mexican fashion with young horses, being wild, his owner, by way of making him *more gentle*, commenced beating him with a stick that might have been selected to kill him; before I had time to know what was going on and interfere for the poor horse, he had looked to his own interests, pulled away, and with a bounding gallop went off, like an escaped prisoner, leading four of our best men and horses some ten miles ahead of the train, and when the runaway was at length overtaken, VAN HORN, PENNYPACKER, Mc. CUSKER, and myself were greatly in advance; the curve we had made from the road was slight, and on reaching it again, no trail told that the company had passed, so we had time to look about us, and loitered to rest our tired horses, when simultaneously we saw the back of a deer or Antelope; its head was hidden by the tall grass in which it was grazing on the soft juicy young shoots at the roots of the old tussocks: VAN HORN, with his unerring aim and Mississippi rifle, the eccentric twist of which, no doubt taken from WESSON's patent, renders these guns superior to all we have tried, was told to kill it. For a few seconds he was lost to our sight, though only a hundred yards from us, so low did he squat in the sparse tufts of dead grass and stinking wormwood. How curious it is to stand waiting the result of the skill and caution of the well tried hunter, at such a time; again and again we saw the back of the Antelope, as he passed one bunch of shrubbery after another, but never saw our hunter: at every moment we expected to see the wary animal with sense of smell so keen as nine times out of ten to save him from his enemies, bound away; but how different was his bound when he did leap, not forward, but straight upward.

And now we saw VAN HORN, a quarter of a mile off, running to where the last leap was made by his prey, and then came on the sluggish air, the crack of his rifle, almost after we had forgotten to listen for it, as a rifle cracks nowhere except on prairies, where neither woods, rocks or hills send back the sound. When I saw this beautiful creature, a most magnificent male, the first I had ever seen in the flesh, though the drawing for the 'Quadrupeds' had been long made and published, how I wished to redraw it! delicate even to the descriptions of the gazelle, muscular and sinewy as the best bred grey hound that Scotland ever produced.

I anticipated a treat, as VAN HORN gave me a hind quarter for our men, which I tied doubly secure to my saddle. But when night came, after ten hours' ride, although we enjoyed our steaks, the deer of the Cordilleras was too fresh in our memories to permit us to say that this Antelope was the best meat we had eaten."

* * * " The eastern spurs of the coast range were just behind us ; the black-tailed deer was scarcely past, for a few miles back, high up on one of the conical velvety hills of this range, we had seen three, looking at us from under one of the dwarf oaks that grow at a certain altitude, in forms peculiar to this country ; above or below, either a different formation or total absence of shrubbery occurring. We were winding along the base of a moderate line of hills of the *Sierra Nevada*, when what we took for a flock of sheep, the trail of which we had been following for three days on the way to the mines from Los Angeles, was discovered, and we hoped for mutton, to say nothing of the company we anticipated ; but our flock of sheep was like the 'Phantom Bark,' for it 'seemed never the nigher,' *au contraire*, turning a hill went out of sight, and we never got another view ; we saw another flock some miles on, and at first, supposing it the same, wondered how they could travel so fast. This was probably another portion of the one we had trailed for so many days. We were gratified by the whole flock running near us, from which we argued we were in the chosen country of the Antelope, the broad Tule valley. The flock ran 'shearing' about, as the formation of the land compelled them to turn to the right or left, showing their sides alternately in light and shade. When they are on the mountain sides and discover a foe, or any object that frightens them, the whole flock rush headlong for the plains, whether the enemy is likely to intercept them or not, and they seem to fly with the single idea, that they are in a dangerous place, and must change it for some other, no matter what ; at times a whole flock would run to within shot of our company, determined as it were to go through the line, and I believe in one or two instances would have done so, if they had not been shot at by our too impatient party. When on

the plains, the same desire possesses them to get to the hills, and back they go a hundred or two in a flock, seldom slackening their speed, except for a few seconds to look again, and be more frightened than ever at what had first startled them. The rolling hills of the western line of the Sierra Nevada were their most favourite locality in this valley, as far as we saw, but Layton and myself met an accidental individual or two, nearly up to Sacramento city, as we travelled through the beautiful, park-like scenes of this portion of California to the diggings of the head waters of the "American Fork."

As to the shedding of the horns of this species, I never was able to ascertain it, but a fine buck we killed, late in November, had a soft space between the head and horn, over the bone, that looked as if it had grown that length in one season. A young Antelope is better eating than a deer, but an old one, is *decidedly goaty.*

GEOGRAPHICAL DISTRIBUTION.

The Prong-horned Antelope is an inhabitant of the western portions of North America, being at no time found to the east of the Mississippi river. Its most northerly range is, according to Richardson, latitude 53° on the banks of the north branch of the Saskatchewan. They range southerly on the plains east of the Rocky Mountains into New Mexico. The precise latitude we have not been able to ascertain, but we have seen specimens that were said to have been obtained along the eastern ridge of the mountains within the tropics in Mexico. The account given by Hernandez, as well as his bad figure of his *Teuthlamacame,* can apply to no other species; this was obtained in Mexico. Lewis and Clarke found it on the plains west of the Columbia River, and it is now known to be an inhabitant of California. It has, therefore, a very extensive geographical range.

GENERAL REMARKS.

We have after much reflection and careful examination, concluded to adopt Mr. Ord's genus *Antilocapra* for this species. It differs in so many particulars from the true Antelopes, that naturalists will be compelled either to enlarge the character of that genus, or place it under one already formed. Its horns are branched, of which no instance occurs among all the species of Antelope; it is destitute of crumens or lachrymal openings, and is entirely deficient in the posterior or accessory hoofs, there being only two on each foot.

Major Hamilton Smith, (Cuv. Animal Kingdom, Vol. v., p. 321,) formed a genus under the name of *Dicranocerus,* under which he placed a second species which he named *A palmata.* Although the generic name given by

Smith is in many respects preferable, as being more classically correct, still, if we were to be governed by the principle that we should reject a genus because the compound word from which it is derived is composed of two languages, or if it does not designate the precise character of the species, we would be compelled to abandon many familiar genera, established by Linnæus himself.

The specific name of Ord, we have also adopted in preference to the more characteristic one "*furcifer*" of Smith, under a rule which we have laid down in this work not to alter a specific name that has been legitimately given.

We have added the *A palmata*, palmated Antelope of Major Smith, as a synonyme. We have compared so many specimens differing from each other in shades of colour and size of horns, that we have scarcely a doubt of his having described a very old male of the Prong-horned Antelope.

CERVUS MACROTIS.—Say.

MULE DEER.

PLATE LXXVIII. Female—Summer Pelage.

C. cornibus sub-dichotomo-ramosis; auriculis longissimis; corpore supra pallide rufescente-fusco, caudâ pallide rufescente cinereâ, apice compresso subtus nudi-osculo nigro.

CHARACTERS.

Horns cylindrical, twice forked; ears very long; body above, brownish grey; tail short, above, pale reddish ash colour, except at the extremity on its upper surface, where it is black. Hair on the body coarse, like that of the Elk; very long glandular openings on the sides of hind legs.

SYNONYMES.

Jumping Deer. Umfreville, Hudson's Bay, p. 164.
Black Tailed or Mule Deer. Gass Journ. p. 55.
Black Tailed Deer, Mule Deer. Lewis and Clarke. Vol. 1, pp. 91, 92, 106, 152, 239, 264, 328. Vol. 2. p. 152. Vol. 3. p. 27, 125.
Mule Deer. Warden's United States. Vol. 1, p. 245.
Cerf Mulet. Desmarest Mam., p. 43.
Black Tailed or Mule Deer. James Long's Exped. Vol. 2, p. 276.
Cervus Macrotis, Say. Long's Expedit. Vol. 2, p. 254.
" " Harlan Fauna, p. 243.
" " Sabine. Franklin's Journey, p. 667.
" " Godman's Nat. Hist. Vol. 2, p. 305.
Great Eared Deer. Griffith's An. King. Vol. 4, p. 133; Vol. 5. p. 794.

DESCRIPTION.

In size this species is intermediate between the Elk and the Virginian Deer, and a little larger than the Columbian Black Tailed Deer, to be noticed hereafter. It is a fine formed animal, bearing a considerable resemblance to the Elk, its long ears constitute its only apparent deformity.

Male.—Antlers slightly grooved, tuberculated at base, a small branch near the base, corresponding to the situation and direction of those of the *C. Virginianus*. The curvature of the anterior line of the antlers, is similar in

Plate LXXVIII

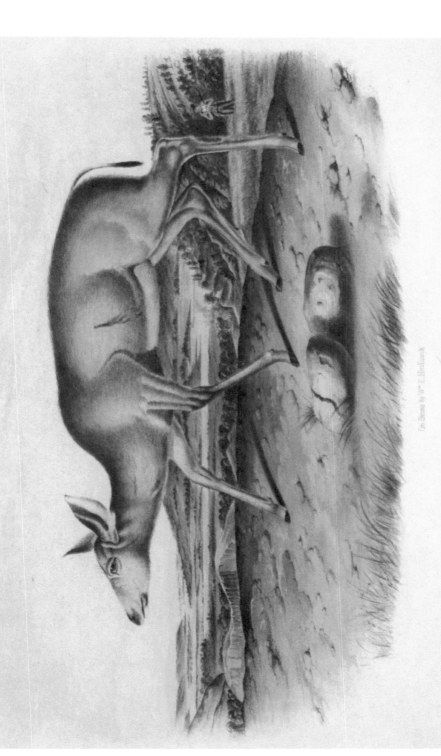

On Stone by Wᵐ E. Hitchcock

Drawn from Nature by J W Audubon

Lith Printed & Colᵈ by J.T. Bowen, Phil

Black Tailed Deer

direction but less in degree than in the Common Deer ; near the middle of the entire length of the antlers they bifurcate equally and each of these processes again divides near the extremity, the anterior of these smaller prongs being somewhat longer than the posterior ones. The lateral teeth are larger in proportion to the intermediate teeth than those of the *Virginianus.* The ears are very long, extending to the principal bifurcation, about half the length of the whole antler. The lachrymal aperture is longer than in the Virginian Deer, the hair is coarser and is undulated or crimped like that of the Elk ; the hoofs are shorter and wider than those of the common Deer, and more like those of the Elk, the tip of the trunk of the tail is somewhat compressed and almost destitute of hair.

Female.—Summer Pelage.—In the length and form of its ears, the animal from which we describe constantly reminds us of the mule, and in this particular may not have been inappropriately named the Mule Deer. The female is considerably larger than the largest male of the Virginian Deer we have ever examined. The head is much broader and longer from the eye to the point of the nose, the eye large and prominent, the legs stouter, and the tail shorter. The gland on the outer surface of the hind legs below the knee, covered by a tuft of hair, is of the unusual length of six inches, whilst in the common deer it is only one inch long. Around the throat, the hair is longer than in the corresponding parts of the Virginian Deer, and near the lower jaw under the throat, it has the appearance of a small tuft or beard. The tail of the summer-specimen is slightly tufted, indicating that in winter it might have a distinct tuft at the end. It is rounded and not broad and flat like that of the Virginian Deer.

The hair on the body is coarse, and lies less compact and smooth, that on the thighs near the buttocks, resembles white cotton threads cut off abruptly.

COLOUR.

Upper portion of nose and sides of face ashy grey ; the forehead is dark brown, and commences a line running along the vertebræ of the back, growing darker till it becomes nearly black. Eyebrows and a few streaks on and along the neck dark brown. Neck, and sides of body, yellowish brown. Outer surface of legs a shade lighter than the sides of the body. Under the chin, inner surface of legs, and belly, greyish white. Belly between the forelegs brownish or yellowish-brown, a line of which colour runs up to the neck. It differs from the Virginian Deer in being destitute of the dark markings under the chin, and has them less conspicuous around the nose. From the root of the tail extending downwards on both buttocks there is a lightish patch seven inches in diameter, making an approach to the yellowish white spot on the buttocks, so characteristic in the elk, rocky mountain sheep, and pronged horned antelope. From the root of the tail to near the extremity the hairs are ashy white. Point of tail for two inches black.

There are no annulations on the hair, which is uniform in colour from the roots.

DIMENSIONS.

Female.

	Ft.	Inches
Nose to anterior canthus of eye		6½
Length of eye		1¼
Nose to opening of ear	1	¼
" end " "	1	8¼
Breadth of ear		3½
Nose to point of shoulder	2	1
Nose to root of tail	4	10
Tail vertebræ		5½
End of hair		10
Tip of shoulder to elbow	1	5
" " " to bottom of feet	3	3
Height to rump	3	6¼
Girth back of shoulder	3	1¾
Round the neck	1	2¾
Nose to angle of mouth		3½
Between eyes at anterior canthus		4
Behind the eyes round the head	1	6

Weight, 132 lbs.

Dimensions of a Male, as given by Say.

	Inches.
Length from base of antlers to origin of basal process,	2
From basal process to principal bifurcations	4½ to 5
Posterior branch	2½ to 3
From anterior base of antlers to tip of superior jaw	9¼
Of the ears	7½
Trunk of the tail	4
Hair at the tip of tail	3 to 4

HABITS.

The first opportunity was afforded us of observing this magnificent animal, on the 12th of May as we were ascending the Missouri, about eleven hundred miles above Fort Leavenworth. On winding along the banks, bordering a long and wide prairie, intermingled with willows and other small brush wood, we suddenly came in sight of four Mule or black-tailed Deer, which after standing a moment on the bank and looking at us, trotted leisurely away, without appearing to be much alarmed. After they had retired a few hundred yards, the two largest, apparently males, elevated themselves on their hind legs and pawed each other in the manner of the horse. They occasionally stopped for a moment, then trotted off again, appearing and disappearing from time to time, when becoming suddenly alarmed, they bounded off at a swift pace, until out of sight. They did not trot or run as irregularly as our Virginian Deer, and they appeared at a distance darker in colour, as the common Deer at this season is red. On the 25th of the same month, we met with four others, which in the present instance did not stop to be examined: we saw them at a distance rapidly and gracefully hurrying out of sight. On the evening of the same day, one of our hunters brought to us a young Buck of this species, the horns of which, however, were yet too small to enable us to judge what would be their appearance in the adult animal. When on the Upper Missouri, near Fort Union, we obtained through the aid of our hunters, the female Black-tailed Deer, from which our figure, description and measurements have been made. We regret exceedingly that we were so unfortunate as not to have been able to procure a male, the delineation of which we must leave to our successors.

The habits of this animal approach more nearly those of the Elk, than of either the long-tailed or Virginian Deer. Like the former they remove far from the settlements, fly from the vicinity of the hunter's camp, and when once fairly started, run for a mile or two before they come to a pause.

The female produces one or two young, in the month of June.

We have figured a female in summer pelage, and have represented the animal in an exhausted state, wounded through the body, and about to drop down, whilst the hunter is seen approaching, through the tall grass, anticipating the moment when she will reel and fall in her tracks.

GEOGRAPHICAL DISTRIBUTION.

The Mule Deer range along the eastern sides of the Rocky Mountains, through a vast extent of country; and according to Lewis and Clarke

are the only species on the mountains in the vicinity of the first falls of the
Columbia River. Their highest northern range, according to RICHARDSON, is
the banks of the Saskatchewan, in about latitude 54° ; they do not come to
the eastward of longitude 105 in that parallel. He represents them as
numerous on the Guamash flats, which border on the Kooskooskie River.
We found it a little to the east of Fort Union on the Missouri River. It
ranges north and south along the eastern sides of the Rocky Mountains
through many parallels of latitude until it reaches north-western Texas,
where it has recently been killed.

GENERAL REMARKS.

Since the days of LEWIS and CLARKE, an impression has existed among naturalists that there were two species of black-tailed Deer ; the one existing
to the east of the Rocky Mountains, and the other, bordering on the Pacific,
and extending through upper California. Although the descriptions of those
fearless and enterprising travellers are not scientific, yet their accounts
of the various species of animals, existing on the line of their travels, have
in nearly every case been found correct, and their description of habits
very accurate. They state that "the black-tailed fallow Deer are peculiar
to this coast (mouth of the Columbia,) and are a distinct species, partaking
equally of the qualities of the Mule and the common Deer (*C. Virginianus.*)
The receptacle of the eye more conspicuous, their legs shorter, their bodies
thicker and larger. The tail is of the same length with that of the common
Deer, the hair on the under side, white ; and on its sides and top of a deep
jetty black ; the hams resembling in form and colour those of the Mule Deer,
which it likewise resembles in its gait. The black-tailed Deer never runs
at full speed, but bounds with every foot from the ground at the same time,
like the Mule Deer. He sometimes inhabits the woodlands, but more often
the prairies and open grounds. It may be generally said that he is of a size
larger than the common Deer, and less than the Mule Deer. The flesh is
seldom fat, and in flavour is far inferior to any other of the species ! It will
be seen from the above, that they regarded the Mule Deer of the plains of
Western Missouri as a distinct species from the black-tailed Deer, which
existed along the Pacific coast near the Columbia river.

SAY gave the first scientific description of the Mule Deer, which he named
" *Cervus Macrotis,*" which having the priority we have retained. RICHARD-
SON, whilst at the Saskatchewan, sought to obtain specimens of this animal
for description, but it being a season of scarcity, the appetites of the hunters
proved superior to their love of gain, and they devoured the Deer they had
shot, even to their skins. When after his return to Europe, in 1829, he

published the animals obtained in the expedition, he very properly added such other species as had been collected by the labours of Douglass, Drum. mond and other naturalists, who had explored the northern and western portions of America. Finding in the Zoological Museum a specimen of black-tailed Deer, procured on the western coast of America, by Douglass, he concluded that it was the species described by Say, *C. macrotis;* at the close of his article, he refers to the animal mentioned by Lewis and Clarke, as the black-tailed Deer of the western coast, of which he states, that he had seen no specimen, designating it (F. B. Am. p. 257) *C. macrotis, var. Columbiana.* We have, however, come to the conclusion that the animal described by Richardson was the very western species to which Lewis and Clarke refer, and that whilst his description of the specimen was correct, he erred in the name, he having described not the Mule Deer of Lewis and Clark and Say, but the Columbian black-tailed Deer, our drawing of which was made from the identical specimen described and figured by Richardson. We have named it, after its first describer, *Cervus Richardsonii.*

The following characters will serve to designate the species.

C. Richardsonii, considerably smaller than *C. macrotis,* the male of the former species being smaller than the female of the latter. The hair of *C. macrotis* is very coarse and spongy, like that of the elk, that of *C. Richardsonii* is much finer and more resembles that of the Virginian Deer. The *C. Richardsonii* has no glandular opening on the outer surface of the hind leg below the knee joint, approaching in this particular the antelopes which are also without such openings, whilst the corresponding portion in *C. macrotis* is longer than that of any known species of Deer, being six inches in length. They differ in the shape of their horns, *C. Richardsonii* having the antlers more slender, much less knobbed, and less covered with sharp points than those of the latter. They are also destitute of the basal process, so conspicuous in *C. macrotis.* We regret exceedingly that from circumstances beyond our control, we have been enabled to give a figure of the female only of *C. macrotis,* and of the male only of *C. Richardsonii.* The former was figured from the specimen we obtained at Fort Union, and for the latter we are indebted to the directors of the Zool. Society of London, who very kindly permitted us to make a drawing from the specimen previously described and figured by Richardson.

Note.—In connection with this subject, we are deeply pained to be compelled to notice the obstructions thrown in the way of our pursuits by the directors of the National Institute at Washington, which city we visited shortly after the return of our exploring expedition, when we were kindly invited by Mr. Peale to an examination of the valuable specimens of Natural History, collected by our adventurous countrymen. We pointed out to him one or two skins of the black-tailed Deer from the Western coast, which we both agreed differed from the *C. Macrotis* of SAY. We proposed to him that he should give a short description of the species, and select the name, which we would afterwards adopt in our work—this is in accordance with the mode usually pursued, and would have only occupied an hour. After the lapse of several years, we made an application by letter to the directors of the Institution for the privilege of making a drawing of the specimen; this we were not only refused, but were even denied the privilege of looking at the specimen, which we were very anxious to see, in order to be enabled to point out in the most satisfactory manner the characteristics by which these two closely allied species of Deer inhabiting our country could be distinguished from each other.

We cannot but contrast the narrow-minded policy pursued towards us in our application at Washington, with the liberality and generosity which was at all times extended to us in Europe under similar circumstances. When we visited England in 1838, the Directors of the Zoological Society opened its museum and assigned to us a private room, of which they gave us the key, and which we occupied for nearly a month—the specimens were taken from the cases by their attendants and brought to us, and when we discovered in the collection undescribed species, we were encouraged and aided in describing them. The same facilities were afforded us in the British museum, and in those of Edinburgh, Paris, Berlin, Dresden, and Zurich. The British Government, as well as our own, gave us all the assistance which could be rendered by either, consistent with other public services, and we derived material advantages from the aid afforded us by the revenue service and the various military stations we have visited in our researches, in Labrador—in Florida— in the far West, and in Texas.

We know not who were the Directors of the National Institute when our reasonable request was so cavalierly rejected, nor have we inquired whether any changes in policy have since taken place in regard to the collection of animals at Washington, but we feel it our duty publicly to protest against a conduct so narrow, selfish, and inconsistent with the liberality of our free institutions and so little adapted to promote one of the objects sought to be gained by the exploring expedition—viz: the advancement of natural history.

When the Hudsons Bay Company received an intimation that we would be glad to obtain any specimens they could furnish us from their trading posts in the arctic regions, they immediately gave orders to their agents and we secured from them rare animals and skins, procured at considerable labour and expense, and sent to us without cost, knowing and believing that in benefitting the cause of natural science they would receive a sufficient reward.

Plate LXXIX

Drawn from Nature by J.J. Audubon, FRSFLS

On Stone by Wm E. Hitchcock

Lith. Printed & Col.d by J. T. Bowen Phil.

Annulated Marmot Squirrel

SPERMOPHILUS ANNULATUS.—Aud. and Bach.

PLATE LXXIX.—Male.

S. Super cervinus, pilis nigris, interspersis, subtus albido. Caudâ corpore longiore, annulis, 17–20 nigris.

CHARACTERS.

Reddish-brown above, speckled with black beneath. Tail, which is longer than the body, annulated, with from seventeen to twenty black bands.

SYNONYME.

Spermophilus Annulatus. Aud. & Bach. Transactions of the Academy of Natural Sciences, Oct. 5th, 1841.

DESCRIPTION.

In size, this species is scarcely larger than the Hudson's Bay Squirrel, (*S. Hudsonius.*) In the shape of the head it resembles *Spermophilus Parryi.* The ears are quite small, being scarcely visible above its short coat of rather coarse, adpressed hairs; they are thickly covered with hair on both surfaces. The nose is sharp; whiskers, (which are numerous,) the length of the head. Eyes of moderate size, situated on the sides of the head. The os-frontis is rounded between the orbits, as in *S. Franklinii.* The cheek pouches are pretty large, and open into the mouth immediately anterior to the grinders. The body is more slender than the spermophiles in general, and in this, and several other peculiarities which will be mentioned, this species approaches the genus *Sciurus.* On the fore-foot, a sharp, conical nail is inserted on the tubercle which represents the thumb. There are four toes, covered to the extremities with a close, smooth coat of hair. The first and the fourth toe are of equal length. The second and third, which are longest, are also uniform in length. The nails are short, crooked and sharp, like those of the Squirrels, and not like those of the Marmots and Spermophili in general, which are long and slender, and but slightly curved. The legs are long and slender. The hair on the back is rather short, and lies close and smooth. The short fur beneath this coarser hair is rather sparingly distributed. On the under surface, the hairs are longer, and so thinly and loosely scattered as to leave the skin visible in many places, especially on the abdomen, and inner surface of the thighs. The hind feet, which are thickly covered with short, smooth hairs, have five toes. The soles, as well as palms, are naked. The tail, by its great length and singular markings, presents a distinguishing peculiarity in this species; it is flattened, and the hairs admit of a distichous arrangement; but the tail is narrower, and less bushy than those of the Squirrels.

COLOUR.

The incisors are deep orange; nails, brown; whiskers, black; nose and sides of the face, chestnut-brown. There is a line of soiled white above and around the eyes. The hairs on the upper surface are yellowish-brown at the roots, barred about the middle with black; then another line of yellowish-brown and tipped with black, giving it a dark, greyish-brown, and in some lights a speckled appearance. The small spots are, however, no where well defined; upper surface of the feet and legs, yellowish-brown; the under parts, chin, throat, belly, and inner surface of the legs and thighs are white. The tail is annulated with about nineteen black, and the same number of cream-coloured bands, giving it a very conspicuous appearance. These annulations commence about three inches from the root of the tail, and continue to be well defined till near the extremity, where the colours become more blended, and the rings are scarcely visible. On the under surface, the tail is pale reddish-brown, irregularly, and not very distinctly barred with black.

DIMENSIONS.

		Inches.	Lines.
Length from point of nose to root of tail,	- ▪ -	8	2
"　　　 tail vertebræ,	- ▪ ▪ ▪ ▪ -	8	0
"　　　 to end of hair,	- ▪ ▪ ▪ ▪ -	9	4
From heel to end of middle hind claw, -	▪ ▪ -	1	10
Height of ear, posteriorly, -	▪ ▪ ▪ ▪ -	0	$1\frac{1}{2}$
Length of longest fore-claw,	▪ ▪ ▪ ▪ -	0	2
Length of longest hind claw,	▪ ▪ ▪ ▪ -	0	$2\frac{1}{2}$

HABITS.

We possess no knowledge of the habits of this species, but presume from its form, that it possesses the burrowing propensities of the genus. All the *Spermophili* avoid thickly wooded countries, and are either found in rocky localities, or burrowing in the prairies.

GEOGRAPHICAL DISTRIBUTION.

The specimen we have described above, was obtained on the Western Prairies, we believe on the east of the Mississippi river ; the locality was not particularly stated. It was politely presented to us by Professor SPENCER F. BAIRD, of Carlisle, Pennsylvania, a young Naturalist of eminent attainments.

GENERAL REMARKS.

In every department of Natural History, a species is occasionally found which forms the connecting link between two genera, rendering it doubtful under which genus it should properly be arranged. Under such circumstances, the Naturalist is obliged to ascertain, by careful examination, the various predominating characteristics, and finally, place it under the genus to which it bears the closest affinity in all its details. The Spermophili are intermediate in character between the Squirrels and Marmots. They have the lightness of form of the former, and burrow in the ground like the latter. By their cheek pouches, of which the true Squirrels and Marmots are destitute, they are distinguished from both. The second inner toe on the forefoot of the Spermophili is the longest, whilst in the Squirrels the third is longest. But in these closely-allied genera, there are species which approach those of another genus. Thus our Maryland Marmot, (*A Monax*,) has a rudimentary cheek-pouch, in which a pea might be inserted, yet in every other particular it is a true *Arctomys*. The downy Squirrel, (*Sciurus lanuginosus*, see Journal Acad. Nat. Science, Vol. 8th, part 1st, p. 67,) by its short ears, broad head, and not very distichous tail, approaches the Spermophili, yet by its being destitute of cheek-pouches, by its soft, downy fur, and its hooked, sharp claws, of which the third, as in the Squirrels, is longest, it is more allied to Sciurus. On the other hand, the species now under consideration has the long legs, slender form, and sharp, hooked claws of the Squirrel. The two middle toes of the fore-feet being of equal length, prove its affinity to both genera; but in the general shape of its body, its cheek pouches, its short ears, and smooth, rigid hair, it must be regarded as belonging to the genus Spermophilus. We consider this species and the downy Squirrel as connecting links between Sciurus and Spermophilus, as we regard *Sciurus Hudsonius* the connecting link between Tamias and Sciurus.

ARVICOLA PINETORUM.—Leconte.

Leconte's Pine-Mouse.

PLATE LXXX.—Male and Female.

A. Capite crasso; naso obtuso; vellere curto; molli bombycino, instar velleri Talpæ; supra fusco-canâ, subtus plumbeo.

CHARACTERS.

Head large, nose blunt; fur short, soft, silky and lustrous, like that of the mole. Colour, above, brown, beneath, plumbeous.

SYNONYMES.

Psammomy's Pinetorum, Le Conte, Annals of the Lyceum of Natural History of New-York, Vol. III. p. 3, p. 2.

Arvicola Scalopsoides, Mole Arvicola. Aud. and Bach. Transactions Acad. Nat. Sciences, October, 1841.

Arvicola Oneida, De Kay, Nat. Hist., N. Y., p. 88.

DESCRIPTION.

This species bears some resemblance to Wilson's Meadow Mouse; it is, however, less in size, and its fur is shorter, more compact and glossy; body rather stout, short and cylindrical; head large and short; nose blunt, and hairy, except the nostrils, which are naked; incisors of moderate size; moustaches, fine, and nearly all short, a few reaching the ear; eyes very small; auditory openings large; ears very short, not visible beyond the fur, thin and membranous, with a few scattered hairs on the upper margin; neck short and thick; legs short and slender, covered with very short, adpressed hairs, not concealing the nails; palms naked. There are four toes on the fore foot, of which the second, on the inner side, is the longest, the first and third nearly equal, and the fourth shortest; in place of a thumb, there is a minute, straight, but not blunt, nail. The hind feet have five toes, the middle longest, the two next on each side being of equal length, and a little shorter than the middle one; the inner toe is considerably shorter, and the fourth, placed far back, is the shortest. The nails are weak, nearly straight, sharp, but not hooked. The fur on the whole body is short, compact and soft, and on the back, glossy.

Plate LXXX

On Stone by W.E. Hitchcock

Leconte's Pine Mouse

COLOUR.

The eyes are black ; nostrils flesh-colour ; incisors light yellowish ; moustaches nearly all white, with a few interspersed of a dark brown colour. Hair from the root plumbeous, tipped on the upper surface with glossy brown. These tips are so broad that they conceal the ashy-grey colours beneath ; cheeks chestnut-brown, upper surface of tail, brown, feet, light-brown, nails, whitish. The hairs on the under surface are shorter than those on the back, and instead of being broadly tipped with brown, like those on the back, are very slightly tipped with very pale brown and whitish, giving the chin, throat, neck and inner surface of legs and whole under surface of body a pale ash colour. The line of demarcation between the colours of the back and under surface, is very distinct in most specimens, commencing on the edges of the mouth, running along the sides of the neck, thence along the shoulder, including the fore legs— along the sides, the two opposite lines meeting near the root of the tail. We have observed in this species a considerable difference in different specimens, both in size and colour, having met some which were but little more than three inches long, whilst others were five. In some, the colours on the back were of a much deeper brown than in others, whilst in others, the brown markings on the cheeks were altogether wanting. It should be observed that in this species, as well as in all our field mice, the colours are much lighter, and inclined to cinereous after the shedding of the hair in summer ; the colours gradually deepen and become brighter toward autumn and winter, and are most conspicuously dark brown in spring.

DIMENSIONS.

	Inches.
From point of nose to root of tail, -	$3\frac{1}{4}$
Tail,	$\frac{1}{8}$
Another Specimen.	
Length of head and body,	$4\frac{1}{4}$
Tail,	$\frac{3}{4}$

HABITS.

The manners of this species do not differ very widely from those exhibited by many other field mice. They however, avoid low grounds, so much the resort of the meadow mice, and prefer higher and drier soils.

This mouse is rather an inhabitant of cultivated fields than of woods, and is seldom found in the forest far removed from the vicinity of plantations, to which it resorts, not only to partake of the gleanings of the fields, but to lay its contributions on the products of the husbandman's labours, claiming a share before the crops are gathered. In the Northern states, it is found in potato fields and in vegetable gardens, gnawing holes into the sides of the potatoes, carrots, ruta-baga, and common turnips, following the rows where green peas and corn have been planted, bringing down threats of vengeance from the farmer on the poor ground mole, which, feeding only on worms, is made a kind of cat's-paw by this mischievous little field mouse, which does the injury in most cases, whilst the other is saddled with the blame. In the South it is, next to the Norway rat, the most troublesome visitant of the cellars and banks in which the sweet potato is stored, destroying more than it consumes, by gnawing holes into the tubers, and causing them to rot. Wherever a bed of Guinea corn, Egyptian millet, or Guinea grass is planted, there you will soon observe numerous holes and nests of this species. We have recently seen an instance where a large bed of kohlrabi was was nearly destroyed by it; the bulbs appearing above the surface were gnawed into holes, which, in some instances, penetrated to the centre. Our friend, the owner, had, as usual, laid the mischief on the broad shoulders of the hated and persecuted ground mole, of whose galleries not a trace could be seen in the vicinity. A number of small holes at the root of a stump, in the garden, indicated the true author of these depredations, and on digging, about a dozen of Leconte's field Mice were captured. This species is particularly fond of the pea or ground nut, (hypogea.) On examining the beds where this nut is cultivated, we have observed the rows on whole acres perforated in every direction by small holes, giving evidence that this troublesome little pest had been at work. In endeavouring to save and collect the seeds of the Gama grass, (*Tripsacum dactyloides*,) we generally found ourselves forestalled by this active and voracious little rat.

This species has young three or four times during the summer. One which we had in confinement, produced young three times, having three, seven, and four, in the different broods. The young were nearly all raised, but, when full-grown, became pugnacious and persecuted each other so much that we were obliged to separate them. They were almost exclusively fed on ground nuts, corn meal and sweet potatoes, but seemed to relish both boiled rice and bread. We have seen nine young taken from one nest.

The nest of this species is generally found under ground, at the distance of about a foot from the surface; it is small, and composed of light, loose materials, collected in the vicinity.

This prolific field rat possesses many enemies to diminish its numbers. The house cat not only watches for it about the fields and gardens, but is fond of devouring it, whilst the bodies of shrews and ground moles are not eaten. The very common Owl, (*Syrnium nebulosum,*) the Barn

Owl (*Strix Americana,*) the Weasel, Ermine, and Mink, all make this species a considerable part of their subsistence.

The only note we have ever heard from this mouse is a low squeak, only uttered when it is either struck suddenly or greatly alarmed. In a state of confinement it was remarkably silent, except when two were engaged in fighting.

GEOGRAPHICAL DISTRIBUTION.

Le Conte's Field Mouse has an extensive geographical range. We have received specimens from our friend, Dr. Brewer, obtained in Massachusetts. It is found in Connecticut, is quite abundant on the farms in Rhode-Island, and in the immediate vicinity of New-York. We found it at Milestown, a few miles from Philadelphia. Mr. Ruffin sent us several specimens from Virginia. We procured it in North Carolina, and received a specimen from Dr. Barritt, Abbeville, South Carolina. It becomes more abundant as you approach the seaboard, in Carolina and Georgia ; and we have specimens sent to us from Alabama, Mississippi and Florida. We have traced it no farther south, have not heard of it to the west of the Mississippi, and are informed that it does not exist in Texas.

GENERAL REMARKS.

From the diminutive figure in Wilson's Ornithology, we might be led to the conjecture that he had this little species in view. The accurate description given by Ord, applies, however, only to the *Arvicola Pennsylvanica.* The first scientific description that appears of this species was given by Le Conte, (Annals of the Lyceum of Nat. Hist. N. Y., Vol. III., p. 3.) Finding that there were some variations in the dentition from the long established genus *Arvicola,* he formed for it a new genus, under the name of *Psamomys.* As this name, however, had been pre-occupied by Ruppel for an Arabian species, the American translator, (Dr. McMurtrie,) of Cuvier's Animal Kingdom, proposed changing the genus to *Pitymis,* Pine Mouse. The variations in the teeth, however, we have found by comparison, do not afford sufficient characters to warrant us in removing it from *Arvicola,* to which, from its shape and habits, it seems legitimately to belong.

We do not feel warranted in changing the specific name of Le Conte, but that name is not expressive of one of its characteristics, as, although it may have been found in the pine woods, we have never, in a single instance, detected it in such localities. We have always found it either in the open fields, or along fences, in the vicinity of gardens and farms.

This species is subject to many changes in colour, and is so variable in size, that it is easy to mistake it ; hence we have added as synonymes, our *A. Scalopsoides,* and the *A. Oneida* of Dr. De Kay.

CERVUS VIRGINIANUS.—Pennant.

COMMON AMERICAN DEER.

PLATE LXXXI.—Fawn.

PLATE CXXXVI.—Male and Female.—Winter pelage.

C. cornibus mediocribus, ramosis, sub-complanatis, retrorsum valde in-
clinatis, dein antrorsum versis; ramo basali-interno retrorso; ramis
plurimis posticis, retrorsum et sursum spectantibus, sinubus suborbitalibus
plicam cutaneam formantibus: vellere aestate fulvo, hyeme canescente-
fusco.

CHARACTERS.

*Horns middle sized, tending to flatten, strongly bent back and then for-
wards; a basal antler on the internal side, pointing backwards; several
snags on the posterior edge, turned to the rear, and upwards; suborbital sinus
making a fold; colour, fulvous in summer, gray-brown in winter.*

SYNONYMES.

Virginian Deer. Penn. Syn., p. 51
    "   "   Penn. Quadrupeds. Vol. 1, p. 104.
    "   "   Shaw's General Zoology. Vol. 2, p. 284.
Amerikanischer Hirsch. Kalm Reise. Vol. 2, p. 326. 3d. p. 482.
Virginischer Hirsch. Zimmerm. Geogr. Gesch. Vol. 2, p. 129.
Cerf de la Louisiane. Cuv. Regn. An., 1ère p. 256.
Cervus Virginianus. Gmel. Vol. 1, p. 179.
Dama Americanus. Erxl. Syst., p. 312.
C. Virginianus. Harlan. Fauna Am., p. 239.
    "   Godm. Am. Nat. Hist. Vol. 2. p. 306.
C. Mexicanus et clavatus. Hamilton Smith, p. 315. Griff. Cuv. Vol. 4. p. 127.
  Vol. 5, p. 315.
C. Virginianus. Dekay's N. Y. Fauna, p. 113.

DESCRIPTION.

Muzzle sharp; head rather long; eyes large and lustrous; lachrymal
pits covered by a fold of the skin. Tail moderately depressed. Legs

Plate LXXXI

On Stone by W^P E. Hitchcock

Drawn from Nature by J.W. Audubon.

Lith. Printed & Col^d by J.T. Bowen Phil.

Common American Deer.

Fawn.

slender. A glandular pouch surrounded by a thick tuft of rigid hairs inside of the hind legs.

COLOUR.

The Virginian Deer varies considerably in colour at different periods of the year. In the spring it is of a dusky reddish or fulvous colour above, extending over the whole head, back, upper surface of the tail and along the sides. In the autumn it is of a bluish or lead colour, and in winter the hairs on the upper surface are longer and more dense and of a brownish dark tint. Beneath the chin, throat, belly, inner surface of legs, and under side of tail, white. There is no perceptible difference in colour between the sexes.

The fawns are at first, bright reddish-brown, spotted with irregular longitudinal rows of white. These spots become less visible as the animal grows older, and in the course of about four months the hairs are replaced by others, and it assumes the colour of the old ones.

DIMENSIONS.

						Feet.	Inches.
Length from nose to root of tail,	-	-	-	-		5	4
" of tail, (vertebræ),	-	-	-	-	-		6
" including hairs,	-	-	-	-	-	1	1
" Height of ear,	-	-	-	-	-		5½

HABITS.

Perhaps no species of wild animal inhabiting North-America, deserves to be regarded with more interest than the subject of our present article, the Common or Virginian Deer; its symmetrical form, graceful curving leap or bound, and its rushing speed, when, flying before its pursuers, it passes like a meteor by the startled traveller in the forest, exciting admiration, though he be ever so dull an observer.

The tender, juicy, savoury, and above all, digestible qualities of its flesh are well known; and venison is held in highest esteem from the camp of the backwoodman to the luxurious tables of the opulent, and, when not kept too long (a common error in our large cities by the way) a fat haunch with jelly and chafing dishes is almost as much relished, as a "hunter's steak," cooked in the open air on a frosty evening far away in the west. The skin is of the greatest service to the wild man, and also useful to the dweller in towns; dressed and smoked by the squaw, until soft and pliable, it will not shrink with all the wettings to which it is exposed. In the form of mocasins, leggings, and hunting shirts, it is the most material part of the dress of many Indian tribes, and in the civilized world is used for breeches, gloves, gaiters, and various other purposes.

From the horns are made beautiful handles for various kinds of cutlery.

The timidity of the Deer is such, that it hurries away, even from the sight of a child, and it is but seldom that the hunter has any danger to apprehend, even from a wounded buck; it does but little injury to the fields of the planter, and is a universal favourite with old and young of both sexes in our Southern States.

The Virginian, or as we wish to designate it, the Common Deer, is the only large animal, if we except the bear, that is not driven from the vicinity of man by the report of the deer-driver's gun, or the crack of the hunter's rifle ; the buffalo and the elk are now rarely seen east of the Mississippi. Hunted by hounds and shot at from day to day, the Deer may retreat from this persecution for a little while, but soon returns again to its original haunts. Although it scarcely ever occupies the same bed on successive nights, yet it is usually found in the same range, or drive as it is called, and often not fifty yards from the place, where it was started before. It is fond of lingering around fences and old fields, that are partially overspread with brush-wood, briar-patches and other cover, to screen it from observation. In the southern States the Deer, especially in summer when they are least disturbed, are fond of leaping the outer fences of plantations, lying through the day in some tangled thicket, overgrown with cane, vines and briars; and in such places you may be so fortunate as to start an old buck in August or September, and many an overgrown denizen of the forest has bowed his huge antlers and fallen a sacrifice to his temerity in seeking a resting-place too near some pea-patch, where his hoofs left traces for many weeks of his nightly depredations.

This habit of resting during the day in the near vicinity of their feeding ground, is however not universal. We during last summer were invited to visit a large cornfield in which a quantity of the Carolina cowpea had been planted among the corn. This had been the nightly resort of the Deer during the whole summer—their tracks of various sizes covered the ground, as if flocks of sheep had resorted to it, and scarcely a pod or even a leaf was remaining on the vines. The Deer, however, were not in the vicinity, where there were several favourable and extensive covers ; they were trailed to some small islands, in a marsh nearly two miles off. We ascertained that the Deer inhabiting the swamps on the east side of the Edisto river, where there are but few cultivated farms, were in the nightly habit of swimming the Edisto and visiting the pea-fields in Barnwell, on the opposite side, returning before day-light to their customary haunts, some four or five miles distant.

The localities selected by Deer as places of rest and concealment during the day are various, such as the season of the year and the nature of the country and climate may suggest to the instincts of the animal. Although we have occasionally in mountainous regions, especially in the higher mountains of Virginia and the Green Mountains of Vermont, detected a Deer lying without concealment on an elevated ledge of bare rock, like the ibex and chamois on the Alps, yet as a general habit, the animal may be said to seek concealment, either among clumps of myrtle or laurel bushes, (*Kalmia*), in large fallen tree-tops, briar-patches, clusters of alder bushes, (*alnus*), or in tall broom-grass, (*Andropogon dissitiflorus*). In cold weather it prefers seeking its repose in some sheltered dry situation, where it is protected from the wind, and warmed by the rays of the sun ; and on these occasions it may be found in briar-patches which face the south, or in tufts of broom-grass in old uncultivated fields. In warm weather it retires during the day to shady swamps, and may often be started from a clump of alder or myrtle bushes near some rivulet or cool stream. To avoid the persecution of moschetoes and ticks, it occasionally, like the moose in Maine, resorts to some stream or pond and lies for a time immersed in the water, from which the nose and a part of the head only project. We recollect an occasion, when on sitting down to rest on the margin of the Santee river, we observed a pair of antlers on the surface of the water near an old tree, not ten steps from us. The half-closed eye of the buck was upon us ; we were without a gun, and he was, therefore, safe from any injury we could inflict on him. Anxious to observe the cunning he would display, we turned our eyes another way, and commenced a careless whistle, as if for our own amusement, walking gradually towards him in a circuitous route, until we arrived within a few feet of him. He had now sunk so deep in the water that an inch only of his nose, and slight portions of his prongs were seen above the surface. We again sat down on the bank for some minutes, pretending to read a book. At length we suddenly directed our eyes towards him, and raised our hand, when he rushed to the shore, and dashed through the rattling canebrake, in rapid style.

The food of the common Deer varies at different periods of the year. In winter, it feeds on buds of several kinds of shrubs, such as the wild rose the hawthorn, various species of bramble, (*Rubus*,) the winter green (*Pyrola*,) the Partridge Berry, (*Mitchella repens*,) the Deer Leaf, (*Hopea tinctoria*,) the bush Honeysuckle, (*Azalea*,) and many others. In spring and summer it subsists on tender grasses, being very select in its choice and dainty in its taste. At these seasons it frequently leaps fences, and visits the fields of the planter, taking an occasional bite at his young wheat and oats, not overlooking the green corn, (*Maize*,) and giving a decided preference to a field planted with cow-peas, which it divests of its young pods and tender leaves; nor does it pass lightly by berries of all kinds, such as the Huckleberry, Blackberry and Sloe, (*Viburnum prunifolium.*) We are informed by a friend that in the vicinity of Nashville, (Tenessee,) there is an extensive park containing about three hundred Deer, the principal food of which is the luxuriant Kentucky blue-grass, (*Poa pratensis.*) In autumn it finds an abundance of very choice food in the chestnuts, chinquepins and beech-nuts strewn over the ground. The localities of the various oaks are resorted to, and we have seen its tracks most abundantly under the Live Oak, (*Quercus virens*,) the acorns of which it appears to prefer to all others. We once observed three deer feeding on these acorns, surrounded by a flock of wild turkeys, all eagerly engaged in claiming their share. The fruit of the Persimmon tree, after having been ripened by the frosts of winter, falls to the ground, and also becomes a favourite food of the Deer.

Possessing such a choice of food, we might suppose this animal would be always fat: this, however, is not the case, and, except at certain seasons of the year, the Deer is rather poor. The bucks are always in fine order from the month of August to November, when we have seen some that were very fat. One which we killed weighed one hundred and seventy-five pounds. We have been informed that some have reached considerably over two hundred pounds. In November, and sometimes a little earlier, the rutting season commences in Carolina, when the neck of the buck begins to dilate to a large size. He is now constantly on foot, and nearly in a full run, in search of the does. On meeting with other males, tremendous battles ensue, when, in some rare instances, the weaker animal is gored to death; generally, however, he flies from the vanquisher, and follows him, crest fallen, at a respectful and convenient distance, ready to turn on his heels and scamper off at the first threat of his victorious rival. In these rencontres, the horns of the combatants sometimes become interlocked in such a manner that they cannot be separated, and the pugnacious bucks are consigned to a lingering and inevitable death by starvation. We have endeavoured to disengage these horns, but found them so completely entwined that no skill or strength of ours was successful. We have several times seen two, and on one occasion, three pairs of horns thus interlocked, and ascertained that the skulls and skeletons of the Deer had always been found attached. These battles only take place during the rutting season, when the horns are too firmly attached to be separated from the skull. Indeed, we have seen a horn shot off in the middle by a ball, whilst the stump still continued firmly seated on the skull. The rutting season continues about two months, the largest and oldest does being earliest sought for, and those of eighteen months at a later period. About the month of January, the bucks drop their horns, when, as if conscious of having been shorn of their strength and honours, they seem humbled, and congregate peaceably with each other, seeking the concealment of the woods, until they can once more present their proud antlers to the admiring herd. Immediately after the rutting season, the bucks begin to grow lean. Their incessant travelling during the period of venery—their fierce battles with their rivals, and the exhaustion consequent on shedding and replacing their horns by a remarkably rapid growth, render them emaciated and feeble for several months. About three weeks after the old antlers have been shed, the elevated knobs of the young horns make their appearance. They are at first soft and tender, containing numerous blood-vessels, and the slightest injury causes them to bleed freely. They possess a considerable degree of heat, grow rapidly, branch off into several ramifications, and gradually harden. They are covered with a soft, downy skin, and are now in what is called "velvet." When the horns are fully grown, which is usually in July or August, the buck shows a restless propensity to rid himself of the velvet covering, which has now lost its heat, and become dry: hence he is constantly engaged in rubbing his horns against bushes and saplings, often destroying the trees by wounding and tearing the bark, and by twisting and breaking off the tops. The system of bony development now ceases altogether, and the horns become smooth, hard, and solid.

The does are fattest from November to January. They gradually get thinner as the season of parturition approaches, and grow lean whilst suckling their young.

The young are, in Carolina, produced in the month of April; young does, however, seldom yean till May or June. In the Northern States, they bring forth a little later, whilst in Florida and Texas the period is earlier. It is a remarkable, but well ascertained fact, that in Alabama and Florida, a majority of the fawns are produced in November. The doe conceals her young under a prostrate tree-top, or in a thick covert of grass, visiting them occasionally during the day, especially in the morning, evening, and at night. The young fawns, when only a few days old, are often found in so sound a sleep that we have, on several occasions, seen them taken up in the arms before they became conscious that they were captives. They are easily domesticated, and attach themselves to their keepers in a few hours. A friend possesses a young deer that, when captured, during the last summer, was placed with a she goat, which reared it, and the parties still live in habits of mutual attachment. We have seen others reared by a cow. A goat, however, becomes the best foster-mother. They breed in confinement, but we have found them troublesome pets. A pair that we had for several years, were in the habit of leaping into our study through the open window, and when the sashes were down they still bounced through, carrying along with them the shattered glasses. They also seemed to have imbibed a vitiated and morbid taste, licked and gnawed the covers of our books, and created confusion among our papers. No shrub in the garden, however valuable to us, was sacred to them; they gnawed our carriage harness, and finally pounced upon our young ducks and chickens, biting off their heads and feet, leaving the body untouched.

The doe does not produce young until she is two years old, when she has one fawn. If in good order, she has two the following year. A very large and healthy doe often produces three, and we were present at Goose Creek when an immense one, killed by J. W. AUDUBON, was ascertained, on being opened, to contain four large and well formed fawns. The average number of fawns in Carolina is two, and the cases where three are produced are nearly as numerous as those in which young does produce only one at a birth.

The wild doe is attached to her young, and its bleat will soon bring her to its side, if she is within hearing. The Indians use a stratagem, by imitating the cry of the fawn, with a pipe made of a reed, to bring up the mother, which is easily killed by their arrows. We have twice observed the doe called up by this imitation of the voice of the young. She is, however, so timid that she makes no effort in defence of her captured offspring, and bounds off at the sight of man.

The common Deer is a gregarious animal, being found on our western prairies in immense scattered herds of several hundred. After the rutting season the males, as we have before stated, herd together and it is only during the season of intercourse that both sexes are found in company. The does, however, although congregating during a considerable portion of the year, are less gregarious than many species of African antelopes, the buffalo, or our domestic sheep; as they are found during the summer separated from the rest of the gang or troop, and are only accompanied by their young.

The Deer is one of the most silent of animals, and scarcely possesses any notes of recognition. The fawn has a gentle bleat that might be heard by the keen ears of its mother at the distance probably of a hundred yards. We have never heard the voice of the female beyond a mere murmur when calling her young, except when shot, when she often bleats loudly like a calf in pain. The buck when suddenly started sometimes utters a snort, and we have at night heard him *emitting* a shrill whistling sound, not unlike that of the chamois of the Alps, that could be heard at the distance of half a mile. The keen sense of smell the Deer possess enables them to follow each other's tracks. We have observed them smelling on the ground and thus following each other's trail for miles. We were on an autumnal morning seated on a log in the pine lands of Carolina when a doe came running past us. In the course of ten minutes we observed a buck in pursuit, with his nose near the ground, following in all the windings of her course. Half an hour afterwards came a second buck, and during another interval a third small buck pursued the same trail. The sense of sight appears imperfect—as we have often, when standing still, perceived the Deer passing within a few yards without observing us, but we have often noticed the affrighted start when we moved our position or when they scented us by the wind. On one occasion we had tied our horse for some time at a stand;—on his becoming restless we removed him to a distance—a Deer pursued by dogs ran near the spot where the horse had originally stood, caught the scent, started suddenly back, and passed within a few feet of the spot where we were standing, without having observed us. Their sense of hearing is as keen as that of smell. In crawling towards them in an open wood, against the wind, you may approach within gun shot, but if you unfortunately break a stick, or create a rustling among the leaves, they start away in an instant.

This animal cannot exist without water, being obliged nightly to visit some stream or spring for the purpose of drinking. During the present year (1850) a general drought prevailed throughout our southern country. On the Hunting Islands between Beaufort and Savannah, the Deer, we were informed, nearly all perished in consequence of the streams on these Islands having dried up. Deer are fond of salt, and like many other wild animals resort instinctively to salt-licks or saline springs. The hunters, aware of this habit, watch at these "licks," as they are called, and destroy vast numbers of them. We have visited some of these pools, and seen the Deer resorting to them in the mornings and evenings and by moon light. They did not appear to visit them for the mere purpose of drinking, but after walking around the sides, commenced licking the stones and the earth on the edges, preferring in this manner to obtain this agreeable condiment, to taking a sudden draught and then retiring. On the contrary they lingered for half an hour around the spring, and after having strayed away for some distance, they often returned a second and even a third time to scrape the sides of it, and renew the licking process. Our common Deer may be said to be nocturnal in its habits, yet on the prairies, or in situations where seldom disturbed, herds of Deer may be seen feeding late in the morning and early in the afternoon. Their time for rest, in such situations, is generally the middle of the day. In the Atlantic States, where constantly molested by the hunters, they are seldom seen after sunrise, and do not rise from their bed until the dusk of the evening. The Deer is more frequently seen feeding in the day time during spring and summer, than in winter; a rainy day, and snowy wintery weather, also invite it to leave its uncomfortable hiding place and indulge in its roaming habits. We have no doubt, that in localities where Deer have been constantly hunted, they, from a sense of fear, allow you to approach much nearer to their place of concealment than in situations where they are seldom disturbed. They continue lying still, not because they are asleep or unaware of your approach, but because they are afraid to expose themselves to view, and hope by close concealment to be passed without being observed. We have seen them lying with their hind legs drawn under them ready for a spring—their ears pressed flat on the sides of the neck, and their eyes keenly watching every movement of the intruder. Under these circumstances your only chance of success is to ride slowly around the animal as if he was not observed, and suddenly fire before he leaps from his bed. This effect of fear, on your near approach, is not confined to our Deer; it may be seen in the common partridge, the snipe, and other game birds. Before being hunted, they are restless—are unwilling to assume the crouching posture called setting, and rise at a distance from their pursuers; but after having been a few times disturbed and shot at, they, in the language of sportsmen, become tame, and permit themselves to be nearly trodden on before they can be induced to rise; this apparent tameness is in reality wildness, and their squatting and hiding the effect of terror to which they are prompted by an instinct of self-preservation.

The gait of this Deer is various. In walking it carries its head very low, and pursues its course cautiously and silently, occasionally moving its ears and whisking its tail; the largest animal is usually the leader of the herd, which travel in what is called Indian file, there seldom being two abreast. Walking is the ordinary pace of the Deer unless frightened, or in some state of excitement. When first started, without being much alarmed, it gives two or three springs, alighting with apparent awkwardness on three feet—and immediately afterwards resting on the opposite side, erecting its white tail and throwing it from side to side. A few high bounds succeed, whilst the head is turned in every direction to enable it to detect the cause of alarm. The leaps and high boundings of the Deer are so graceful, that we have never witnessed them without excitement and admiration. When, however, the Deer observes you before it is routed from its bed, it bolts off with a rush, running low to the ground, with its head and tail on a line with the body, and for a few hundred yards rivalling the speed of a race horse. But this rattling pace cannot be kept up for any length of time—after the first burst its speed slackens, it foams at the mouth, and exhibits other evidences of fatigue. We have sometimes seen it overtaken and turned by an active rider in the open wood, and under other favourable circumstances, and on one occasion a fat buck was headed by a fearless driver, lashed with his whip, brought to bay, and finally knocked in the head and taken without having been shot. We have witnessed a few instances where a pack of hounds, after a four hours' chase, succeeded in running down a Deer. These cases are, however, rare, nor would we give any encouragement to this furious Sylvan race, in which the horse and his mad rider are momentarily exposed to the danger of a broken neck from the many holes in the pine lands. The Deer, after an attempt at bringing it to bay, frequently succeeds in escaping from the hunter and the hounds, by dashing into a swamp or crossing a river, and even should it be captured, after a long chase the venison is found to be insipid and of no value.

In riding through the woods at night in the vicinity of Deer, we have often heard them stamp their feet, the bucks on such occasions giving a loud snort, then bounding off for a few yards and again repeating the stamping and snorting, which appear to be nocturnal habits.

Deer take the water freely, and swim with considerable rapidity ; their bodies are on such occasions submerged, their heads only being visible above the surface. We have witnessed them crossing broad rivers and swimming the distance of two miles. When thus under way, they cleave the water with such celerity that a boat can scarcely overtake them.

Along our southern sea-board the Deer, when fatigued by the hounds, plunge into the surf and swim off for a mile or two, floating or swimming back with the returning tide, when they ascend the beach near the same place where they entered the water.

As already remarked, the flesh of our common Deer is the best flavoured and most easy of digestion of all the species with which we are acquainted, except the black-tailed Deer; it is superior to the Elk or Moose of our country, or the red Deer or Roebuck of Europe. It is, however, only a delicacy when it is fat, which is generally the case from the beginning of

August to the month of December. In Carolina, the haunch and loin only are served up on the tables of the planters, the shoulders and skin are the perquisites of the driver, or negro huntsman. The Indians eat every part of the Deer, not omitting the entrails and the contents of the stomach—the latter many of the tribes devour raw, without subjecting them to any cooking or roasting process. It is stated, even by white men, that the stomach, with all its half-digested ingredients, is very palatable. Hunger and hardships seldom fail to give a zest to the appetite. Vegetable food is scarce in the wilderness or on the prairies. The traveller who has long been obliged to sleep in a tent and make his toilet in the woods, soon becomes indifferent to the etiquette of civilized life, and does not inquire whether his dish has been prepared according to the recipe of the cookery-books. A Deer paunch contains a mixture of many ingredients, picked up from various shrubs, seeds, and grasses, and may become a substitute for vegetables where the kitchen-garden has not yet been introduced. According to a northern traveller (Lyon's Narrative, p. 242), who referred, however, to another animal, the reindeer of our continent, it is "acid and rather pungent, resembling a mixture of sorrel and radish leaves," its smell like "fresh brewer's grains." As we have never been subjected to the necessity of testing the virtues of this primitive chowder, we are unable to pronounce it a delicacy, and must leave the decision to those who may be disposed to make the experiment.

The capture of the common Deer exercised the ingenuity and patience of the Indian, ages before the pale faces intruded on his hunting-grounds, with their rifles, their horses, and hounds. He combatted with the wolf and the cougar for their share of the prey, leaving on our minds a melancholy impression of the near approach of the condition of savage life to that of the brute creation. Different modes of hunting were suggested by the peculiar face of the localities of the country, and the degrees of intelligence or native cunning of the several tribes. The bow and arrow evidently must have been in common use throughout the whole length and breadth of our land, as the numerous arrow-heads still every where turned up by the plough abundantly attest.

The Rein Deer, inhabiting the extensive, cold, and inhospitable regions of the British possessions to the north of Quebec, were caught in snares manufactured from the hide, and sometimes of the sinews, of the animal. During the season of their annual migrations, rude fences of brush-wood were constructed, which were a mile or two apart at the entrance, narrowing down to nearly a point at the other end, in which the snares were placed, and at the termination of this "cul de sac" was erected a high fence or pound, secured by stakes, stones, and other strong materials, in which the Deer that escaped from the snares were finally enclosed and shot with arrows. The common Deer, however, is more suspicious and timid, and will seldom suffer itself to be circumvented in this manner.

The American Rein Deer is also brought near to the hunter lying in wait behind the concealment of a clump of bushes, or heap of stones. by the waving of a small flag of cloth, or a deer's tail, which, exciting its atten-tion, it falls a sacrifice to its curiosity. This stratagem is also successfully practised on our western prong-horned Antelope.

The Common Deer is frequently brought within bow-shot by the Indians, who call up the does, as we have already mentioned, by imitating, with a pipe made of a reed, the bleating of the fawn, and also the bucks, by an imitation of the shrill, whistling sound which they emit during the rutting season. The wily savage often clothes himself in the hide of a Deer, with the horns and ears attached—imitating the walk and other actions of the animal, by which means he is enabled to approach and almost mingle with the herd, and kill several with his arrows before they take the alarm. Since the introduction of fire-arms, however, many tribes of Indians have laid aside the bow and arrow, and adopted the gun. The traders who visit them, usually supply them with an inferior article, and we have never seen any considerable number of Indians expert in the use of the rifle. The late Dr. Leitner informed us that the Florida Indians seldom shot at a Deer beyond twenty-five or thirty yards, exercising great patience and caution before they ventured on firing ; the result, however, under these favourable circumstances, was usually successful. We believe the Indians of North America never used poisoned arrows in the destruc-tion of game, like the natives of Caffraria and other portions of Africa, or the aborigines of Brazil and the neighbouring regions of South America.

The white man conducts his hunting excursions in various modes suited to his tastes and adapted to the nature of the country in which he resides. In mountainous, rocky regions, where horses cannot be used with advan-tage, he goes on foot, armed with a rifle, carries no dog, and seeks for the Deer in such situations as his sagacity and experience suggest. He either espies him in his bed, or silently steals upon him behind the covert of the stem of a large tree whilst he is feeding, and leisurely takes a steady and fatal aim. On the contrary, in situations adapted to riding, where the woods are thickly clothed with underbrush. where here and there wide openings exist between briar-patches, and clumps of myrtle-bushes, as in the Southern States, the Deer are almost universally chased with hounds, and instead of the rifle, double-barrelled deer-guns, of different sizes, carrying from twelve to twenty buck-shot, are alone made use of by the hunters.

It may not be uninteresting to our readers if we point out the different modes in which Deer hunts are conducted.

In the early settlement of our country, when men hunted for food, and before they accustomed themselves to study their ease and comfort even in the chase, "still hunting," as it is termed, was universally practised. The wolves and other depredating animals, by which the colonists were surrounded, as well as the proximity of hostile Indians, almost precluded them for many years from raising a sufficient supply of sheep, hogs, and poultry. The cultivation of a small field furnished them with bread, while for meat they were chiefly dependent on the gun. Hence a portion of their time was from a kind of necessity devoted to the chase. The passion for hunting seems however to be innate with many persons, and we have observed that it often runs in families and is transmitted to their posterity, as is known to be the case with the descendants of the hunters in the Alps. There are even now many persons in our country, who devote weeks and months to the precarious employment of Deer hunting, when half the industry and fatigue in regular labour would afford their families every necessary and comfort. Hunting is a pleasant recreation, but a very unprofitable trade ; it often leads to idleness, intemperance, and poverty.

For success in still-hunting it is essential that the individual who engages in it, should be acquainted with the almost impenetrable depths of the forest, as well as the habits of the Deer. He must be expert in the use of the rifle, possess a large stock of patience, and be constitutionally adapted to endure great fatigue. Before the dawn of day, he treads the paths along which the animal strays in returning from its nightly rambles to the covert usually its resting-place for the day. He ascends an elevation, to ascertain whether he may not observe the object of his search feeding in the vallies. If the patience and perseverance of the morning are not attended with success, he seeks for the Deer in its bed—if it should be startled by his stealthy tread and spring up, it stops for a moment before bounding away, and thus affords him the chance of a shot ; even if the animal should keep on its course without a pause, he frequently takes a running, or what is called a chance shot, and is often successful.

There is another mode of deer hunting we saw practised many years ago in the Western parts of the State of New-York, which we regard as still more fatiguing to the hunter, and as an unfair advantage taken of the unfortunate animals. The parties sally out on a deep snow, covered by a crust, which sometimes succeeds a rain during winter. They use light snow-shoes and seek the Deer in situations where in the manner of the moose of Nova Scotia, they have trampled paths through the snow in

the vicinity of the shrubs on which they feed. When started from these retreats they are forced to plunge into the deep snow; and breaking through
the crust leave at every leap traces of blood from their wounded legs;
they are soon overtaken, sometimes by dogs, at other times by the hunters,
who advance faster on their snow-shoes than the exhausted Deer, which
fall an easy prey either to the hunter's knife or his gun. In this manner
thousands of Deer were formerly massacred in the Northern States.

We have ascertained that our common Deer may be easily taken by the
grey-hound. A pair of the latter, introduced into Carolina by Col. CATTEL,
frequently caught them after a run of a few hundred yards. The Deer
were trailed and started by beagles—the grey-hounds generally kept in
advance of them, making high leaps in order to get a glimpse of the Deer
which were soon overtaken, seized by the throat, and thrown down. The
nature of the country, however, from its swamps and thick covers often prevented the huntsmen from coming up to the captured animal before it was
torn and mutilated by the hounds, and many Deer could not be found, as
the pack becomes silent as soon as the Deer is taken. We predict, however,
that this will become the favourite mode of taking Deer on the open wes
tern prairies, where there are no trees or other obstructions, and the whole
scene may be enacted within view of the hunters.

Some hunters, who are engaged in supplying the salt and red Sulphur
Springs of Virginia with venison during summer, practise a novel and an
equally objectionable mode in capturing the Deer. A certain number of
very large steel-traps made by a blacksmith in the vicinity, are set at night
in the waters of different streams at the crossing-places of the Deer.
The animal when thus captured instead of tearing off its leg by violent
struggles is said to remain standing still, as passive as a wolf when similarly entrapped. Another and still more cruel mode is sometimes prac-
ticed in the South: The Deer have particular places where they leap the
fences to visit the pea-fields; a sharpened stake is placed on the inside of
the fence—the Deer in leaping over is perforated through the body by
this treacherous spike, and is found either dead or dying on the following
morning. It is also a frequent practice in the South for the hunter during
clear nights to watch a pea-field frequented by Deer. To make sure of
this game he mounts some tree, seats himself on a crotch or limb which is
above the current that would convey the scent to the keen olfactories of
the Deer, and from this elevation leisurely waits for an opportunity to
make a sure shot.

In some parts of the Northern and Middle States the Deer are captured
by the aid of boats. We observed this mode of hunting pursued at Sara-
toga and other lakes, and ascertained that it was frequently attended with

success.   The hounds are carried to the hills to trail, and start the Deer before day light. Some of the hunters are stationed at their favourite crossing places to shoot them should they approach within gun shot. After being chased for an hour or two the Deer pushes for the lake. Here on some point of land a party lie in wait with a light and swift boat ; after the Deer has swam to a certain distance from the shore he is headed and approached by the rowers, a noose is thrown over the head, and the unfortunate animal drawn to the side of the boat, when the captors proceed to cut its throat in violation of all the rules of legitimate sporting.

Fire hunting is another destructive mode of obtaining Deer. In this case two persons are essential to success. A torch of resinous wood is carried by one of the party, the other keeps immediately in front with his gun. The astonished Deer instead of darting off seems dazzled by the light, and stands gazing at this newly kindled flame in the forest. The hunter sees his eyes shining like two tapers before him ; he fires and is usually successful ; sometimes there are several Deer in the gang, who start off for a few rods at the report of the gun, and again turn their eyes to the light. In this manner two or three are frequently killed within fifty yards of each other. This kind of hunting by firelight is often attended with danger to the cattle that may be feeding in the vicinity, and is prohibited by a law of Carolina, which is however frequently violated. The eyes of a cow are easily mistaken for those of a deer. We conversed with a gentleman who informed us that he had never indulged in more than one fire-hunt, and was then taught a lesson which cured him of his passion for this kind of amusement. He believed that he saw the eyes of a Deer and fired, the animal bounded off, as he was convinced, mortally wounded. In the immediate vicinity he detected another pair of eyes and fired again. On returning the next morning to look for his game, he found that he had slaughtered two favourite colts. Another related an anecdote of a shot fired at what was supposed to be the shining eyes of a Deer, and ascertained to his horror that it was a dog standing between the legs of a negro, who had endeavoured to keep him quiet. The dog was killed and the negro slightly wounded.

There is still another mode of Deer hunting which remains to be decribed. It is called "driving," and is the one in general practice, and the favourite pastime among the hospitable planters of the Southern States. We have at long intervals, occasionally joined in these hunts, and must admit that in the manner in which they were conducted, this method of Deer hunting proved an exciting and very agreeable recreation. Although we regret to state that it is pursued by some persons at all seasons of the year, even when the animals are lean and the venison of no value, yet the more thoughful and judicious huntsmen are satisfied to permit the Deer to rest and multiply for a season, and practice a little self-denial, during summer when the oppressive heats which usually prevail—the danger of being caught in heavy showers—and the annoyance of gauzeflies, mosquetoes, and ticks, present serious drawbacks to its enjoyment. The most favourable season for this kind of amusement is from the beginning of October to January. The Deer are then in fine order ; the heats of summer are over ; the crops of rice gathered, and the value of the planter's crop can be calculated. The autumn of the Southern States possesses a peculiar charm ; high winds seldom prevail, and the air is soft and mellow ; although many of the summer warblers have migrated farther to the south, yet they have been replaced by others : The blue-bird, cat-bird, and mocking-bird have not yet lost their song, and the swallows and nighthawks are skimming through the air in irregular and scattered groups on their way to the tropics. Vegetation has been checked, but not sufficiently destroyed to give a wintry aspect to the landscape. The *Gentians Gerardias* and other autumnal flowers are still disclosing a few lingering blossoms and emitting their fragrance. The forest trees present a peculiar and most striking appearance. A chemical process has been going on among the leaves, since the first cool nights have suspended the circulation, giving to those of the maple and sweet gum, a bright scarlet hue, which contrasted with the yellow of the hickory, and the glossy green of the magnolia grandiflora, besides every shade of colour that can be imagined, render an American forest, more striking and beautiful than that of any other country. It is the season of the year that invites to recreation and enjoyment. The planters have been separated during the summer ; some have travelled from home—others have resided at their summer retreats ;—they are now returning to their plantations, and the intercourse of the neighbourhood, that has been suspended for a season, is renewed. We recall with satisfaction some past scenes of pleasureable associations of this kind. The space already taken up by this article will preclude us from entering into minute detail, and restrict us to a few incidents which will present the general features of a Carolina Deer hunt. We comply with the oft-repeated invitation to make our annual visit to our early and long-tried friend Dr. Desel at his hospitable residence some twenty miles from the city, which his friends have named Liberty Hall. The mind requires an occasional relaxation as well as the body. We have resolved to fly for a day or two, from the noise and turmoil of the city—to leave books and cares behind us—to break off the train of serious thought—to breathe the fresh country air, and mingle in the innocent sports of the field and the forest. Reader, you will go with us and enter into our feelings and enjoyments. As we approach the long avenue a mile from the residence of the companion of thirty-five years, we are espied by his domestics who welcome us with a shout, and inform us that their "Boss" is looking out for us. Our friend soon perceives us, and hurries to the gate. How pleasant are the greetings of friendship—the smiling look of welcome, the open hand, and the warm heart of hospitality.

The usual invitation is sent to a neighbour, to lunch, dine, and meet a friend. The evening is spent in social converse and closed with the family bible, and offerings of gratitude and praise to the Giver of all good. The sleep of him, who has escaped from the din of the city to the quiet of the country, is always refreshing. The dawn of day invites us to a substantial breakfast. The parties now load their double-barrelled guns, whilst the horses are being saddled. The horn is sounded, and the driver, full of glee, collects his impatient hounds. The party is unexpectedly augmented by several welcome guests. Our intelligent friend HARRIS, from New-Jersey, has come to Carolina, to be initiated into the mysteries of Deer hunting, as a preparation to farther exploits on the Western prairies, among the elk and the buffalo; with him comes AUDUBON, the Nestor of American ornithology, and his son, together with Dr. WILSON. After the first greetings are over, we hasten to saddle additional horses for those of our guests, who are disposed to join us. The old ornithologist, having no relish for such boyish sports, sallies to the swamps in search of some rare species of woodpecker. We proceed to the drives, as they are called, viz., certain woods, separated by old fields and various openings, in some parts of which the Deer have their usual run, where the parties take their stands. These drives are designated by particular names, and we are familiar with Crane pond, Gum thicket, the Pasture, the Oak swamp, and a number of bays, one of which we would be willing to forget, for there we missed a Deer, and the bay was named after us, to our mortification. The driver is mounted on a hardy, active, and sure-footed horse, that he may be enabled to turn the course of the Deer, if he attempts to run back, or to stop the dogs. We were carried round to our stands by our host, when a Deer bounced up before us; in an instant a loud report is heard waking the echoes of the forest—the animal leaps high into the air, and tumbles to the ground. Thus, our venison is secured, and we carry on our farther operations from the mere love of sport. Anxious to give our friend HARRIS an opportunity of killing his first Deer, we place him at the best stand. Our mutual wishes are soon gratified. He is stationed at the edge of a bay—a valley overgrown with bay-trees (*Magnolia glauca*)—which from that day received the cognomen of Harris' bay. The hounds after considerable trailing rouse two noble bucks, one of them bounds out near our friend. He is obliged to be ready in a moment, before the Deer comes in the line with another hunter. At the report of his gun we perceive that the buck is wounded. "Mind," cries out friend WILSON, "your shot have whistled past me." Friend H. grows pale at the thought of having endangered the life of another, but we comfort him by stating, that his shot had not reached within fifty yards of the nervous hunter, and moreover, that the old buck was wounded and would soon be his. We observed where he had laid down in the grass, and was started up again by the dogs. Now for a chase of a wounded buck. He takes through an old field once planted with cotton, now full of ruts and ditches, and grown up with tall broom-grass. We agree to let the boys have the pleasure of the chase whilst we are the silent spectators. They bound over ditches and old corn-fields, firing as they run. Suddenly the hounds become silent, and then the loud sounding of the horn is heard mingled with the whoops of the hunters, which inform us, that the game is secured; it proves to be a majestic buck. The successful hunter is now obliged to submit to the ordeal of all who have fleshed their maiden sword, and killed their first Deer. "I submit," he said good naturedly, "but spare my spectacles and whiskers." So his forehead and cheeks were crossed with the red blood of the buck, and the tail was stuck in his cap. The hunt proceeded merrily and successfully. Young AUDUBON, however, had not yet obtained a shot. At length a Deer was started near our host. He would not shoot it, but strove to drive it to his neighbour. He ran after it, and shouted, stumbled over a root, and in the fall threw off his spectacles; but as he was groping for them among the leaves, he ascertained that his generous efforts had been successful; the Deer had been turned to Mr. AUDUBON. One barrel snapped—then came a sharp report from the other—a loud whoop succeeded, and we soon ascertained that another Deer had fallen. We now conceived that we had our wishes for a successful hunt fully gratified; the dinner hour had arrived. Five noble Deer were strung upon the old pecan-nut tree in sight of our festive hall. The evening passed off in pleasant conversation—some of those present displayed their wit and poetical talents by giving the details of the hunt in an amusing ballad, which however has not yet found its way into print. Thus ended a Carolina Deer hunt.

We regret to be obliged to state, that the Deer are rapidly disappearing from causes that ought not to exist. There are at present not one-fifth of the number of Deer in Carolina that existed twenty years ago. In the Northern and Middle States, where the farms have been sub-divided, and the forests necessarily cleared, the Deer have disappeared because there was no cover to shelter them. In the Southern States, however, where there are immense swamps subject to constant inundations and pine barrens too poor for cultivation, they would remain undiminished in numbers were it not for the idle and cruel practice of destroying them by firelight, and hunting them in the spring and summer seasons by overseers and idlers. There is a law of the State forbidding the killing of Deer during certain months in the year. It is, however, never enforced, and Deer are exposed for sale in the markets of Charleston and Savannah at all seasons. In some neighbourhoods, where they were formerly abundant, now none exist, and the planters have given up their hounds. In New-Jersey and Long Island, where the game laws are strictly enforced, Deer are said to be on the increase. In some parts of Carolina, where the woods are enclosed with fences, not sufficiently high to prevent the Deer from straying out, but sufficient to prevent the hunters from persecuting them in summer, they have greatly multiplied and stocked the surrounding neighbourhoods. If judicious laws were framed and strictly enforced the Deer could be preserved for ages in all our Southern States, and we cannot refrain from submitting this subject to the consideration of our southern legislators.

GEOGRAPHICAL DISTRIBUTION.

This animal is found in the State of Maine; north of this it is replaced by larger species, the moose and reindeer. It exists sparingly in Upper Canada. In all the Atlantic States it is still found, although in diminished numbers. Where care has been used to prevent its being hunted at unseasonable periods of the year, as in New-York and New-Jersey, it is said to be rather on the increase. In the mountainous portions of Virginia it is hunted with success. It is still rather common in North and South Carolina, Georgia and Florida, especially in barren or swampy regions, of which vast tracts remain uncultivated. In Mississippi, Missouri, Arkansas, and Texas, it supplies many of the less industrious inhabitants with a considerable portion of their food. It is very abundant in Texas and New Mexico, and is a common species in the northern parts of Mexico. We cannot say with confidence that it exists in Oregon, and in California it is replaced by the black tailed Deer.—*C. Richardsonii.*

GENERAL REMARKS.

This species has been given under different names, and we might have added a long list of synonymes. The specimens we saw in Maine and at Niagara were nearly double the size of those on the hunting islands in South Carolina. The Deer that reside permanently in the swamps of Carolina are taller and longer legged than those in the higher grounds. The deer of the mountains are larger than those on the sea-board, yet these differences, the result of food or climate, will not warrant us in multiplying them into different species.

CANIS LUPUS.—Linn: Var. Rufus.

RED TEXAN WOLF.

PLATE LXXXII.—Male.

C. Colore supra inter fulvum nigrum variante, subtus dilutior; cauda apice nigro.

CHARACTERS.

Varied with red and black above, lighter beneath.   End of tail black.

DESCRIPTION.

In shape the Red Texan Wolf resembles the common gray variety. It is more slender and lighter than the white Wolf of the North West, and has a more cunning fox-like appearance. The hairs on the body are not woolly like those of the latter but lie smooth and flat. Its body and legs are long, nose pointed, and ears erect.

COLOUR.

The body above is reddish-brown mixed up with irregular patches of black; the shorter hairs being light yellowish-brown at the roots, deepening into reddish at the tips; many of the longer hairs interspersed are black from the roots through their whole extent. Nose, outer surface of ears, neck, and legs, chestnut-brown, a shade paler on the under surface. There is a brown stripe on the fore-legs extending from the shoulders to near the paws. Moustaches few and black; inner surfaces of ears soiled-white; nails black; along the upper lip, under the chin, and on the throat, grayish-white. Upper surface and end of tail, as well as a broad band across the middle portion, black.

DIMENSIONS.

	Ft.	Inches.
From point of nose to root of tail, - - - -	2	11
Tail, - - - - - - - -	1	1

Plate LXXXII

Drawn from Nature by J.W. Audubon.

On stone by W.E. Hitchcock

Lith.ª Printed & Col.ª by J.T. Bowen, Philad.ª

Red Texan Wolf.

HABITS.

This variety is by no means the only one found in Texas, where Wolves, black, white and gray, are to be met with from time to time. We do not think, however, that this Red Wolf is an inhabitant of the more northerly prairies, or even of the lower Mississippi bottoms, and have, therefore, called him the Red Texan Wolf.

The habits of this variety are nearly similar to those of the black and the white Wolf, which we have already described, differing somewhat, owing to local causes, but showing the same sneaking, cowardly, yet ferocious disposition.

It is said that when visiting battle-fields in Mexico, the Wolves preferred the slain Texans or Americans, to the Mexicans, and only ate the bodies of the latter from necessity, as owing to the quantity of pepper used by the Mexicans in their food, their flesh is impregnated with that powerful stimulant. Not vouching for this story, however, the fact is well known that these animals follow the movements of armies, or at least are always at hand to prey upon the slain before their comrades can give them a soldier's burial, or even after that mournful rite; and if anything could increase the horrors displayed by the gory ensanguined field, where man has slain his fellows by thousands, it would be the presence of packs of these ravenous beasts disputing for the carcasses of the brave, the young, and the patriotic, who have fallen for their country's honour!

No corpse of wounded straggler from his troop, or of unfortunate traveller, butchered by Camanches, is ever "neglected" by the prowling Wolf, and he quarrels in his fierce hunger in his turn over the victim of similar violent passions exhibited by man!

The Wolf is met on the prairies from time to time as the traveller slowly winds his way. We will here give an extract from the journal kept by J. W. AUDUBON while in Texas, which shows the audacity of this animal, and gives us a little bit of an adventure with a hungry one, related by POWELL, one of the gallant Texan Rangers.

" Like all travellers, the ranger rides over the wide prairie in long silences of either deep thought or listless musings, I have never been able to decide which; but when, riding by the side of WALKER or HAYS, who would like to say that a vacant mind was ever in the broad brow or behind the sparkling eye either of him with the gray, or of him with the brown? but at times when watching closely I have thought I could trace in the varying expression, castle after castle mounting higher and higher, till a creek ' to water at,' or a deer which had been sound asleep and to

windward of us, started some 30 or 40 yards off our path to wake up the
dreamers of our party.   No one is certain that his queries will be welcome to the backwoodsman on a march through a strange country, any
more than would be those of a passenger, put to the captain of a vessel as he
leans over the weather-rail looking what the wind will be, or thinking of
the disagreeable bustle he will have, when he gets into port, compared
to his lazy luxury on shipboard : but as I rode by the side of POWELL we
started no deer, nor came to a ' water hole,' but a Red Wolf jumped up
some two or three hundred yards from us, and took to the lazy gallop so
common to this species ; ' Run you ———,' cried POWELL, and he sent a
yell after him that would have done credit to red or white man for its
shrill and startling effect, the Wolf's tail dropped lower than usual, and
now it would have taken a racer to have overtaken him in a mile ; a
laugh from POWELL, and another yell, which as the sound reached the Wolf
made him jump again, and POWELL turned to me with a chuckle, and
said, ' I had the nicest trick played me by one of those rascals you ever
heard of.' The simple, how was it, or let's have it, was all that he wanted,
and he began at the beginning. ' I was out on a survey about 15 miles
west of Austin, in a range that we didn't care about shooting in any more
than we could help, for the Camanches were all over the country ; and
having killed a deer in the morning, I took the ribs off one side and wrap-
ping them in a piece of the skin, tied it to my saddle and carried it all day,
so as to have a supper at night without hunting for it ; it was a dark, dismal
day, and I was cold and hungry when I got to where I was to camp to wait
for the rest of the party to come up next day ; I made my fire, untied my
precious parcel, for it was now dark, with two sticks put up my ribs
to roast, and walked off to rub down and secure my horse, while they
were cooking ; but in the midst of my arrangements I heard a stick crack,
and as that in an Indian country means something, I turned and saw, to
my amazement, for I thought no animal would go near the fire, a large
Red Wolf actually stealing ' my ribs' as they roasted ; instinct made me
draw a pistol and ' let drive' at him ; the smoke came in my face and I saw
nothing but that my whole supper was gone. So not in the most
philosophical manner I lay down, supperless, on my blanket ; at daylight
I was up to look out for breakfast, and to my surprise, my half-cooked ribs
lay within twenty feet of the fire, and the Wolf about twenty yards off,
dead ; my ball having been as well aimed as if in broad daylight."

    We have represented a fine specimen of this Wolf, on a sand-bar, snuff-
ing at the bone of a buffalo, which, alas ! is the **only fragment of " animal matter" he has in prospect for breakfast.**

GEOGRAPHICAL DISTRIBUTION.

In all species of quadrupeds that are widely diffused over our continent, it has often appeared to us that toward the north they are more subject to become white—toward the east or Atlantic side gray—to the south black—and toward the west red. The gray squirrel, (*S. migratorius*), of the Northern and Eastern States presents many varieties of red as we proceed westwardly towards Ohio. In the south, the fox squirrel in the maritime districts is black as well as gray, but not red. On proceeding westwardly, however, through Georgia and Alabama, a great many are found of a rufous colour. In Louisiana, there are in the southern parts two species permanently black as well as the foxsquirrel, which in about half the specimens are found black, and the remainder reddish. The same may be said in regard to the Wolves. In the north there is a tendency towards white—hence great numbers are of that colour. Along the Atlantic coast, in the Middle and Northern States, the majority are gray. To the south, in Florida, the prevailing colour is black, and in Texas and the southwest the colour is generally reddish. It is difficult to account, on any principles of science, for this remarkable peculiarity, which forms a subject of curious speculation.

This variety of Wolf is traced from the northern parts of the State of Arkansas, southerly through Texas into Mexico ; we are not informed of its southern limits.

GENERAL REMARKS.

The Wolves present so many shades of colour that we have not ventured to regard this as a distinct species ; more especially as it breeds with those of other colours, gangs of Wolves being seen, in which this variety is mixed up with both the gray and black.

GENUS LAGOMYS.—Geoff.

DENTAL FORMULA.

$$Incisive \; \frac{2-2}{1-1} \quad Canine \; \frac{0-0}{0-0}; \quad Molar \; \frac{5-5}{5-5} = 26.$$

Teeth and toes similar to those of the genus Lepus, upper incisors in pairs, two in front and two immediately behind them, the former large and the latter small.

Ears moderate ; eyes, round ; hind legs not much longer than fore legs ; fur under the feet ; no tail ; mammæ four or six ; clavicles nearly perfect.

Native of cold and Alpine regions. They lay up stores for winter provision which is never done by the true hares. They have a call-note resembling that of some species of *Tamiæ*.

The name of this sub-genus, *Lagomys*, is derived from the Greek words λαγως, (*lagos*), a Hare, and μυς, (*mus*), a Mouse.

Four species of this genus are described ; one, the *Pika*, exists in the northern mountains of the Old World, one in Mongolian Tartary, one in the south eastern parts of Russia, and one in the Rocky Mountains of North America.

LAGOMYS PRINCEPS.—Richardson.

Little-Chief Hare.

PLATE LXXXIII.—Males.

L. Ecaudatus, fuscus, latere pallidior, subtus griseus, capite brevi ; auriculis rotundatis.

CHARACTERS.

Tailless ; colour blackish brown, beneath gray ; head short and thick ; ears rounded.

Plate LXXXIII

Drawn from Nature by J.J.Audubon, F.R.S.F.L.S.

On Stone by Wm E. Hitchcock

Lith Printed & Col.d by J.T. Bowen, Phil.

Little Chief Hare.

LEPUS (LAGOMYS PRINCEPS).   Rich. Fauna B. Am. p. 227.
  "   "   "   Fischer's Mamalium. p. 503.

DESCRIPTION.

"On comparing the skull of this animal with that of a true Hare, there appears a larger cavity in proportion to its size, for the reception of the brain. The breadth of the skull, too, behind, is increased by very large and spongy processes. The bone anterior to the orbit is not cribriform as in the Hares, although it is thin, and there is no depression of the frontal bone between the orbits.

The upper anterior incisors are marked with a deep furrow near their anterior margins, and have cutting edges which present conjointly three well marked points, the middle one of which is common to both teeth, and is shorter than the exterior one. These incisiors are much thinner than the incisors of the Hare, and are scooped out like a gouge behind. The small round posterior or accessary upper incisors, have flat summits. The lower incisors are thinner than those of the Hares, and are chamfered away toward their summits, more in the form of a gouge than like the chisel-shaped-edge of the incisors of a Hare.

Grinders.—The upper grinders are not very dissimilar to those of the Hare, on the crowns, but the transverse plates of enamel are more distinct. They differ in each tooth having a very deep furrow on its inner side, which separates the folds of enamel. This furrow is nearly obsolete in the Hares, whilst in the *lagomys* it is as conspicuous as the separation betwixt the teeth. The small posterior grinder which exists in the upper jaw of the adult Hare is entirely wanting in the different specimens of the Little-Chief Hare which I have examined. The lower grinders, from the depths of their lateral grooves, have at first sight a greater resemblance to the grinders of some animal belonging to the genus *Arvicola* than those of a Hare ; their crowns exhibit a single series of acute-triangles with hollow areas. The first grinder has three not very deep grooves on a side, and is not so unlike the corresponding tooth of a Hare as those which succeed it. The second, third, and fourth, have each a groove in both sides so deep as nearly to divide the tooth, and each of the crowns exhibits two triangular folds of enamel. The posterior grinder forms only one triangle."—(RICHARDSON).

In size this species is a little smaller than the alpine *pika* of Siberia. The body is thick ; the head broad and short, and the forehead arched. The ears are ovate, and do not appear to have any incurvations on their inner margins. The eyes are small, resembling those of the *arvicolæ ;* there is a marked prominent tubercle at the root of each claw.

COLOUR.

The Little-Chief Hare is, on the upper surface dark brown, varied with irregular bands of brownish-black running from the sides across the back. There are slight variations in different specimens, some having these blackish markings more distinct than others. The fur is, for three-fourths of its length, of a grayish-black colour, then partly yellowish-brown and white; on the sides of the head and fore shoulders this yellowish-brown colour prevails more than in other parts. The ears are bordered with white; the whole under surface is yellowish-gray, and the small protuberance, which represents the tail, light coloured.

DIMENSIONS.

	Inches.
Length of head and body - - - - -	$6\frac{1}{2}$
" from nose to eye - - - - -	$\frac{3}{4}$
Breadth of ear - - - - - -	$\frac{3}{4}$
Fur on the back - - - - - -	$\frac{3}{4}$
Length of head - - - - - -	$2\frac{1}{4}$
Height of ear - - - - - - -	1
Length of heel - - - - - -	$1\frac{1}{8}$

HABITS.

Little is known with regard to the habits of this animal.
The following extract is made from the Fauna Boreali Americana:

"Mr. DRUMMOND informs me, that the Little-Chief Hare frequents heaps of loose stones, through the interstices of which it makes its way with great facility. It is often seen at sunset, mounted on a stone, and calling to its mate by a peculiar shrill whistle. On the approach of man, it utters a feeble cry, like the squeak of a rabbit when hurt, and instantly disappears, to reappear in a minute or two, at the distance of twenty or thirty yards, if the object of its apprehension remains stationary. On the least movement of the intruder, it instantly conceals itself again, repeating its cry of fear; which, when there are several of the animals in the neighbourhood, is passed from one to the other. Mr. DRUMMOND describes their cry as very deceptive, and as appearing to come from an animal at a great distance, whilst in fact the little creature is close at hand; and if seated on a grey limestone rock, is so similar, that it can scarcely be discovered. These animals feed on vegetables. Mr. DRUMMOND never found their burrows, and he thinks they do not make any, but that they construct their nests among the stones. He does not know whether they store up hay for winter or not, but is certain, that they "do not come abroad during that season."

To the above account, it affords us pleasure to annex the extract of a letter, which we received from Mr. NUTTALL on the same subject.

Of this curious species of Lepus, (*L. princeps* of RICHARDSON), we were not fortunate enough to obtain any good specimens. I found its range to be in that latitude (42°) almost entirely alpine. I first discovered it by its peculiar cry, far up the mountain of the dividing ridge between the waters of the Columbia and Colorado, and the Missouri, hiding amongst loose piles of rocks, such as you generally see beneath broken cliffs. From this retreat I heard a slender, but very distinct bleat, so like that of a young kid or goat, that I at first concluded it to be such a call; but in vain trying to discover any large animal around me, at length I may almost literally say, the mountain brought forth nothing much larger than a mouse, as I discovered that this little animal was the real author of this unexpected note."

GEOGRAPHICAL DISTRIBUTION.

Dr. RICHARDSON states, that this animal inhabits the Rocky Mountains from latitude 52° to 60° The specimen of Mr. TOWNSEND was procured in latitude 42°, and therefore within the limits of the United States.

GENERAL REMARKS.

Until recently it was not supposed, that we had in America any species of this genus. We have compared it with the Pika, (*Lagomys alpinus*), of the Eastern continent, described by PALLAS. Our animal is not only of smaller size, but differs from it in the formation of the skull and several other particulars.

SPERMOPHILUS FRANKLINII.—Sabine.

Franklin's Marmot Squirrel.

PLATE LXXXIV.—Male and Female.

S. corpore super cervino ferrugineave creberrimè nigro maculato subter albido, vultu ex nigro canescenti, caudâ elongata cylindricâ pilis albis nigro ter quatorve torquatis vestita.

CHARACTERS.

Cheek pouches, the upper surface of the body spotted thickly with black, on a yellowish-brown ground, under surface grayish-white; face black and white, intimately and equally mixed; tail long, cylindrical, and clothed with hairs which are ringed alternately with black and white.

SYNONYMES.

Arctomys Franklinii. Sabine. Linnean Transactions, Vol. 13, p. 19.
" " Franklin's Journey, p. 662.
" " Harlan's Fauna, p. 167.
" " Godman, Nat. Hist. Vol. 2d p. 109.
" " Richardson, F. B. Am. p. 168. pl. 12.

DESCRIPTION.

Franklin's Marmot is about the size of the Carolina Gray Squirrel, and resembles it in form, its ears however are shorter, and its tail, which is narrower, presents a less distichous appearance. The ears have an erect rounded flap, and although not as large as those of *S. Douglassii*, are prominent, rising above the fur considerably more than those of *S. Richardsonii* or *S. Annulatus*. The body is rather slender for this genus; eyes large and rather prominent; cheek pouches small; moustaches few and short.

The legs are shorter than those of the squirrels, and stouter than those of *S. Annulatus*. The thumb has one joint, with a small nail; the second toe from the inside is the longest; the palms are naked. The soles of the hind feet are hairy for about two-thirds of their length from

Plate LXXXIV.

On stone by W.E Hitchcock

Drawn from Nature by J.J.Audubon F.R.S.F.L.S

Franklin's Marmot Squirrel.

Lith⁴ Printed & Col⁴ by J.T.Bowen Phila⁵

the heels. The claws are nearly straight being much less hooked than those of *S. Annulatus.*

The hair is rather coarse, and the under fur not very dense.

The tail is clothed with hair, but has on it no under fur. It is capable of a somewhat distichous arrangement, but as we are informed by Sir JOHN RICHARDSON, when this animal is pursued, the tail is cylindrical, the hairs standing out in every direction. The hind feet, when stretched out, reach to the middle of the tail.

COLOUR.

Incisors orange ; eyes and whiskers, black ; nails, dark-brown ; the septum and naked margins of the nostrils, and margins of the lips are of a light flesh-colour ; eyelids, white : below the nostrils, sides of face, chin, and throat, yellowish-white. Upper parts of the head to beyond the ears and neck, light brindled-gray, composed of blackish hairs tipped with white, without any admixture of brown. The hairs on the back, are at the roots, plumbeous, then brown, succeeded by a line of black, and finally tipped with brown, giving it on the back a brownish-speckled appearance. On the chest and inner surfaces of legs white, with a slight brownish tinge. The hairs on the tail are barred with black and white ; they are light-coloured at the roots, then twice barred with black and white, and broadly tipped with white. Towards the extremity of the tail there is a broader black bar, the apical portion being white. When the tail is distichously arranged it presents two indistinct longitudinal stripes of black.

DIMENSIONS.

							Inches.
From point of nose to insertion of tail,	-	•	-			-	9¾
Tail (vertebræ),	-	•	-	-	•	•	4⅜
To end of hair,	-	•	-	•	•	•	5¾
From heel to end of middle claw,	-	•	-	•		-	2
Height of ear,	-	•	-	-	•	•	¼

HABITS.

We possess but little information of the habits of several of the Spermophili of America. None of the species are found in the settled portions of our country, where opportunities are afforded the naturalist to observe and note down their habits ; every one has undoubtedly an interesting history attached to its life, which yet remains to be collected and written.

RICHARDSON observes of this species, that it lives in burrows in the sandy soil amongst the little thickets of brushwood that skirt the plains. That it is about three weeks later in its appearance in the spring than the *Arctomys Richardsonii*, probably from the snow lying longer on the shady places it inhabits, than on the open plains frequented by the latter. It runs on the ground with considerable rapidity, but has not been seen to ascend trees. It has a louder and harsher voice than the *A. Richardsonii*, more resembling that of *Sciurus Hudsonius* when terrified. Its food consists principally of the seeds of liguminous plants, which it can procure in considerable quantity as soon as the snow melts and exposes the crop of the preceding year. Mr. TOWNSEND, who observed it in Oregon, does not refer particularly to any habit differing from the above.

GEOGRAPHICAL DISTRIBUTION.

This is a northern and western species ; Dr. RICHARDSON having obtained it in the neighbourhood of Carlton House, and TOWNSEND near the Columbia River.

GENERAL REMARKS.

Although several different Spermophiles bear a strong resemblance to each other, we have not observed that this species has as yet been mistaken for any other, and it has as far as we can ascertain retained its name without change in the works of all new describers.

Plate LXXV.

On Stone by W.ᵐ E. Hitchcock

Lith. Printed & Col.ᵈ by J.T. Bowen, Phil.ᵃ

Jumping Mouse.

GENUS MERIONES.—Illiger.

DENTAL FORMULA.

$$Incisive \; \frac{2}{2}; \quad Canine \; \frac{0-0}{0-0}; \quad Molar \; \frac{3-3}{3-3} = 16.$$

Cheek-teeth tuberculous, the first with three, the second with two, and the third with one, tubercle.

Nose sharp, ears moderate; fore-feet short, with the rudiment of a thumb; hind legs long, terminated by five toes with nails, each with a distinct metatarsus. Tail, very long and slender; mammæ. from two to four pectoral, and from two to four abdominal.

Habits nocturnal, many hibernate.

There have been eleven species described as belonging to this genus, as it is now restricted; one well determined species has been discovered in North America, the rest are found in sandy and elevated regions, in parts of Asia and Africa.

The word Meriones is derived from the Gr. μηριον, (mĕriŏn), the thigh.

MERIONES HUDSONICUS.—Zimmerman.

Jumping Mouse.

PLATE LXXXV.—Male and Female.

M. Supra saturate fuscus, infra albus, lineâ laterali flava inter colorem fuscom albumque intermedia; caudâ corpore longiore.

CHARACTERS.

Dark reddish-brown above, with white underneath; sides yellow, separating the colours of the back from the white beneath; tail much longer than the body.

SYNONYMES.

Dipus Hudsonicus. Zimmerman. Geogr. Geschich., II. p.
" Americanus. Barton, Am. Phil. Trans., 4. vol. p. 358—262. A. D. 1782.
" Canadensis. Davies' Linn. Trans., 4. 155.

GERBILLE DU CANADA. Desm. Mammal., p. 132.
       "   "   Fr. Cuvier in Dict. des Sc. Nat., 18. p. 464.
MERIONES LABRADORIUS. Sabine, Franklin's Journ., p. 155 and 157.
G. CANADENSIS ET LABRADORIUS. Harlan, Fauna, p. 155 and 157.
       "   "   Godman, vol. 2. p. 94 and 97.
MERIONES LABRADORIUS. Richardson, Fau. Bore. Am., p. 144.
       "   AMERICANUS. De Kay. Nat. Hist. N. Y., p. 71. pl. XXIV., fig. 2d.

DESCRIPTION.

Head, narrow and conical.   Nose, tolerably sharp, with an obtuse tip
projecting a little beyond the incisors.   Nostrils small, facing sideways
and protected anteriorly by a slight ventricose arching of their naked
inner margins.   The mouth is small and far back.   Whiskers, long,
extending to the shoulder; eyes. small; ears, semi-oval, rounded at
the tips, clothed on both surfaces with short hair.   Fore feet small, nail in
place of a thumb; hind legs long and slender; there are five hind-toes,
each with a long slender tarsal bone; the toes, when expanded, resembling
those of some species of birds.   The soles are naked to the heels; upper
surface of hind-feet covered with short adpressed hairs; tail, long, scaly,
has a velvety appearance, soft to the touch, is thinly covered with such
soft short hairs, that without a close examination it would appear naked.
The hair on the body is of moderate fineness, and lies smooth and
compact.

COLOUR.

Upper surface of nose, forehead, neck, ears, and a broad line on the
back, dark-brown; the hairs being plumbeous at their roots, tipped with
yellowish-brown and black; under the nose, along the sides of the face,
outer surface of the legs, and along the sides, yellowish; lips, chin, and
all the under surface white; as is also the under surface of the tail in some
specimens, though in others brownish-white.   The colours between the
back and sides, as well as between the sides and belly, are in most specimens separated by a distinct line of demarcation.   This species is subject
to considerable variations in colour.   We have seen some young animals, in which the dark reddish-brown stripe along the back was wholly
wanting; others where the line of demarcation between the colours was
very indistinct : nearly all are pure white on the under surface; but we
possess two specimens that are tinged on those parts with a yellowish
hue.

DIMENSIONS.

						Inches.
Length of head and body	-	-	-	-	-	$2\frac{5}{8}$
do of tail	-	-	-	-	-	$4\frac{3}{4}$
Height of ear posteriorly	-	-	-	-	-	$\frac{1}{4}$
From heel to longest nail	-	-	-	-	-	$1\frac{7}{8}$

HABITS.

This species was familar to us in early life, and we possessed many opportunities of studying its peculiar and very interesting habits. We doubt whether there is any quadruped in the world of its size, that can make its way over the ground as rapidly, or one that can in an open space so quickly evade the grasp of its pursuers. The ploughman in the Northern and Middle States, sometimes turns up this species from under a clod of earth, when it immediately commences its long leaps. He drops his reins and hurries after it; whilst the little creature darts off with great agility, pursuing an irregular zig-zag direction, and it requires an active runner to keep pace with it, as it alternately rises and sinks like the flying-fish at sea, and ere the pursuer is aware, is out of sight, hidden probably behind some clod, or concealed under a tuft of grass. We have frequently seen these mice start from small stacks of wheat, where the bundles had been temporarily collected previous to their being removed to the barn. In such cases they usually effect their escape among the grass and stubble. A rapid movement seems natural to this animal, and is often exhibited when it is not under the influence of fear, and apparently for mere amusement. Our kind friend Maj. Le Conte, now of New-York, informs us, that he has seen it in former times, near the northern end of the Island of New-York, springing from the ground and passing with the velocity of a bird, until its momentum being exhausted it disappeared in the tall grass, apparently with ease and grace, again springing forth in the same manner. It must not, however, from hence be believed that the Jumping Mouse walks on its hind feet only, and progresses at all times by leaps, without using its fore-feet. We have frequently seen it walking leisurely on all its feet, in the manner of the white-footed mouse. It is chiefly when alarmed, or on special occasions, that it makes these unusual leaps; the construction of the body proves that this species could not for any length of time be sustained on its tarsi. In its leaps we have always observed that it falls on all its four feet.

We experienced no difficulty in capturing this species in box-traps, and preserved a female in a cage from spring to autumn ; she produced two young a few days after being caught ; she reared both of them, and they had become nearly of full size before autumn, when by some accident our pets escaped. We placed a foot of earth at the bottom of the cage, in this they formed a burrow with two outlets. They used their feet and nails to advantage, as we observed them bury themselves in the earth, in a very short time. They were usually very silent, but when we placed a common mouse in the cage, squeaked with a loud chattering noise, like some young bird in pain. They skipped about the cage, were anxious to make their escape from the mouse, and convinced us that this species is very timid. They were in their habits strictly nocturnal, scarcely ever coming out of their holes during the day, but rattling about the wires of the cage throughout the night.

We observed that every thing that was put into their cage, however great might be the quantity, was stored away in their holes before the next morning. We fed them on wheat, maize, and buckwheat. They gave the preference to the latter, and we observed that when they had filled their store-house with a quart of buckwheat, they immediately formed a new burrow in which they deposited the surplus.

We are inclined to believe that this species produces several times during the summer, as we have seen the young on several occasions in May and August ; They are from two to four ; we have usually found three.

The fact of the females being frequently seen with the young attached to their teats, carrying them along in their flight when disturbed, is well ascertained. We have also observed this in several other species ; in the white-footed mouse, the Florida rat, and even the common flying squirrel. We are not, however, to argue from this that the young immediately after birth become attached to the teats in the manner of the young opossoms, and are incapable of relaxing their hold ; on the contrary the female we had in confinement, only dragged her young along with her, when she was suddenly disturbed, and when in the act of giving suck ; but when she came out, of her own accord, we observed that she had relieved herself from this incumbrance. This was also the case with the other species referred to.

Dr. DEKAY, regards it as a matter of course that in its long leaps, it is aided by the tail. We doubt whether the tail is used in the manner of the kangaru ; the under surface of it is never worn in the slightest manner, and exhibits no evidence of its having been used as a propeller. Its long heel and peculiarly long slender tarsal bones on each toe, seem in themselves sufficient to produce those very long leaps. We have often watched this species, and although it moves with such celerity as to render an examination very difficult, we have been able to decide, as we think, that the tail is not used by the animal in its surprising leaps and rapid movements.

The domicil of the Jumping Mouse in summer, in which her young are produced, we have always found near the surface, seldom more than six inches under ground, sometimes under fences and brushwood, but more generally under clods of earth, where the sward had been turned over in early spring, leaving hollow spaces beneath, convenient for the summer residence of the animal. The nest is composed of fine grass, mixed with which we have sometimes seen feathers, wool, and hair.

We are, however, under an impression that the Jumping Mouse in winter resorts to a burrow situated much deeper in the earth, and beyond the influence of severe frosts, as when fields were ploughed late in autumn, we could never obtain any of this species. It may be stated as a general observation, that this animal is a resident of fields and cultivated grounds; we have, however, witnessed two or three exceptions to this habit, having caught some in traps set at night in the woods, and once having found a nest under the roots of a tree in the forest, occupied by an old female of this species with three young two-thirds grown; this nest contained about a handful of chestnuts, which had fallen from the surrounding trees.

It is generally believed, that the Jumping Mouse, like the Hampster of Europe, (*Cricetus vulgaris*), and the Marmots, (*Arctomys*), hibernates, and passes the winter in a profound lethargy. Although we made some efforts many years ago, to place this matter beyond a doubt by personal observation, we regret that our residence, being in a region where this species does not exist, no favourable opportunity has since been afforded us.

Naturalists residing in the Northern and Middle States could easily solve the whole matter, by preserving the animal in confinement through the winter.

To us the Jumping Mouse has not been an abundant species in any part of our country. Being, however, a nocturnal animal, rarely seen during the day unless disturbed, it is in reality more numerous than is generally supposed. We have frequently caught it in traps at night in localities where its existence was scarcely known.

This species, feeding on small seeds, does very little injury to the farmer; it serves, like the sparrow, to lessen the superabundance of grass seeds, which are injurious to the growth of wheat and other grains; it is fond of the seeds of several species of *Amaranthus*, the pigweed, (*Ambrosia*), burr-marygold, beggar or sheep ticks, (*Bidens*), all of which are regarded as pests, he therefore should not grumble at the loss of a few grains of wheat or buckwheat.   Its enemies are cats, owls, weasels, and foxes, which all devour it.

GEOGRAPHICAL DISTRIBUTION.

If there is no mistake in regarding all the varieties of Jumping Mice in the northern parts of America as one species, this little animal has a range nearly as extensive as that of the white-footed Mouse. It exists, according to RICHARDSON, as far to the North as great Slave Lake, Lat. 62°.   It is found in Labrador and Nova Scotia, and in Upper and Lower Canada. We have seen it in the Eastern and Middle States, and obtained a specimen on the mountains of Virginia, but have not traced it farther to the South ; although we are pretty sure that it may, like the *Sciurus Hudsonius* be found on the whole range of the Alleghanies. SAY observed it on the base of the Rocky Mountains, and Mr. TOWNSEND brought specimens from Oregon, near the mouth of the Columbia River.   We can scarcely doubt, that it will yet be discovered on both sides of the mountains in California and New-Mexico.

GENERAL REMARKS.

On looking at our synonymes our readers will discover that this species has been described under an endless variety of names.   We have omitted a reference to RAFINESQUE, who indicated several new species in the American Monthly Magazine.   We have concluded, that a writer exhibiting such a want of accuracy, who gives no characters by which the species can be known, and who has involved the science in great confusion, and given such infinite trouble to his successors, does not deserve to be quoted.

We had attached to our plate the specific name given by Dr. BAR-TON, (*M. Americanus*), this we would have preferred to either of the others, especially as it now seems probable, that this is the only species in North America.   The names *Hudsonius, Labradorius,* and *Canadensis,* are all exceptionable, as it appears to be as abundant in the Northern and Eastern States, as it is in Hudson's Bay, Labrador, or Canada.   There is an evident impropriety, although we confess when hard pressed for a name we have often committed the error ourselves, in naming species after localities where they have been found.   The *Meles Labradoria* of SABINE, and the *Lepus Virginianus* of HARLAN, are both familiar examples. Having recently had an opportunity of consulting the original description of ZIMMERMAN, published between the years 1778 and 1783, we are convinced that he was the first scientific describer, and we have accordingly adopted his name. BARTON, at a little later period, published a good description with a figure. DAVIES shortly afterwards published it under the name of *Dipus Canadensis*. SABINE published a specimen with a mutilated tail, which he named *M. Labradorius*, and RICHARDSON a specimen from the North, which he referred to the northern species, under the name of *M. Labradorius*, supposing there was still another species, which had been described as *G. Canadensis*. We have compared many specimens from all the localities indicated by authors. There is a considerable variety in colour, young animals being paler and having the lines of demarcation between the colours less distinct. There is also a great difference between the colour of the coat of hair in the spring, before it is shed, and that of the young hair which replaces the winter pelage. The tail varies a little, but is always long in all the specimens. The ears, size, and habits of all are similar. We have thus far seen no specimen that would warrant us in admitting more than one species into our American Fauna.

GENUS FELIS.—Linn

$$Incisive \; \tfrac{6}{6}; \; Canine \; \tfrac{1-1}{1-1} \; ; \; Molar \quad \tfrac{4-4}{3-3} = 30.$$

There are two conical teeth, or false molars, in the upper jaw, which are wanting in the genus *Lynx* ; a large carnivorous tooth with three lobes; the fourth cheek-tooth in the upper jaw nearly flat, and placed transversely ; the two anterior cheek-teeth in the lower jaw false.

Head, round ; ears, short and generally triangular, not tufted ; in many species a white spot on their outer surfaces ; no mane ; tail, long ; tongue roughened with prickles ; anterior extremities with five toes, posterior, with four ; nails curved, acute, and retractile.

Habit savage, feeding in a state of nature on living animals only, which they seize by surprise, and not by the chase, as is the habit of the dog wolf, &c.; leaping and climbing with facility ; speed moderate ; sense of sight good ; that of smell imperfect.

There are 33 species of *Long-tailed Cats* described, inhabiting the four quarters of the world. Four species only are positively known to exist north of the tropics in America.

The generic name is derived from the latin word *Felis*—a cat.

FELIS PARDALIS.—Linn.

Ocelot, or Leopard-Cat.

PLATE LXXXVI.—Male.—Winter Pelage.

F. Magnitudine. Lynx rufus. Cana. (*s. potius flava*), maculis ocellaribus magnis fulvis nigro-limbatis, in lateribus facias oblequas formantibus ; fronte striis 2 lateribus nigricantibus caudâ corporis longitudine dimedia.

CHARACTERS.

Size of the Bay Lynx ; general colour gray, marked with large fawn-coloured spots, bordered with black, forming oblique bands on the flanks ; two black lines bordering the forehead laterally.

Plate LXXXVI

On Stone by Wm E Hitchcock

Drawn from Nature by J. W. Audubon

106 Printed & Colᵈ by J. Bowen Philᵃ

Ocelot or Leopard Cat

FELIS PARDALIS. Linn., p. 62.
   "   "   Harlan's Fauna, p. 96.
   "   "   Cuv. An. King., vol. 2, p. 476.
   "   "   Griffith's An. King., vol. 5, p. 167.
   "   "   Shaw's Zoology, vol. 2d, p. 356.

DESCRIPTION.

Head, short ; neck, long and thin ; body, long and slender; tail, rather thick, and of moderate size ; hair, rather soft, and not very dense.

COLOUR.

The outer surface of the ear is black, with a white patch beneath ; chin and throat white, with a black bar immediately beneath the chin, and another under the neck. On the chest and under surface, white, with irregular black patches. There are small black spots disposed on the head, surrounded by reddish-brown, a black line runs longitudinally on the sides of the head to the neck. The whole back is marked with oval figures, and in some specimens with longitudinal black stripes edged with fawn-colour. Upper surface of the tail irregularly barred with black and white, the extremity black.

Specimens vary much in their markings, and we have not found two precisely alike.

DIMENSIONS.

Male, procured by Col. HARNEY in Texas, seven miles from San Antonio, December, 1845.

	Feet.	Inches.
From point of nose to root of tail,	2	11
Tail,	1	3
Height from nails to shoulder,	1	2
" of ear posteriorly,		$1\frac{3}{4}$

Female.

	Feet.	Inches.
Length of head and body	2	4
" tail	1	1
From nose to shoulder.	1	1

HABITS.

Before describing the habits of this beautiful species, we must enter into the difficult task of separating it from several other spotted, leopard-like cats, that have been confounded with it. Of these, the most similar in appearance is perhaps the *Felis mitis*, which is found in the tropical portions of North America, and in the warmer parts of South America.

The *Felis mitis* has in fact been figured, and described by Shaw, Vol. 2, p. 356, (unless we deceive ourselves), as the Ocelot, (our present species) while his figure of the *Jaguar*, (opposite p. 354), is probably drawn from the Ocelot, although, so poor a figure as to be hardly recognisable. The descriptions and figures of the Ocelot, that we find in old works on natural history, are so confusing, and unsatisfactory, that we are obliged to throw aside all reference to them in establishing any one of the feline tribe as our animal, and leave the reader to decide whether Buffon, speaking of the Ocelot, as two feet and a-half high and about four feet in length, meant the subject of our article, which is only two feet-six inches long from nose to root of tail, the *Felis mitis*, or the Jaguar ; and whether Pennant referred to the same animal, which he describes, when speaking of the Ocelot, "as about four times the size of a large cat," (about the size of our specimen of the Ocelot).

The description of this species in Linnæus is so short, that it is almost equally applicable to either the Jaguar, the Ocelot, or Felis mitis : "*Felis cauda elongata, corpore maculis superioribus virgatis, inferioribus orbiculatis.*" Sys. Nat. Gmel. p. 78. Brisson is also very concise in giving the character of the Ocelot ; *F. rufa, in ventre exalbo flavicans, maculis nigris in dorso longis, in ventre orbiculatis variegata.*" Quadr. 169. We are on the whole inclined to consider the species described by Pennant as the Mexican Cat, the Ocelot or Leopard-Cat of the present article, and the larger animal described by other authors, as the *Felis mitis*, as young of the Jaguar, or perhaps females of this last named species, and we have not yet met with the *Felis mitis* within our range, although we have seen such an animal alive in New-York, one having been brought by sea from Yucatan.

Our animal is quite well known in Texas as the Leopard-Cat, and in Mexico is called the Tiger-Cat, it is in the habit of concealing itself in hollows in trees, and also by squatting upon the larger branches. It is rather nocturnal, and preys upon the smaller quadrupeds, and on birds, eggs, &c., when they can be seized on the ground.

The activity and grace of the Leopard-Cat, are equal to the beauty of its fur, and it leaps with case amid the branches of trees, or runs with swiftness on the ground. These Cats seldom stray far from woods, or thickets bordering on rivers, streams, or ponds, very rarely lying on the hill-sides, or out on the plains.

They run like foxes, or wild-cats, when chased by the hunters with hounds or other dogs, doubling frequently, and using all the stratagems of the gray fox, before they take a straight course, but when hard pressed and fatigued, they always ascend a tree, instead of running to earth.

Like all the cat tribe, the Ocelot is spiteful when confined in a cage, and snarls and spits at the spectator when he draws near; but we have never seen it strike through the bars like the leopard, which sometimes inflicts severe wounds on the incautious or fool-hardy person, who, to see it better, approaches too closely its prison.

According to our information, the Ocelot only has two young at a litter, but we have not had an opportunity of ascertaining this point ourselves.

The specimen from which our figure was drawn, was procured by Gen. HARNEY, who sent it fresh killed to J. W. AUDUBON, then at San Antonio on an expedition in search of the quadrupeds of Texas, for our work. We here give an extract from his journal.

" But for the kindness of Col. HARNEY, I might never have made the drawing of this most beautiful of all the North American feline race. Col. HARNEY sent for my trunks, and while I waited the return of the sergeant's guard, who went to fetch them, I saw him daily. He introduced me to Mrs. BRADLY, where he and Capt. MYERS, afterwards my friend, boarded, and the lady of the house made it a home to me.

I was invited out to the camp, and as I talked of the animals I was most anxious to procure, all seemed desirous to aid me. Col. HARNEY, fond of field sports, as active and industrious as he was tall and magnificent-looking, waked at day light the lone prairies and swamps with shouts of encouragement to his small pack of well-chosen dogs, till they in turn burst forth in full cry on the hot trail of a magnificent specimen of this most interesting species. I had just returned from an examination of all my steel-traps; some were sprung, yet nothing but fur was left, showing that a strong wolf or lynx had been caught, but had pulled away ; thus preventing perhaps, the capture of some smaller animal that I wanted ; and rats, mice, skunks, or other little quadrupeds, were eaten nightly whilst fast in the steel teeth, by these prowlers. I sat down, to think of spring-guns, and long for means to prevent this robbery of my traps, when a sergeant came in, with the result of Col. HARNEY's morning's chase, the beautiful Ocelot, from which my drawing was made.

This was a new animal to me, as, though I knew of its existence, I had never seen one, so that my delight was only equalled by my desire to paint a good figure of it. Its beautiful skin makes a most favourite bullet pouch, and its variegated spots are only surpassed by the rich glossy coat and fur of the far famed ' black otter.' "

In his many long hunts, Col. HARNEY must have often and often past the lurking Wako and Camanche, who quailed at his soldierly bearing, while any other man would have had perchance a dozen arrows shot at him.

GEOGRAPHICAL DISTRIBUTION.

We have heard of an occasional specimen of this cat having been obtained in the southern parts of Louisiana. NUTTALL saw it in the State of Arkansas; our specimens were procured in Texas. It is common in Mexico; its southern range has not been accurately determined.

GENERAL REMARKS.

Much confusion still exists among writers in reference to the spotted cats of Mexico and South America, which can only be removed by the careful observations of naturalists in the native regions of these closely allied species.

Plate LXXXVII

American Red Fox.

VULPES FULVUS.—Desm

AMERICAN RED FOX.

PLATE LXXXVII.—Male.

V. Rufo-fulvoque varius ; collo subtus ventreque imo albis ; pectore cano ; antibrachiis antice prodiisque nigris ; digitis fulvis ; caudâ apice albâ.

CHARACTERS.

Fur reddish or fulvous ; beneath the neck and belly white ; chest gray ; front part of the fore legs and feet, black ; toes fulvous ; tip of the tail white.

SYNONYMES.

CANIS FULVUS. Desm. Mamm. p. 203.
    "   "   Fr. Cuvier, in Dict. des. Sc. Nat. VIII. p, 568.
RENARD DE VIRGINIE. Palesot de Beauvois Mem. Sur.
LE RENARD. Bullet, Soc. Phil.
RED FOX. Sabine, Franklin's Journ. p. 656.
CANIS FULVUS. Harlan, 89.
    "   "   Godman, vol. 1, p. 280.
VULPES FULVUS. Rich. Fauna, B. A. p. 91.
    "   "   De Kay, Nat. Hist. N. Y., p. 44, fig. 1, pl. 7.

DESCRIPTION.

This animal bears so strong a resemblance to the European Fox. (*v. vul_ garis*), that it was regarded as the same species by early naturalists. No one, however, who will compare specimens from both countries, can have a doubt of their being very distinct. Our Red Fox is a little the largest, its legs are less robust, its nose shorter and more pointed, the eyes nearer together, its feet and toes more thickly clothed with fur, its ears shorter, it has a finer and larger brush, and its fur is much softer, finer, and of a brighter colour.

It stands higher on its legs than the Gray Fox, and its muzzle is not so long and acute, as in that species. It is formed for lightness and speed, and is more perfect in its proportions than any other species in the genus with which we are acquainted.

The hair on the whole body is soft, silky, and lustrous ; the ears are clothed with short hairs on both surfaces, and the feet and toes are so clothed with hair, that the nails are concealed. The body of this species has a strong musky smell, far less disagreeable, however, than that of either the skunk or mink.   It becomes less offensive in a state of domestication.

COLOUR.

Point of nose, outer extremity of ears, and outer surfaces of legs below the knees, black; forehead, neck, flanks, and back, bright-reddish, and a little deeper tint on the back and fore-shoulders; around the nostrils, margins of the upper jaw, and chin, pure white; throat, breast and a narrow space on the under surface, dingy-white; extreme end of brush slightly tipped with white; inner surface of ears, and base of the outer surface, yellowish. The hair on the body is of two sorts: long hairs interspersed among a dense coat of softer, brighter, and more yellowish fur; on the tail the longer interspersed hairs are more numerous, and many of them are quite black, giving the tail a more dusky appearance than rest the of the body.

In addition to the distinct varieties of this species, the black and cross Fox, we have seen some shades of difference in colour in the red variety. In some the colours on the back are considerably darker than in others We have seen several with the nose and chin nearly black, and in others the white tip at the tail is replaced with black.

DIMENSIONS.

							Feet.	Inches.
From point of nose to root of tail,	-	-	-	-			2	6
Tail (vertebræ)	-	-	-	-	-	-	1	1
" to end of hair,	-	-	-	-	-	-	1	5
Height at shoulders,	-	-	-	-	-	-	1	1
" of ears posteriorly	-	-	-	-	-			2¾

HABITS.

This Fox, in times gone by, was comparatively rare in Virginia, and farther south was unknown. It is now seldom or never to be met with beyond Kentucky and Tennessee. Its early history is not ascertained, it was probably for a long time confounded with the Gray Fox, (which is in many parts of the country the most abundant species of the two,) and afterwards was supposed to have been imported from England, by some Fox-hunting governor of one of the "colonies." It was first distinguished from the Gray Fox and hunted, in Virginia; but now is known to exist in all the Northern States, and we are somewhat surprised that it should so long have been overlooked by our forefathers. No doubt, however, the cultivation and improvement of the whole country, is the chief reason why the Red Fox has become more numerous than it was before the Revolution, and it will probably be found going farther south and west, as the woods and forests give place to farms, with hens, chickens, tame turkeys, ducks, &c., in the barn-yards.

The Red Fox is far more active and enduring than the Gray, and generally runs in a more direct line, so that it always gives both dogs and hunters a good long chase, and where the hounds are not accustomed to follow, it will frequently beat-out the whole pack, and the horses and huntsmen to boot.

In some parts of the country, however, it is chased and killed with dogs, in fine style. The following account of the mode of taking the Red Fox, at the sea side in New-Jersey, near Cape May, is from an interesting letter written to us in December, 1845, by our friend EDWARD HARRIS, Esq., of Moorestown, in the neighbourhood of Philadelphia; it is quite different from the ordinary mode of hunting the Red Fox. He begins thus:

" On Saturday, a week ago, I went to Cape May Court-house, where I spent Monday and Tuesday among the quails, (perdrix virginianus), which I found exceedingly abundant, but the ground so bad for shooting, that in both days two of us shot but thirty-three birds. On Wednesday my friend Mr. HOLMES took me to BEASLEY's Point at the northern extremity of the county; here I was sorry to learn that young BEASLEY, who was to have returned from Philadelphia on the Saturday previous, had not yet made his appearance; his father, however, showed a great desire to forward my views in regard to "Monsieur Reynard." The next day it rained cats and dogs, and TOM BEASLEY did not arrive in the stage. In the afternoon it cleared off sufficiently to make a " a drive" in the point, where we started a noble specimen in beautiful pelage, but alas! he would not come near the standers.

The next morning, we drove the same ground, being the only place on the main land where there was any prospect of driving a Fox to standers without dogs, (of which there are none in the vicinity). This time we saw none. After dinner I took my pointer, and bagged eight brace and a half of quails, having this time found them on good ground. The next day, Saturday, with three drivers, and three standers, we drove the beach for five and a-half miles, without seeing a fox, and so ended this unsuccessful expedition. I had great hopes of this beach, (PECK's), as it had not been hunted since the winter before the last, although some of the gunners told me they had seen but few " signs" since that time.

The mode of driving, which requires no dogs, is for the drivers to be fur nished with two boards, or shingles, which they strike together, or with

what is better, a rattle, similar to a watchman's. The standers are sent
ahead to a narrow part of the beach, where the creeks of the salt-marshes
approach nearest to the sand-hills : when they are supposed to have reached their stands, the drivers enter, and walk abreast among the bushes,
between the sand-hills and the marshes, making all the noise they can,
with their lungs, as well as their boards or rattles ; and these unusual noises are almost sure to drive the Foxes to the standers, where
if they pass harmless, they have again to run the gauntlet to the
end of the beach, at the inlet, where, Mr. Beasley assures me, he
has known seven Red Foxes cornered, out of which four were killed, and
three escaped from bad shooting. We made four drives in the five and
a-half miles.

The facts in regard to the history of the Red Fox on the Jersey coast
that I have been able to collect, are few ; such as they are I will give
them to you.

Certain it is that they frequent the beaches in great numbers, and so far
as I can learn, the Gray Fox is not found in the same places, nor is the
raccoon, which we know to be so abundant on the sea islands and beaches
of our southern coast. They pass to the beaches on the ice, in the winter
season, when the " sounds" are frozen, and have frequently been seen in the
day time, making their passage, though doubtless it is more frequently performed in the night. Their means of subsistence there are ample, consisting
of wild fowl of various kinds, upon which they spring while they are asleep
upon the ponds and creeks, but more particularly upon the wounded fowl
which escape from the numerous gunners, also crabs and fish, which are
thrown up dead by the surf, and rabbits and wading birds, in the summer.
A marvellous story is told of their sagacity in selecting the food they like
best, which is vouched for by Mr. Beasley, and all the gunners along
shore, but which I think requires confirmation, at least so far as to have
the fish in question, seen by some naturalist in the state described by the
narrators, in order to ascertain its name, or describe it, if new, before its
publication is ventured on. The story is, that a certain fish, called the cramp-
fish, from its supposed power of paralizing the hand which touches it while
living, is thrown ashore dead, by the surf in the winter season, that every
one of these fishes contains a bird, such as the coot, (either *fusca* or *per-
spicillata*), or a gull, which appears to have destroyed the fish, by its prov-
ing rather hard to digest, without having been plucked. Mr. Fox finds
the fish that has come to this deplorable end, and either in the vain hope
of restoring animation to the unfortunate defunct, or for the gratification of
a less noble impulse, he makes a longitudinal incision into the peritonæum
of the subject, and extracts the bird, of which he makes a meal ; but, mind

you, Mr. Fox has profited by the awful example before him—he picks the bird before he eats it. Moral—never swallow what you cannot digest. But, to be serious, I do not mean to ridicule the fact, which I cannot but believe with the testimony which accompanies it, but if it be new, which I cannot answer for, it might in its plain, unvarnished form, without being announced in pedantic Latin, afford too tempting a morceau for the snarling critic. The fish are said to reach sometimes the length of four-feet, with a mouth twenty-two inches wide, they are scaled, and are said to resemble, somewhat, the sea cat-fish, with which I am not acquainted. The Fox on the beach when hunted by hounds, resorts to his usual trick of taking the water, to throw the dogs off the scent, by following the retreating surf, so that its return may efface his trail, then lying down among the sand hills to rest, while the dogs are at fault. In the woods on the main land both Red and Gray Foxes are abundant, the latter rather predominating. The Foxes are abundant on some of the beaches, and generally may be procured. Mr. Spencer, of Mount Holly, has been on a party when five were killed, but I do not know where, nor whether it was this season or before."

We have not been able to procure the fish which is alluded to in the foregoing, but have no doubt of the correctness of the account. The Red Fox will eat fish as well as birds, and when hard pressed does not refuse even carrion. It is, therefore, probable that the discovery of the bird within the dead fish, may be the result of accident rather than of instinct, reason, or keenness of smell on the part of the Fox; for when he begins to devour a fish he must soon find the more savoury bird in its stomach, and being fonder of fowl than of fish, he would of course eat the bird and leave the latter. A Fox after having in this way discovered coots, gulls, or any other bird, would undoubtedly examine any dead fish that he came across, in hopes of similar good luck. Hence the foxes on the beaches have, we suppose, acquired the habit of extracting birds from the stomachs of such fish as have swallowed them, and are cast ashore dead by the storms on the coast; and they also at times get a plentiful meal from the dead birds that float ashore. We received a beautiful specimen of the Red Fox, in the flesh, from our friend Mr. Harris, not long after the foregoing letter, and our figure was drawn from it. We represented the animal just caught in a steel-trap.

The Red Fox brings forth from four to six young at a litter, although not unfrequently as many as seven. The young are covered, for some time after they are born, with a soft woolly fur, quite unlike the coat of the grown animal, and generally of a pale rufous colour. Frequently, however, the cubs in a litter are mixed in colour, there being some red and some black-cross Foxes together: when this is the case it is difficult to tell which are the red and which the cross Foxes until they are somewhat grown. In these cases the parents were probably different in colour.

This animal feeds upon rats, rabbits, and other small quadrupeds, and catches birds, both by lying in wait for them, and by trailing them up in the manner of a pointer dog, until watching an opportunity he can pounce or spring upon them. In our article on the Gray Fox, (vol. 1., p. 164) we have described the manner in which this is done by that species, and the Red Fox hunts in the same way.

The Red Fox also eats eggs, and we have watched it catching crickets in an open field near an old stone wall. It is diverting to witness this—the animal leaps about and whirls round so quickly as to be able to put his foot on the insect, and then gets hold of it with his mouth; we did not see him snap at them; his movements reminded us of a kitten playing with a mouse.

We once knew a Red Fox that had been chased frequently, and always escaped at the same spot, by the hounds losing the track: the secret was at last found out, and proved to be a trick somewhat similar to the stratagem of the Gray Fox related in our first volume, p. 171; the Red Fox always took the same course, and being ahead of the dogs so far that they could not see him, leaped from a fallen log on to a very sloping tree, which he ascended until concealed by the branches, and as soon as the dogs passed he ran down and leaping on to his old track ran back in his former path. So dexterously was this "tour" performed that he was not suspected by the hunters, who once or twice actually whipped their dogs off the trail, thinking they were only following the "back track."

The Red Fox is in the habit of following the same path, which enables the fox hunters to shoot this species from "stands," even in a country where the animal has room enough to take any course he may choose to run. The "hunters" who go out from the city of New-York, are a mixed set, probably including Germans, Frenchmen, Englishmen, and Irishmen, and each one generally takes his own dog along, (on the speed and prowess of which he is ready to bet largely,) and the hunt is organized on the height beyond Weehawken in "the Jerseys," where a good many Red Foxes are to be found, as well as more Gray ones.

The men are all on foot, and station themselves along ridges, or in gaps in the rocky hilly country, now running to a point, to try and get a shot, now yelling to their dogs, and all excitement and hubbub. If the Fox doubles much, he is very apt to get shot by some one before he passes all the "standers," and the hunters then try to start another;

but the Fox often gets away, as the underbrush is thick and a good deal of the ground swampy, and in that case he makes for a large rocky hill which stands in the Newark marshes, familiarly known as *Rattle-snake* hill. When running across the low level to this strong - hold the Fox is frequently seen by the whole company of hunters, and the chase is lengthened out to a run of many miles, as Reynard will turn again toward the high ridges nearer the Hudson River.

We will give an an account of one of these hunts as related by some young friends, who having two fine harriers (to contribute their share of dogs to the pack,) were gladly hailed by the other gentlemen in the field.

"After some beating about among the thickets and ravines, we found the dogs had strayed away down the side of the hills nearly to the level of the marshes, and raising our horn to call them up, observed that they were running toward a cur-dog that appeared to have come from somewhere in them; we immediately gave a loud halloo, and urged all the hounds to the chase. The cur turned tail at once, the whole pack "opened" after him in full cry, and all the hunters came running forth from the woods to the brow of the hill, whence we had a view of the whole scene. The cur looked a good deal like a Fox, at a distance, and most of the hunters thought he was one " certain," he shewed good bottom, took several leaps over the stone walls and fences, and dodged about and round patches of briars and rocks with extraordinary agility, until he got fairly off towards his home, when he positively "streaked it," until, to the utter amazement of the hunters, he jumped on to a wall enclosing a small farm yard, and disappeared within, immediately setting up a loud bark of defiance, while some of the hunters who had expressed most confidence, were loudly laughed at by their comrades, who banteringly asked what they would take for their dogs, &c., and broke out in fresh roars of merriment."

The Red Fox is taken in traps, but is so very wary that it is necessary to set them with great nicety.

Dr. RICHARDSON tells us that the best fox hunters in the fur countries use *assafœtida*, *castoreum*, and other strong smelling substances, with which they rub their traps and the small twigs set up in the neighbourhood, alleging that Foxes are fond of such perfumes.

The same author informs us that their flesh is ill tasted, and is eaten only through necessity.

Red Foxes have gradually migrated from the Northern to the Southern States. This change of habitation may possibly be owing to the more extensive cultivation to which we have alluded, (at p. 265, in this article,) as a reason for this species having become more numerous than it was before the Revolution. This idea, however, would seem to be over-thrown by the continued abundance of Gray Foxes in the Eastern States. In the early history of our country the Red Fox was unknown south of Pennsylvania, that State being its Southern limit. In process of time it was found in the mountains of Virginia, where it has now become more abun-dant than the Gray Fox. A few years afterwards it appeared in the more elevated portions of North Carolina, then in the mountains of South Carolina, and finally in Georgia ; where we have recently observed it.

This species was first seen in Lincoln County, Georgia, in the year 1840, since then it has spread over the less elevated parts of the country, and is not rare in the neighbourhood of Augusta. We are informed by Mr. BEILE, an intelligent observer of the habits of animals, that on one occasion near Augusta, as he was using a call for wild turkeys, a little before sunrise, in the vicinity of Augusta, two Red Foxes came to the call, suppos-ing it to be that of a wild turkey, and were both killed by one discharge of his gun

In order to ascertain whether the speed of the Red Fox was as great in the south as in the colder regions of the north, several gentlemen near Augusta, in the winter of 1844, resolved to test the question by a regular Fox chase. They congregated to the number of thirty, with one hun-dred hounds, many of them imported dogs, and all in fine running order. They started a Fox at two o'clock on a moonlight morning. He took to a pretty open country on the west bank of the Savannah river. A number of gentlemen were mounted on fleet horses. Mr. BEILE rode in succession three horses during the chase, two of which were good hunters. The pursuit of the flying beast was kept up till three o'clock in the afternoon, having continued thirteen hours, when the horses and the whole pack of hounds were broken down, and the hunt was abandoned. This account does not accord with that given by RICHARDSON, who states (Fauna Boreali. Am. p. 93,) "The Red Fox does not possess the wind of its English congener. It runs for about a hundred yards with great swiftness, but its strength is exhausted in the first burst, and it is soon overtaken by a wolf or a mounted huntsman." It is quite evident that our estimable friend never had an opportunity of participating in the chase of the American Red Fox.

Whilst the Gray Fox seldom is known to dig a burrow, concealing its young usually beneath the ledges of rocks, under roots, or in the hollow of some fallen tree, the Red Fox on the contrary, digs an extensive burrow with two or three openings. To this retreat the Fox only flies after a hard chase and as a last resort. If, as often happens, the burrow is on level ground it is not very difficult by ascertaining the direction of the galleries and sinking a hole at intervals of seven or eight feet, to dig out and capture the animal. When thus taken he displays but little courage—sometimes, like the Opossum, closing his eyes and feigning death.

The young, from four to six at a birth, are born in February and March, they are blind when born, and are not seen at the mouth of the den for about six weeks.

It is at this period, when the snows in the Northern States are still on the ground, that the Fox, urged by hunger and instinct, goes out in search of prey. At a later period, both the parents hunt to provide food for their young. They are particularly fond of young lambs, which they carry off for miles to their burrows. They also kill geese, turkeys, ducks, and other poultry, and have a bad reputation with the farmer. They likewise feed on grouse and partridges, as well as on hares, squirrels, and field-rats of various species, as we have previously mentioned.

GEOGRAPHICAL DISTRIBUTION.

The Red Fox exists in the fur countries to the North, is found in Labrador to the East, and in the Russian settlements on the West of our continent. Its Southern limit at present is Abbeville, in South Carolina, and Augusta, in Georgia; a few individuals have been seen in those States, near the sea-board. It also appears in Tennessee, Kentucky, and Missouri. We have not heard of its existence in Florida, Louisiana, or Texas.

GENERAL REMARKS.

It is now so generally admitted that the Red Fox of America is a distinct species from the European Fox ; that a comparison seems unnecessary. We have seen no specimen in this country that can be referred to *Canis vulpes.*

LEPUS ARTEMISIA.—Bach.

Worm-wood Hare.

PLATE LXXXVIII.—Males and Female.

L. Parvus, canescens, nucha et cruribus dilute ferugineis, cauda supra canescens, subtus alba, gula et ventre albis, vellere toto ad basin cano; auriculis longitudine capitis, tarsus dense vestitis.

CHARACTERS.

Small; of a gray colour, pale rufus on the back of the neck and legs; tail, above, the colour of the body; beneath, white; under parts of the neck, and lower surface of the body, white; all the fur gray at the base; ears as long as the head; tarsus, well clothed.

SYNONYMES.

Lepus Artemisia. Bach, Worm-wood Hare. Journal Acad. Nat. Sciences, vol. 8, p. 1, p. 94.

DESCRIPTION.

This small Hare is a little less than our common gray Rabbit, the ears are longer and more conspicuous. The head is much arched, and the upper incisors deeply grooved.

COLOUR.

This species is grayish-black and brownish-white above; the fur is soft, pale-gray at the base, shaded into brownish externally, annulated with brownish-white near the apex, and black at the tips; under parts, and inner sides of the limbs, white; the hairs pale-gray at the base; neck, with the hairs on the sides, and under parts gray, tipped with brownish-white, having a faint yellow hue; chin and throat grayish-white, the hairs being gray at their base, and white at their tips. The whole back of the neck and limbs exteriorly of a pale rusty-fawn colour; hairs on the neck uniform to the base; soles of the feet, very pale soiled yellowish-brown; tail, coloured above as the back, with an admixture of grayish-black hairs, beneath, white; ears, externally on the anterior

Plate LXXXVIII

Drawn from Nature by J.J.Audubon. F.R.S.F.L.S

On Stone by W E Hitchcock

Lith.d Printed & Col.d by J.T.Bowen,Philad.a

Worm=wood Hare.

part, coloured as the crown of the head; posteriorly, asny white; at the apex margined with black; internally, nearly naked, excepting the posterior part, where they are grizzled with grayish black and white; in the apical portion they are chiefly white.

DIMENSIONS.

	Inches.	Lines.
Length from nose to root of tail, -	12	0
From heel to point of longest nail, -	3	2
Height of ears externally, - - -	2	8
From ear to point of nose, - -	2	7
Tail (vertebræ) about, - - -	1	1
To end of fur, - - - - -	1	9

HABITS.

Mr. Townsend, who procured this species at Fort Walla-walla, remarks, " it is here abundant but very shy and retired, keeping constantly in the densest wormwood bushes, and leaping with singular speed from one to another when pursued. I have never seen it dart away and run to a great distance like other Hares. I found it very difficult to shoot this animal, for the reasons stated. I had been residing at Fort Walla-walla for two weeks, and had procured only two, when at the suggestion of Mr. Pambrun, I collected a party of a dozen Indians armed with bows and arrows, and sallied forth. We hunted through the wormwood within about a mile of the Fort, and in a few hours returned bringing eleven Hares. The keen eyes of the Indians discovered the little creatures squatting under the bushes, where to a white man they would have been totally invisible. This Hare, when wounded and taken, screams like our common species.

GEOGRAPHICAL DISTRIBUTION.

" This small Hare," we are informed by Mr. Townsend, " inhabits the wormwood plains near the banks of the streams in the neighbourhood of Fort Walla-walla. I cannot define its range with any degree of certainty, but I have every reason to believe that it is very contracted, never having met with it many miles from this locality."

SCIURUS SAYII.—Aud. and Bach.

Say's Squirrel.

PLATE LXXXIX.—Males.

S. Sciurus cinereus magnitudine sub æquans. Corpore supra lateribus-que cano-nigroque variis ; capitis lateribus orbitis que pallide cano-ferrugineis ; genis auriculusque saturate fuscis ; caudâ supra ferrugineo-nigroque varia, infra splendide ferrugineâ.

CHARACTERS.

About the size of the cat-squirrel (S. cinereus) ; body above, and on the sides mixed with gray and black ; sides of the head and orbits, pale ferruginous ; cheek and under the eye, dusky ; tail, above, mixed with ferruginous and black, beneath, bright ferruginous.

SYNONYMES.

Sciurus Macrourus. Say, Long's Exped. vol. 1., p. 115.
S. Magnicaudatus. Harlan, Fauna, p. 178.
S. Macroureus. Godman's Nat. Hist. vol. 2, p. 134.

DESCRIPTION.

In size and form this species bears a considerable resemblance to the Cat-Squirrel (*S. cinereus*). It is a little longer in body, not quite as stout, and has shorter ears. In length and breadth of tail, they are about equal. The first molar tooth in the upper jaw, which in some of the species is deciduous and in others permanent, was wanting in the six specimens we examined ; we presume, however, it exists in very young animals ; mammæ, 8, placed equi-distant on the sides of the belly ; palms, as is usual in this genus, naked, the rudimental thumb protected by a short blunt nail ; the feet are covered with hair, which extends between the toes, half concealing the nails; hair on the body, of moderate length, not as coarse as that of the Fox-Squirrel, (*S. capistratus*), but neither as fine or woolly as that of *S. cinereus*. Our specimens were obtained in summer.—Say has remarked :

" The fur of the back in the summer dress, is from three-fifths to seven-tenths of an inch long ; but in the winter dress, the longest hairs of the middle of the back are from one inch to one and three-fourths in length.

Plate LXXXIX.

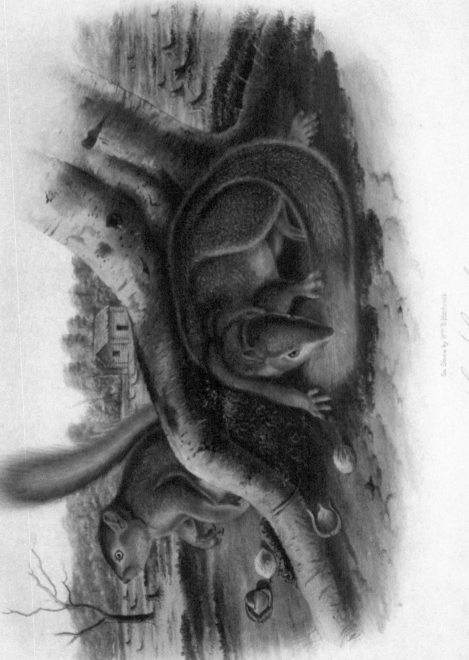

Drawn from Nature by J.J. Audubon, F.R.S.F.L.S.

On Stone by Wm B. Hitchcock

Lith. Printed & Col.d by J.T. Bowen, Philad.a

Say's Squirrel.

He also remarks that it is only in winter that the ears are fringed, which is the necessary consequence of the elongation of the hair ; in our summer specimens, the ears are thinly clothed with hair, not rising above the margins.

COLOUR.

The fur on the back, is for one half its length from the base plumbeous, then pale cinnamon, then a narrow line of black, then cinereous, and broadly tipped with black, giving it what is usually termed an iron-gray colour ; the hairs on the under surface are of a light-ash colour at base, and without any annulations brighten into ferruginous at apex, the paler colours beneath giving way to the broader markings on the extremities; the eyes and moustaches are black ; nails, dark-brown ; sides of face, around the eyes, both surfaces of ears, feet, chin, neck, inner surfaces of legs, and under surface of tail, bright ferruginous ; the hairs on the tail, are at their roots reddish-yellow, with three black annulations, and are broadly tipped with reddish-yellow.

DIMENSIONS.

	Feet.	Inches.
From point of nose to root of tail - - -	1	0
Tail (vertebræ) - - - - - -		10¼
" to end of fur - - - - -		13
Height of ear posteriorly - - - -		⅝

HABITS.

The habits of this Squirrel are not very different from those of the Cat Squirrel, to which it is most nearly allied. It does not run for so great a distance on the ground before taking a tree as the southern Fox Squirrel, nor does it leap quite as actively from tree to tree as the northern Gray Squirrel, (*S. migratorius*,) but appears to possess more activity, and agility than the Cat Squirrel.

The forests on the rich bottom lands of the Wabash, the Illinois, and the Missouri rivers are ornamented with the stately pecan-tree (*Carya olivæformis*), on the nuts of which these squirrels luxuriate ; they also resort to the hickory and oak trees, in the vicinity of their residence, as well as to the hazel bushes, on the fruits of which they feed

They are becoming troublesome in the corn-fields of the farmer, who has commenced planting his crops in the remote but rapidly improving states and territories west of the Ohio.

The flesh is represented by all travellers as delicate, and is said to be equal in flavour to that of any of the species.

GEOGRAPHICAL DISTRIBUTION.

This squirrel is found along the shores of the Missouri, and in the wooded portions of the country, lying east and north of that river ; we have received several specimens, from Michigan, and it seems to be observed west and north of that State.

GENERAL REMARKS.

This species was first discovered by Mr. THOMAS SAY, and by him described and named *Sciurus Macrourus*. This name, unfortunately, was pre-occupied, the Ceylon Squirrel having been so designated : (vide PENNANT, Hist. Quad. ii. p. 140, No. 330.)

Dr. HARLAN and Dr. GODMAN in their respective works, seeing this, applied other names. The former calls it (*Sciurus magnicaudatus,*) the latter (*Sciurus macroureus.*) Authors copied Mr. SAY's description almost literally. Dr. HARLAN gives SAY's name (*S. macrourus,*) as a synonyme, and Dr. GODMAN gives his name (*Sciurus macroureus*) as SAY's name : giving in a note intimation that he has taken the liberty of changing the name by the addition of a single letter, which he considers sufficient to render further change unnecessary. Neither of these gentlemen claimed the discovery of this species, gave original descriptions, or appear to have ever seen the animal ; and, according to all rules which should govern naturalists, they had no right to name it. We, therefore, having procured a good many specimens, and having from them identified, and described this species, have used the grateful privilege of naming it in honour of its discoverer, Mr. SAY, and have given Dr. HARLAN's and Dr. GODMAN's names as synonymes.

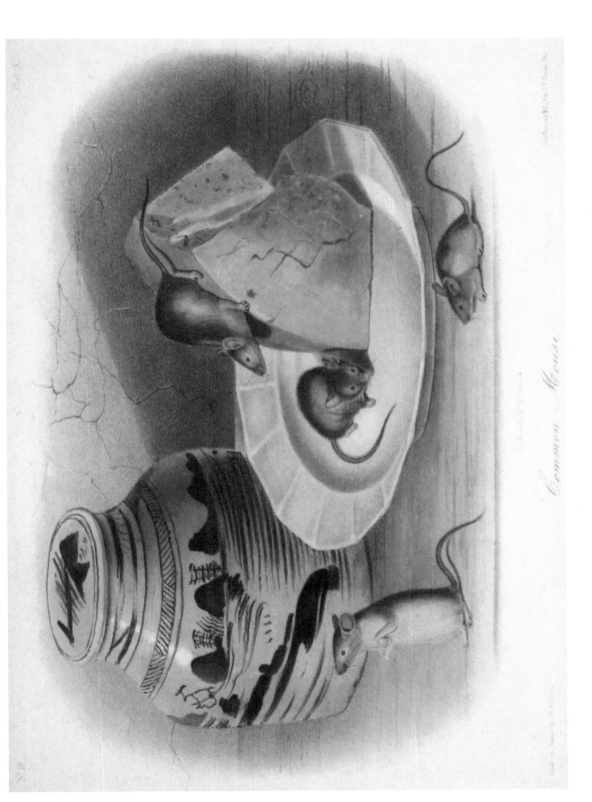

Common Mouse

MUS MUSCULUS.—Linn.

Common Mouse.

PLATE XC.—Male, Female, and Young.

M. Corpore fusco ; subtus ciner ascenti.

CHARACTERS.

Dusky gray above, cinereous beneath.

SYNONYMES.

Mus Musculus. Linn., 12 Ed., p. 83.
Mouse. Pennant, Arct. Zool. vol. 1, p. 131.
Mus Musculus. Say, Long's Expedition, vol. 1, p. 262.
"   "   Harlan, p. 149.
"   "   Godman, vol. 2, p. 84.

DESCRIPTION.

The Common Mouse is more generally and familiarly known than any other species, and therefore requires no very minute description. It is small in size ; head, elongated ; nose, sharp ; ears, large, erect, ovate, and nearly naked on both surfaces ; legs, slender ; nails, sharp, slightly hooked ; tail, round, nearly as long as the body, scaly, and slightly covered with short hair.

COLOUR.

Eyes, black ; incisors, yellowish ; whiskers, mostly black ; fur on the back, plumbeous at the roots, slightly tipped with brownish, giving it a dusky grayish colour ; ears a shade lighter ; under surface, and beneath the tail, obscure ash-colour.

There are some varieties :—very rarely one is found black, others spotted white and black ; one variety is an albino, white with red eyes, breeds in confinement, and produces young with white colour, and the red eyes of the parents.

DIMENSIONS.

		Inches.
Length of head and body	- - - -	$3\frac{1}{4}$
" Tail	- - - - - - -	$3\frac{1}{8}$
Height of ear	- - - - - - -	$4\frac{1}{10}$

HABITS.

We have attempted to shew a portion of a shelf in a pantry, on which stands a china jar, with its indigo-blue peaked mountains, its fantastic trees and its (take them altogether) rather remarkable landscapes, reminding us more of the sweetmeats it contains than of aught in the way of nature; and we have also portrayed a plate, with a piece of hard old cheese in it, on which a Mouse is standing in the act of listening, while another in the plate, and two more on the shelf, likewise appear a little startled, and are expecting to be disturbed ere they can make their intended meal; the little rascals have reason to fear, for the careful housekeeper has heard them of late, squealing in their squabblings with each other, has found the marks of their teeth on the bread and butter, and is determined to get rid of them instanter, if possible; she is calling now to her faithful pussy cat, and inquiring for the trap.

But although the thievish Mouse is often frightened, and may be said to eat his dinner with " a cat " over his head, although he is assailed with pokers, broomsticks, &c., whenever he unluckily runs across the floor, and in fact is killed as often as his death can be compassed by the ingenuity of man, or the cunning and quickness of his ally the cat, the Mouse will not retire from the house, and even where the supply of food for him is small, or in rooms that have long been shut up, he may be found; and would he let our drawings and books alone, we should willingly allow him the crumbs from our table; but he will sometimes gnaw into shreds valuable papers, to make a bed behind some bureau or old chest. He in his turn frightens man at times, and should the hard-hearted hoarding wretch who has made gold his God, while with aged, trembling hands, locked in his inmost chamber, he counts his money-bags, but hear a little Mouse; what a feeling of terror shoots through his frame; despair seems for an instant to be written on his face, and he clutches convulsively the metal to which he is a slave; another moment, and he recovers, but he is still agitated, and hastily secures with locks and bolts the treasure which is to him more precious than the endearments of a wife, the love of children, the delights of friendship and society, the blessings and prayers of the poor, or the common wants of humanity in his own person.

Many a young lady will scream at sight of a poor little Mouse, and many a brave young man might be startled in the stillness of the night by the noise made by this diminutive creature, especially if given to the reading of the " Mysteries of Udolpho" or the " Castle of Otranto," late in the hours of darkness, alone in a large old lumbering house.

The Common Mouse is a graceful, lively little animal—it is almost omnivorous, and is a great feeder, although able to live on but little food if the supply is scanty. This species has from four to ten young at a litter, and the female suckles her young with tender care. When first born, they are very small, almost naked, and of a pinkish colour. The Mouse has several litters every year. We kept a pair in confinement, which produced four times, having from four to nine in each litter. Dr. GODMAN quotes Aristotle, who says that " a pregnant female being shut up in a chest of grain ; in a short time a hundred and twenty individuals were counted."

On examining our corn-crib in the spring, and cleaning it out ; although it was constructed with a special view to keep off rats and vermin, being on posts, and the floor raised from the ground some three feet, with boards outside inclining downwards all round, we found and killed nearly fifty Mice. A basket in the crib, hanging by a rope from a cross-beam, in which we had put some choice corn for seed, had been entered by them, and every grain of corn in it devoured. We found in the basket nothing but husks, and the remains of a Mouse's nest. The animal must therefore have climbed up to the roof of the crib, and then descended the cord by which the basket of corn was suspended.

The activity, agility, and grace of the Mouse, have made it a favourite pet with the prisoner in his solitary cell, and it has been known to answer his call, and come out of its hiding places to play with the unfortunate captive, showing the greatest fondness for him, and eating out of his hand without fear.

Of late years, white Mice have been in request in London, where they are taught various tricks, and are exhibited by boys in the streets. It is stated that in order to increase the number of this variety, persons exclude them from the light, this they pretend causes a great many of them to be born albinos. We are however satisfied from personal experience that a pair of albinos, accidentally produced, would continue to propagate varieties of the same colour without the aid of darkness ; as is the case in the albino variety of the English rabbit.

GEOGRAPHICAL DISTRIBUTION.

The Common Mouse is not a native of America, but exists in all countries where ships have landed cargo, and may be said to tread closely on the heels of commerce. It was brought to America in the vessels that conveyed to our shores the early emigrants.

Plate XCI.

Polar Bear

GENUS URSUS.—Linn.

DENTAL FORMULA.

$$Incisive \; \frac{6}{6}; \; Canine \; \frac{1-1}{1-1} \; ; \; Molar \quad \frac{6-6}{7-7} = 42.$$

Head, large; body, stout, and covered with a coat of thick hair; ears, large, slightly acuminated.

Legs, stout; five toes, furnished with strong curved claws, fitted for digging.

Tail, short; mammæ, six, two pectoral and four ventral; no glandular pouch under the tail.

Omnivorous, nocturnal, but frequently seen wandering about during the day.

The generic name is derived from the Latin *ursus*, a Bear.

Eight species of this genus have been described, three existing in Europe, one of which, the Polar Bear, is common also to America, one in the mountainous districts of India, one in Java, one in Thibet, and three in North America.

URSUS MARITIMUS.—Linn.

Polar Bear.—White Bear.

PLATE XCI.—Male.

U. Capite elongata; cranio applanato; collo longo; pilis longis mollibus, albis.

CHARACTERS.

Head, elongated ; skull, flat ; neck, long ; hair, long, soft, and white

SYNONYMES.

White Bear. Marten's Spitz. Trans., p. 107. An. 1675.
Ursus Maritimus. Lin. Syst.
Ursus Albus. Brisson, Regne, an. p. 260.
L'Ours Blanc. Buffon, vol. 15, p. 128. An. 1767.
Ursus Marinus. Pallas, vol. 3, p. 69.
Polar Bear. Penn. Arct. Zool., p. 53.

URSUS MARITIMUS. Parry's 1st voyage, Supp., p. 183.
    "   "   Franklin's 1st voyage, p. 648.
    "   "   Parry's 2nd voyage, Appendix, p. 288.
    "   "   Richardson, Fauna, p. 30.
    "   "   Scoresby's Account of the Arctic Regions.

DESCRIPTION.

Head and muzzle narrow, prolonged on a straight line with the fore-head, which is flattened ; snout, naked ; ears, short ; neck, long ; body, long in proportion to its height ; soles of the hind feet equal to one-sixth of the length of the body ; hair, rigid, compact and long on the body and limbs, is from two to three inches in length, with a small quantity of fine and woolly hair next the skin. The whole animal wears the appearance of great strength without much agility.

COLOUR.

The naked extremity of the snout, the tongue, margins of the eyelids, and the claws, are black ; lips, purplish black ; eyes, dark-brown ; interior of the mouth pale violet. The hairs on every part of the body are of a yellowish-white colour.

DIMENSIONS.

Specimen in the Charleston Museum :—

	Feet	Inches.
Head and body, - - - - -	6	9
Tail, (vertebræ), - - - - -		10
" to end of hair, - - - - -	1	1
Height of ear, - - - - - -		3¾
Height from shoulder, - - - - -	3	3
Girth around the body, - - - - -	6	3
" around the hind leg, - -	1	7
Length of canine teeth, - -		1⅝
" of incisors, - - - -		0¾

We append the following measurements taken from specimens in the flesh, by Capt. J. C. Ross, R.N., F.R.S., &c. :—

	MALE.	FEMALE.
	Inches.	Inches.
Length from snout to end of tail, - -	94	78
Snout to shoulder, - - - - -	33.5	26.3
Snout to occiput, - - - - -	18.4	15.6
Circumference before the eyes, -	20.4	15.8

		MALE. Inches.	FEMALE. Inches
At broadest part of the head, -	-	32.2	28
At largest part of the abdomen, -	-	65.2	57.6
Length of alimentary canal. - -	-	61	52
Weight,		900lbs.	700lbs.

The weight varies very much according to the season and condition of the animal.

The largest measured 101.5 inches in length, and weighed 1028 lbs., although in poor condition.

HABITS.

We have journeyed together, friend reader, through many a deep dell, and wild wood, through swamp and over mountain ; we have stemmed the current of the Mississippi, sailed on our broad lakes, and on the extended sea coast, from Labrador to Mexico ; we have coursed the huge buffalo over the wide prairies, hunted the timid deer, trapped the beaver. and caught the fox ; we have, in short, already procured, figured, and described, many of our animals ; and now, with your permission, we will send you with the adventurous navigators of the Polar Seas, in search of the White Bear, for we have not seen this remarkable inhabitant of the icy regions of our northern coast amid his native frozen deserts ; and can therefore give you little more than such information as may be found in the works of previous writers on his habits. During our visit to Labrador in 1833, we coasted along to the north as far as the Straits of Belleisle, but it being midsummer, we saw no Polar Bears, although we heard from the settlers that these animals were sometimes seen there ; (on one occasion, indeed, we thought we perceived three of them on an ice-berg. but the distance was too great for us to be certain), although the abundance of seals and fish of various kinds on the shores, would have afforded them a plentiful supply of their ordinary food. They are doubtless drifted far to the southward on ice-bergs from time to time, but in our voyages to and from Europe we never saw any, although we have been for days in the ice.

The Polar Bear is carnivorous, in fact omnivorous, and devours with equal voracity the carcases of whales, abandoned, and drifted ashore by the waves ; seals, dead fish, vegetable substances, and all other eatable matters obtainable, whether putrid or fresh. Dr. RICHARDSON, in the Fauna Boreali Americana, has given a good compiled account of this animal, and we shall lay a portion of it before our readers. The Dr. says :— " I have met with no account of any Polar Bear, killed of late years, which exceeded nine feet in length, or four feet and a-half in height. It is possible that larger individuals may be occasionally found; but the greatness of the dimensions attributed to them by the older voyagers has, I doubt not, originated in the skin having been measured after being much stretched in the process of flaying."

The great power of the Polar Bear is portrayed in the account of a disastrous accident which befel the crew of BARENTZ's vessel on his second voyage to Waigat's Straits.   "On the 6th of September, 1594, some sailors landed to search for a certain sort of stone, a species of diamond.   During this search, two of the seamen lay down to sleep by one another, and a White Bear, very lean, approaching softly, seized one of them by the nape of the neck.   The poor man, not knowing what it was, cried out "who has seized me thus behind ?" on which his companion, raising his head, said, " Holloa, mate, it is a Bear," and immediately ran away.   The Bear having dreadfully mangled the unfortunate man's head, sucked the blood. The rest of the persons who were on shore, to the number of twenty, immediately ran with their match-locks and pikes, and found the Bear devouring the body ; on seeing them, he ran upon them, and carrying another man away, tore him to pieces.   This second misadventure so terrified them that they all fled.   They advanced again, however, with a reinforcement, and the two pilots having fired three times without hitting the animal, the purser approached a little nearer, and shot the Bear in the head, close by the eye.   This did not cause him to quit his prey, for, holding the body, which he was devouring, always by the neck, he carried it away as yet quite entire.   Nevertheless, they then perceived that he began himself to totter, and the purser and a Scotchman going towards him, they gave him several sabre wounds, and cut him to pieces, without his abandoning his prey.

In BARENTZ's third voyage, a story is told of two Bears coming to the carcass of a third one that had been shot, when one of them, taking it by the throat, carried it to a considerable distance, over the most rugged ice, where they both began to eat it.   They were scared from their repast by the report of a musket, and a party of seamen going to the place, found that, in the little time they were about it, they had already devoured half the carcase, which was of such a size that four men had great difficulty in lifting the remainder.   In a manuscript account of Hudson's Bay, written about the year 1786, by Mr. Andrew Graham, one of PENNANT's ablest correspondents, and preserved at the Hudson's Bay house, an anecdote of a different description occurs.   "One of the Company's servants who was tenting abroad to procure rabbits, (*Lepus*

Americanus), having occasion to come to the factory for a few necessaries, on his return to the tent passed through a narrow thicket of willows, and found himself close to a White Bear lying asleep. As he had nothing wherewith to defend himself, he took the bag off his shoulder and held it before his breast, between the Bear and him. The animal arose on seeing the man, stretched himself and rubbed his nose, and having satisfied his curiosity by smelling at the bag, which contained a loaf of bread and a rundlet of strong beer, walked quietly away, thereby relieving the man from his very disagreeable situation."

Dr. RICHARDSON says, "They swim and dive well, they hunt seals and other marine animals with great success. They are even said to wage war, though rather unequally, with the walrus. They feed likewise on land animals, birds, and eggs, nor do they disdain to prey on carrion, or, in the absence of this food, to seek the shore in quest of berries and roots. They scent their prey from a great distance, and are often attracted to the whale vessels by the smell of burning *kreng*, or the refuse of the whale blubber."

The Dr. quotes Captain LYONS, who thus describes the mode in which the Polar Bear surprises a seal:—"The Bear, on seeing his intended prey, gets quietly into the water, and swims to the leeward of him, from whence, by frequent short dives, he silently makes his approaches, and so arranges his distance, that, at the last dive, he comes to the spot where the seal is lying. If the poor animal attempts to escape by rolling into the water, he falls into the bear's clutches; if, on the contrary, he lies still, his destroyer makes a powerful spring, kills him on the ice, and devours him at leisure." Captain LYONS describes the pace of the Polar Bear, at full speed, as "a kind of shuffle, as quick as the sharp gallop of a horse."

The Polar Bear is by no means confined to the land, on the contrary he is seldom if ever seen far inland, but frequents the fields of ice, and swims off to floating ice or to ice-bergs, and is often seen miles from shore.

It is said that these animals "are often carried from the coast of Greenland to Iceland, where they commit such ravages on the flocks that the inhabitants rise in a body to destroy them." Captain SABINE saw one about midway between the north and south shores of Barrow's Straits, which are forty miles apart, although there was no ice in sight to which he could resort to rest himself upon. The Polar Bear is said to be able to make long leaps or springs in the water.

This species is found farther to the north than any other quadruped, having been seen by Captain PARRY in his adventurous boat-voyage beyond 82 degrees of north latitude.

PENNANT, who collected from good authorities much information relative to their range, states that they are frequent on all the Asiatic coasts of the Frozen Ocean, from the mouth of the Obi eastward, and abound in Nova Zembla, Cherry Island, Spitzbergen, Greenland, Labrador, and the coasts of Baffin's and Hudson's Bays. Dr. RICHARDSON says,—"They were seen by Captain PARRY within Barrow's Straits, as far as Melville Island; and the Esquimaux to the westward of Mackenzie river, told Captain FRANKLIN that they occasionally, though very rarely, visited that coast. The exact limit of their range to the westward is uncertain, but they are said not to be known on the islands in Behring's Straits, nor on the coast of Siberia to the eastward of Tchutskoinoss. They are not mentioned by LANGSDORFF and other visitors of the Northwest Coast of America; nor did Captain BEECHEY meet with any in his late voyage to Icy Cape. None were seen on the coast between the Mackenzie and Copper-Mine River; and PENNANT informs us, that they are unknown along the shores of the White Sea, which is an inlet of a similar character."

Dr. RICHARDSON does not think that the Polar Bear is under the same necessity for hibernating that exists in the case of the Black Bear, which feeds chiefly on vegetable matters, and supposes that although they may all retire occasionally to caverns in the snow, the pregnant females alone seclude themselves for the entire winter. In confirmation of this idea the Dr. mentions that "Polar Bears were seen in the course of the two winters that Capt. PARRY remained on the coast of Melville Peninsula; and the Esquimaux of that quarter derive a considerable portion of their subsistence, not only from the flesh of the female Bears, which they dig together with their cubs from under the snow, but also from the males, that they kill when roaming at large at all periods of the winter. To this statement is added HEARNE's account; he says:—"The males leave the land in the winter time and go out on the ice to the edge of the open water in search of seals, whilst the females burrow in deep snow-drifts from the end of December to the end of March, remaining without food, and bringing forth their young during that period; that when they leave their dens in March, their young, which are generally two in number, are not larger than rabbits, and make a foot-mark in the snow no bigger than a crown piece."

"In winter," says Mr. GRAHAM, "the White Bear sleeps like other species of the genus, but takes up its residence in a different situation, generally under the declivities of rocks, or at the foot of a bank, where the snow drifts over it, to a great depth; a small hole, for the admission of fresh air, is constantly observed in the dome of its den. This, however, has regard solely to the she Bear, which retires to her winter-quarters in November, where she lives without food, brings forth two young about Christmas, and leaves the den in the month of March, when the cubs are as large as a shepherd's dog. If, perchance, her offspring are tired, they ascend the back of the dam, where they ride secure either in water or ashore. Though they sometimes go nearly thirty miles from the sea in winter, they always come down to the shores in the spring with their cubs, where they subsist on seals and sea-weed. The he Bear wanders about the marshes and adjacent parts until November, and then goes out to the sea upon the ice, and preys upon seals."

The Esquimaux account of the hibernation of the Polar Bear is curious: it was related to Capt. Lyons by one of their most intelligent men, rejoicing in the euphonious name of (Mr.) Ooyarrakhioo! and is as follows :—" At the commencement of winter the pregnant bears are very fat, and always solitary. When a heavy fall of snow sets in, the animal seeks some hollow place in which she can lie down and remain quiet, while the snow covers her. Sometimes she will wait until a quantity of snow has fallen, and then digs herself a cave : at all events, it seems necessary that she should be covered by, and lie amongst, the snow. She now goes to sleep, and does not wake until the spring sun is pretty high, when she brings forth two cubs. The cave by this time has become much larger by the effect of the animal's warmth and breath, so that the cubs have room to move, and they acquire considerable strength by continually sucking. The dam at length becomes so thin and weak, that it is with great difficulty she extricates herself, when the sun is powerful enough to throw a strong glare through the snow which roofs the den." The Esquimaux affirm that during this long confinement the Bear has no evacuations, and is herself the means of preventing them by stopping all the natural passages with moss, grass, or earth. The natives find and kill the Bears during their confinement by means of dogs, which scent them through the snow, and begin scratching and howling very eagerly. As it would be unsafe to make a large opening, a long trench is cut of sufficient width to enable a man to look down and see where the bear's head lies, and he then selects a mortal part, into which he thrusts his spear. The old one being killed, the hole is broken open, and the young cubs may be taken out by the hand, as, having tasted no blood, and never having been at liberty, they are then very harmless and quiet. Females, which are not pregnant, roam throughout the whole winter in the same manner as the males.

The Polar Bear is at certain seasons and under peculiar circumstances a dangerous animal. Like the Grizzly Bear it possesses both strength and activity enough to render it at all times formidable. Although, like all Bears, it appears clumsy, can run with great swiftness either on the ground or on the ice, and it can easily ascend the slippery sides of icebergs by the assistance of its claws, being in the habit of mounting on their ridges and pinnacles to look out for food or survey the surrounding fields of ice.

When in confinement, the great strength of this Bear is sometimes manifested to the terror of the spectators. One that was secured in a cage fronted with rods of inch iron, bolted into a horizontal flat plate of the same metal, several inches wide, near the bottom, and well fastened at top, in the stout oak boarding of which the cage was constructed, one day when we were present became enraged by the delay of his keeper in bringing his food, and seized two of the rods with such a furious grip that one of them bent and instantly came out, when the huge beast nearly made his escape, and was only prevented from succeeding by the promptness of the attendants, who instantly placed the wooden front, used when travelling, on the open part of the broken cage and closed it effectually. This Bear, like all others we have seen caged, was very restless, and would walk backwards and forwards in his prison-house for hours together, always turning his head toward the bars in front, at each end of this alternating movement, and occasionally tossing his head up and down as he walked to and fro.

Many anecdotes are related of accidents to the crews of boats detached from whaling vessels to kill the White Bear, and by all accounts it appears to be exceedingly dangerous to attack this animal on the ice. One of these accounts, with others of a different character, we will repeat here, although they have been published by several authors.

Dr. Scoresby tells us, that " a few years ago, when one of the Davis's Strait whalers was closely beset among the ice at the ' South-west,' or on the coast of Labrador, a Bear that had been for sometime seen near the ship, at length became so bold as to approach alongside, probably tempted by the offal of the provision thrown overboard by the cook. At this time the people were all at dinner, no one being required to keep the deck in the then immovable condition of the ship. A hardy fellow, who first looked out, perceiving the Bear so near, imprudently jumped upon the ice, armed only with a handspike, with a view, it is supposed, of gaining all the honour of the exploit of securing so fierce a visitor by himself. But the bear, regardless of such weapons, and sharpened probably by hunger, disarmed his antagonist, and seizing him by the back with his powerful jaws, carried him off with such celerity, that on his dismayed comrades rising from their meal and looking abroad, he was so far beyond their reach as to defy pursuit."

An equally imprudent attack made on a Bear by a seaman employed in one of the Hull whalers, was attended with a ludicrous result. "The ship was moored to a piece of ice, on which, at a considerable distance, a large Bear was observed prowling about for prey. One of the ship's company, emboldened by an artificial courage derived from the free use of rum, which in his economy he had stored for special occasions, undertook to pursue and attack the Bear that was within view. Armed only with a whale-lance, he resolutely, and against all persuasion, set out on his adventurous exploit. A fatiguing journey of about a half a league, over a yielding surface of snow and rugged hummocks, brought him within a few yards of the enemy, which, to his surprise, undauntedly faced him, and seemed to invite him to the combat. His courage being by this time greatly subdued, partly by evaporation of the stimulus, and partly by the undismayed and even threatening aspect of the Bear, he levelled his lance, in an attitude suited either for offensive or defensive action, and stopped. The Bear also stood still; in vain the adventurer tried to rally courage to make the attack; his enemy was too formidable, and his appearance too imposing. In vain, also, he shouted, advanced his lance, and made feints of attack; the enemy, either not understanding, or despising such unmanliness, obstinately stood his ground. Already the limbs of the sailor began to quiver; but the fear of ridicule from his messmates had its influence, and he yet scarcely dared to retreat. Bruin, however, possessing less reflection, or being regardless of consequences, began, with audacious boldness, to advance. His nigh approach and unshaken step subdued the spark of bravery, and that dread of ridicule that had hitherto upheld our adventurer; he turned and fled. But now was the time of danger; the sailor's flight encouraged the Bear in turn to pursue, and being better practised in snow travelling, and better provided for it, he rapidly gained upon the fugitive. The whale-lance, his only defence, encumbering him in his retreat, he threw it down, and kept on. This fortunately excited the Bear's attention; he stopped, pawed, bit it, and then renewed the chase. Again he was at the heels of the panting seaman, who, conscious of the favourable effects of the lance, dropped one of his mittens; the stratagem succeeded, and while Bruin again stopped to examine it, the fugitive improving the interval, made considerable progress ahead. Still the Bear resumed the pursuit with a most provoking perseverance, except when arrested by another mitten, and finally, by a hat, which he tore to shreds between his teeth and paws, and would, no doubt, soon have made the incautious adventurer his victim, who was now rapidly losing strength.

but for the prompt and well-timed assistance of his shipmates—who, observing that the affair had assumed a dangerous aspect, sallied out to his rescue. The little phalanx opened him a passage, and then closed to receive the bold assailant. Though now beyond the reach of his adversary, the dismayed fugitive continued onwards, impelled by his fears, and never relaxed his exertions, until he fairly reached the shelter of his ship. The Bear once more came to a stand, and for a moment seemed to survey his enemies with all the consideration of an experienced general ; when, finding them too numerous for a hope of success, he very wisely wheeled about, and succeeded in making a safe and honourable retreat."

Several authors speak of the liver of the Polar Bear as being poisonous. This is an anomaly for which no reason has yet been assigned ; the fact seems, however, well ascertained. All the other parts of the animal are wholesome, and it forms a considerable article of food to the Indians of the maritime Arctic regions.

The skin of the Polar Bear is a valuable covering to these tribes, and is dressed by merely stretching it out on the snow, pinning it down, and leaving it to freeze, after which the fat is all scraped off. It is then generally hung up in the open air, and " when the frost is intense, it dries most perfectly ; with a little more scraping it becomes entirely dry and supple, both skin and hair being beautifully white." " The time of the year at which the sexes seek each other is not positively known, but it is most probably in the month of July, or of August. HEARNE, who is an excellent authority, relates that he has seen them killed during this season, when the males exhibited an extreme degree of attachment to their companions. After a female was killed, the male placed his fore-paws over her, and allowed himself to be shot rather than relinquish her dead body."

" The pregnant females during winter seek shelter near the skirt of the woods, where they excavate dens in the deepest snow-drifts, and remain there in a state of torpid inaction, without food, from the latter part of December or early in January till about the end of March ; they then relinquish their dens to seek food on the sea-shore, accompanied by their cubs."—GODMAN, Vol. I., pp. 152, 153.

The affection of the female Polar Bear for her young is exemplified by several stories in the Polar voyages. SCORESBY says, " a she Bear with her two cubs, were pursued on the ice by some of the men, and were so closely approached, as to alarm the mother for the safety of her offspring. Finding that they could not advance with the desired speed, she used various artifices to urge them forward, but without success. Determined to save them, if possible, she ran to one of the cubs, placed her nose under it, and threw it forward as far as possible ; then going to the other, she performed the same action, and repeated it frequently, until she had thus conveyed them to a considerable distance. The young Bears seemed perfectly conscious of their mother's intention, for as soon as they recovered their feet, after being thrown forward, they immediately ran on in the proper direction, and when the mother came up to renew the effort, the little rogues uniformly placed themselves across her path, that they might receive the full advantage of the force exerted for their safety."

The sagacity of the Polar Bear is said to be great, and it is very difficult to entrap this animal, as he scents the ground, and cautiously approaches even when the snare is concealed by the snow. SCORESBY relates an instance of a Bear which, having got his fore-foot in a noose, very deliberately loosened the slip-knot with the other paw, and leisurely walked off to enjoy the bait which he had abstracted.

Capt. J. C. Ross states in regard to this species :—" During our stay at Fury Beach many of these animals came about us, and several were killed. At that time we were fortunately in no want of provisions, but some of our party, tempted by the fine appearance of the meat, made a hearty meal off the first one that was shot. All that partook of it soon after complained of a violent headache, which with some continued two or three days, and was followed by the skin peeling off the face, hands, and arms ; and in some who had probably partaken more largely, off the whole body. On a former occasion I witnessed a somewhat similar occurrence, when, on Sir Edward Parry's Polar journey, having lived for several days wholly on two Bears that were shot, the skin peeled off the face, legs, and arms of many of the party. It was then attributed rather to the quantity than the quality of the meat, and to our having been for sometime previous on very short allowance of provisions. The Esquimaux eat its flesh without experiencing any such inconvenience, but the liver is always given to the dogs, and that may possibly be the noxious part. The Esquimaux of Boothia Felix killed several during their stay in our neighbourhood in 1830, all males."

GEOGRAPHICAL DISTRIBUTION.

The Polar Bear inhabits the north of both continents, having been found in the highest latitudes ever reached by navigators. It was seen by Capt. Parry in latitude 82°. It exists on all the Asiatic coasts of the Frozen Ocean, from the mouth of the Obi, eastward, and abounds in Nova Zembla and Spitzbergen. In America it is found in Greenland, Labrador, and on the coasts of Baffin's and Hudson's Bays. They seem not to be found on the islands in Behring's Straits.

McKENSIE informs us that these animals are unknown in the White Sea, or on the coast of Siberia to the eastward of Tchutskoinoss. They have been seen on floating icebergs from fifty to a hundred miles at sea. Capt. Ross states that this species was found in greater numbers in the neighbourhood of Port Bowen and Batty Bay in Prince Regent's Inlet, than in any other part of the Polar Regions that were visited by the several expeditions of discovery. This he supposed was owing to the food they were enabled to procure in that vicinity, Lancaster Sound being but seldom covered by permanently fixed ice, and therefore affording them means of subsistence during the severity of an arctic winter, and also from its being remote from the haunts of the Esquimaux.

On Stone by Wᵐ E. Hitchcock

Texan Lynx

LYNX RUFUS—VAR. MACULATUS.—Horsfield and Vigors.

Texan Lynx.

PLATE XCII.—Female.—Winter pelage.

L. rufo-grisea, dorso saturatiore, corporis lateribus memberisque externe bruneo-maculatis, gulâ, corpore infra, membrisque internè albis, bruneo latius maculati auribus pencillatis.

CHARACTERS.

Brownish-gray on the upper surface, sides of body and outer surface of legs, with small brown spots; under surface of body and inner surface of legs, white, broadly spotted with brown; ears, pencilled.

SYNONYMES.

Felis Maculatus. Horsfield and Vigors.
      "   "   Zoological Journal, vol. 4, p. 380.
      "   "   Reichenbach, Regnum Animale, vol. 1, p. 6, pl. 37.

DESCRIPTION.

In size, in shape, in its naked soles—in the form of the skull—the disposition and character of its teeth, and in all its habits, this species is so much like the Bay Lynx, (*L. rufus,*) that were it not for the different shades of colour, and the peculiar markings of some parts of the body, no naturalist would have ventured to describe it as a new species. One of the characters given to this supposed species by its original describers is that of pencilled ears; this character, however, exists also in the Bay Lynx; in both cases these hairs drop out when the other hairs are shed in spring, and are not replaced till the following autumn. The same peculiarity exists in many of our American squirrels. There is, as in *L. rufus*, a short ruff under the throat of the male. The hair is of two kinds: the inner, fine, and the outer and longer, not very coarse, and the fur, although much shorter, is fully as fine as that of specimens of the Bay Lynx obtained in Pennsylvania and New-York.

COLOUR.

The hairs on the back are at their roots yellowish-white, gradually becoming light-yellow, which colour continues for three-fourths the length, when they are barred with brownish-black, then yellowish-brown, tipped with black; on the sides, the hairs are tipped with white; on the under surface, they are white throughout, with a shade of pale-yellow at the base. Where black spots exist on the body, the hairs are less annulated —are dark-brown at the roots, deepening into black ; and in some spots on the sides, and the bands on the tail, the hairs are pure black from the roots.

Moustaches, white ; around the nose, around the eye, and cheeks, pale fawn colour ; lips white ; forehead, obscurely and irregularly marked with longitudinal stripes of dark-brown on a light-yellowish ground-colour. There are two black lines commencing at a point on a line with the articulation of the lower jaw, where they form an acute angle, diverging from thence to the sides of the neck, and unite with the ruff, where it is an inch broad. The ears are yellowish-white on the inner surface, black on the outer, with a broad white patch in the middle, including nearly their whole breadth. The slight pencil of hairs at the extremity of the ear is black ; on the back the colours are waved, and blended with obscure yellowish and brown spots—assuming on the dorsal line slight indications of narrow longitudinal stripes. The feet, on the upper surface are dotted with small brown spots ; on the under surface the ground colour is whitish, with irregular patches of black. This is more especially the case on the inner surfaces of the thighs and fore legs, which present long stripes and patches of black, somewhat irregularly disposed. The tail is white on the under surface, barred above with rufous and black ; towards the extremity there is first a bar of black about one-third of an inch wide, then brownish-gray, then an inch of black ; the white on the under surface rises above the black, making the tip of the tail white.

DIMENSIONS.

	Male.—Weight 25 lb.		Female.—Weight 20 lb.	
	Feet.	Inches.	Feet.	Inches.
End of nose to eye, - - -		2	- - - -	$1\frac{3}{4}$
" " to burr of ear, -		$4\frac{3}{4}$	- - - -	$3\frac{5}{8}$
Between ears, - - - -		$3\frac{1}{8}$	- - - -	3
Nose to crown of head, - -		$5\frac{1}{2}$	- - - -	$5\frac{1}{4}$
" to root of tail, - - -	2	9	- - - -	2 5
Tail (vertebræ) - - -		7	- - - -	6
" " to end of hair, -		$7\frac{1}{2}$	- - - -	$6\frac{1}{2}$
Hind legs (stretched) beyond tail,		$11\frac{1}{2}$	- - - -	10
Fore " " beyond nose,		$6\frac{1}{2}$	- - - -	6
Height of shoulder from ground,	1	$7\frac{1}{2}$	- - - -	1 $5\frac{1}{2}$
Round body behind shoulder,	1	$4\frac{1}{2}$	- - - -	1 $2\frac{1}{4}$
" " at the loin, -	1	$4\frac{1}{2}$	- - - -	1 0

HABITS.

This variety of Lynx may be called the Common Wild-Cat of Texas, where it is occasionally found even on the prairies, although it generally confines itself to the neighbourhood of woods and chaparal.

The Texan Wild-Cat is, like the *Lynx rufus*, a wily and audacious depredator—he steals the fowls from the newly-established rancho, or petty farm ; follows the hares, rats, and birds, and springs upon them in the tall rank grass, or thick underbrush, and will sometimes even rob the ranger of a fine turkey ; for should the Wild-Cat be lurking in the dense thicket, when the crack of the rifle is heard, and the wild gobbler or hen falls slanting to the earth, he will, instead of flying with terror from the startling report of the gun, dart towards the falling bird, seize it as it touches the ground, and bear it off at full speed, even if in sight of the enraged and disappointed marksman who brought down the prize. In general, however, the Southern Lynx (as this species is sometimes called) will fly from man's presence, and will only come abroad during the day when very hard pressed by hunger, when it may be occasionally seen near little thickets, on the edges of the prairies, or in the open ground, prowling with the stealthy sneaking gait observed in the domestic cat, when similarly employed. This species of Wild-Cat is better able to escape from an ordinary pack of dogs, than the Common Lynx, being accustomed to the great distances across the high dry prairies, which it must frequently cross at full speed. We have known one chased, from 11 o'clock in the morning till dark night, without being "treed." The animal, in fact, prefers running, to resorting to a tree at all times, and will not ascend one unless it be nearly exhausted, and hard pressed by the hounds.

GEOGRAPHICAL DISTRIBUTION.

This variety of the Bay Lynx is believed to exist throughout Mexico ; we have seen specimens, obtained in that country, in several Museums of Europe, especially those of Berlin and Dresden ; in the latter, the specimen described and figured by REICHENBACH is preserved. His figure, however, which we have compared with the original, is likely to mislead : the legs and tail being much too long. It exists in New Mexico, and we have heard that a Wild-Cat, supposed to be the present variety, is found in California. The specimen from which our drawing was made, was procured with several others by JOHN W. AUDUBON, in the vicinity of Castroville, on the head waters of the Medina, in Texas : we possess a specimen nearly of the same markings, procured by our deceased friend, the late lamented Dr. WURDEMANN.

GENERAL REMARKS.

We have admitted this as a variety of the Bay Lynx with some doubt and hesitation, and not without misgivings that it might yet be proved to be a distinct species. The permanency of its colours, together with the smaller size of our specimens, and their softer fur, may afford sufficient characters to entitle it to the name of *Maculatus*, as given by HORSEFIELD and VIGORS. Aware, however, of the many varieties in the Bay Lynx, we have not felt authorised to regard it as positively distinct.

Plate XLIX.

Black Footed Ferret

PUTORIUS NIGRIPES.—Aud. and Bach.

Black-Footed Ferret.

PLATE XCIII.—Male.

P. Magnitudine mustelam martem equans, fronte, caudæ, apice, pedibusque nigris; supra e flavido fuscus   infra albus.

CHARACTERS.

Size of the pine marten; forehead, feet, and extremity of tail, black; yellowish-brown above, white beneath.

SYNONYME.

Putorius Nigripes. Aud. and Bach, Quadrupeds of North America. vol. 2, pl. 93.

DESCRIPTION.

In its dentition this species possesses all the characteristics belonging to putorius and from the number and disposition of the teeth, cannot be placed in the genus, *mustela*. The canine teeth are stout and rather long, extending beyond the lips; they are slightly arched and somewhat blunt; the two outer incisors in the upper jaw are largest, the remainder are smaller, but regular and conspicuous. The first false molar is small but distinctly visible, it is without a lobe; the second is larger and has a slight lobe on each side. The great tuberculous tooth has two points and an external lobe; the last molar is rather small. In the lower jaw the incisors are small, and much crowded together. The three false molars on each side increase in size from the first, which is smallest and simple, to the third, which is largest and tuberculated. The great internal tooth has three lobes but no tubercle on the inner side, as is the case in the genus *mustela*; the last, or back tooth, is small but simple.

Body, very long; head, blunt; forehead, arched and broad; muzzle, short; eyes, of medium size; moustaches, few; ears, short, erect, broad at base, and triangular in shape, clothed on both surfaces with short hair; neck, long; legs, short and stout; toes, armed with sharp nails, very slightly arched; the feet on both surfaces covered with hair even to the soles, concealing the nails.

The pelage is of two kinds of hair, it is short soft and very fine, the outer and interspersed hairs are not so fine, but are not long and very coarse. The fur is finer than that of the mink or pine marten, and even shorter than that of the ermine. The hairs below the ears, under the forearms and belly are the coarsest; the tail is cylindrical, and less voluminous than that of the mink, containing more coarse hair, and less fine fur, than in that animal.

COLOUR.

The long hairs on the back are at the roots whitish, with a yellowish tinge, broadly tipped with reddish-brown; the soft under fur is white, with a yellowish tinge, giving the animal on the back a yellowish-brown appearance, in some parts approaching to rufous; on the sides and rump the colour is a little lighter, gradually fading into yellowish-white. Whiskers, white and black; nose, ears, sides of face, throat, under surface of neck, belly, and under surface of tail, white, a shade of brownish on the chest between the forelegs. There is a broad black patch commencing on the forehead, enclosing the eyes, and running down within a few lines of the point of the nose; outer and inner surfaces of the legs, to near the shoulders and hips, black, with a tinge of brownish; the tip of the tail is black, for two inches from the extremity.

DIMENSIONS.

					Feet.	Inches.
From point of nose to root of tail,	-	-	-	-	1	7
" " Tail, (vertebræ)	-	-	-	-		4
" head to end of hairs	-	-	-	-		5½
Height of ear posteriorly,	-	-	-	-		½
From shoulder to end of fore leg,	-	-	-	-		4

HABITS.

It is with great pleasure that we introduce this handsome new species; it was procured by Mr. CULBERTSON on the lower waters of the Platte River, and inhabits the wooded parts of the country to the Rocky Mountains, and perhaps is found beyond that range, although not observed by any travellers, from LEWIS and CLARK to the present day. When we consider the very rapid manner in which every expedition that has crossed the Rocky Mountains, has been pushed forward, we cannot wonder that many species have been entirely overlooked, and should rather be surprised at the number noticed by LEWIS and CLARK, and by

NUTALL, TOWNSEND, and others. There has never yet been a Government expedition properly organized, and sent forth to obtain *all* the details, which such a party, allowed *time* enough for thorough investigation, would undoubtedly bring back, concerning the natural history and natural resources of the regions of the far west. The nearest approach to such an expedition having been that so well conducted by LEWIS and CLARK. Nor do we think it at all probable that Government will attend to such matters for a long time to come. We must therefore hope that private enterprise will gradually unfold the zoological, botanical, and mineral wealth of the immense territories we own but do not yet occupy.

The habits of this species resemble, as far as we have learned, those of the ferret of Europe. It feeds on birds, small reptiles and animals, eggs, and various insects, and is a bold and cunning foe to the rabbits, hares, grouse, and other game of our western regions.

The specimen from which we made our drawing was received by us from Mr. J. G. BELL, to whom it was forwarded from the outskirts or outposts of the fur traders on the Platte river, by Mr. CULBERTSON. It was stuffed with the wormwood so abundant in parts of that country, and was rather a poor specimen, although in tolerable preservation. We shall have occasion in a future article to thank Mr. BELL for the use of other new specimens, this being only one of several instances of his kind services to us, and the zoology of our country, in this way manifested.

GEOGRAPHICAL DISTRIBUTION.

As before stated, the specimen which we have figured and described was obtained on the lower waters of the Platte river. We are not aware that another specimen exists in any cabinet

LEPUS NUTTALII.—Bachman.

Nuttal's Hare.

PLATE XCIV.—Males.

L. parvus, supra fuscus cum aureo mistus subtus dilute flavo-canescens, auriculis amplis rotundatisque, cauda longiuscula, subtus albus.

CHARACTERS.

Small ; tail of moderate length, general colour above, a mixture of light buff and dark brown, beneath, light yellowish grey ; ears, broad and rounded ; lower surface of the tail white.

DESCRIPTION.

The anterior upper incisors are more rounded than those of the American Hare, but in the deep longitudinal furrows, and in other particulars they bear a striking resemblance to those of that species ; the accessory, upper incisors resemble those of the Hares in general. The lower incisors are rather thinner than those of the American Hare, and like the upper, more of an oval shape. The upper grinders are furrowed longitudinally, like those of other Hares, and have a slight furrow on the inner side, but not more apparent than in *Lepus aquaticus ;* indeed, all the American Hares have this furrow, which differs considerably in individuals belonging to the same species.

This Hare bears some resemblance to the young of *Lepus sylvaticus ;* the forehead is more arched, and there is no depression in the frontal bone, as in the American Hare ; its fur is also much softer, and differs in colour ; the whiskers are nearly the length of the head. The ears appeared rather short and shrivelled in the dried specimen, but when moistened for the purpose of having a drawing of them made became much distended ; the incurvation on their outer margin was as distinct as in other Hares, bearing no resemblance to the funnel-shaped ears of the *pika.* The tail in the living animal must be conspicuous, although in the dried specimen it is concealed by the long fur of the posteriors. The feet are thickly clothed with soft hair, completely covering the nails. There are five toes on the fore and four on the hind feet.

COLOUR.

Teeth, yellowish white ; whiskers, white and black ; the former colour

Plate XCIV.

Drawn from Nature by J.W. Audubon.

On Stone by Wm E. Hitchcock.

Nuttall's Hare

Printed & Cold by J.T. Bowen, Phila.

predominating; the whole of the upper surface of the body, a mixture of buff and dark brown; under surface light buff-grey. The fur on the back is, for three-fourths of its length from the roots, plumbeous, then light ash mixed with buff; and the long interspersed hairs are all tipped with black. The ears are pretty well clothed, internally and externally, with hairs of an ash colour, bordered with a line of black anteriorly, and edged with white. From behind the ears to the back, there is a very broad patch of buff, and the same colour, mixed with rufus, prevails on the outer surface of the legs, extending to the thighs and shoulders. The soles of the feet are yellowish brown. The claws, which are slightly arched, are light brown for three-fourths of their length, and are tipped with white; under surface of the tail, white.

DIMENSIONS.

	Inches.
Length from point of nose to insertion of tail,	$6\frac{1}{4}$
" of Heel,	2
" Fur on the back,	$\frac{3}{4}$
" of Head,	$2\frac{1}{8}$
Height of ear,	$1\frac{1}{2}$
Tail vertebræ,	$\frac{3}{4}$
Including fur,	$1\frac{1}{4}$

HABITS.

The only information which we have been able to obtain of the habits of this diminutive species is contained in the following note from Mr. NUTTAL, which accompanied the specimen.

"This little Hare we met with west of the Rocky Mountains, inhabiting thickets by the banks of several small streams which flow into the Shoshonee and Columbia rivers. It was frequently seen, in the evening, about our encampment, and appeared to possess all the habits of the *Lepus Sylvaticus*."

GEOGRAPHICAL DISTRIBUTION.

We have not heard of the existence of this Hare in any part of California, or New Mexico; and although it is doubtless found in other localities than those mentioned above, we cannot venture to assert that it is widely distributed

GENERAL REMARKS.

We described this species from the only specimen we have had an opportunity of examining. It would be satisfactory to be able to investigate further, as it needs more information than we have been able to obtain, to pronounce decidedly upon its characters, and give its true geographical distribution.

Plate XCV

On Stone by Wᵐ E. Hitchcock.

Drawn from Nature by J. W. Audubon.

Lith. Printed & Colᵈ by J. T. Bowen. Phil.

Orange Colored Mouse

MUS (CALOMYS) AUREOLUS.—Aud. and Bach.

ORANGE-COLOURED MOUSE.

PLATE XCV.—Male and Females.

M. supra saturate luteus infra pallide flavus ; auriculis longis, cauda corpore curtiore.

CHARACTERS.

Ears long ; tail shorter than the body ; bright orange-coloured above, light buff beneath.

DESCRIPTION.

This species bears a general resemblance in form to the white-footed Mouse, (*Mus leucopus.*) It is, however, a little larger, and its ears rather shorter. Head, long ; nose, sharp ; whiskers, extending beyond the ears. Fur, very soft and lustrous. The legs, feet, and heels, clothed with short, closely adpressed hairs, which extend beyond the nails ; ears, thinly covered with hairs, which do not entirely conceal the colour of the skin ; mammæ, four ; situated far back.

COLOUR.

Head, ears, and whole upper surface, bright orange ; the fur being for three-fourths of its length from the roots, dark plumbeous ; whiskers, nearly black, with a few white hairs interspersed ; tail, above and beneath, dark brown ; throat, breast, and inner surface of the forelegs, white ; belly, light buff. There are no very distinct lines of separation between these colours.

DIMENSIONS.

	Inches.	Lines.
Length of head and body, - -	4	3
" Tail, - - - - -	3	1
" Head, - - - - -	1	3
" Ear posteriorly, - - - -	0	3
" Tarsus, including nail, - -	0	9

HABITS.

In symmetry of form and brightness of colour, this is the prettiest species of *Mus* inhabiting our country. It is at the same time a great climber. We have only observed it in a state of nature in three instances in the oak forests of South Carolina ; it ran up the tall trees with great agility, and on one occasion concealed itself in a hole (which apparently contained its nest,) at least thirty feet from the ground. The specimen we have described, was shot from the extreme branches of an oak, in the dusk of the evening, where it was busily engaged among the acorns. It is a rare species in Carolina, but appears to be more common in Georgia, as we received from Major Le Conte, three specimens obtained in the latter State.

GEÒGRAPHICAL DISTRIBUTION.

We found this species in Carolina, where it is rather rare ; we also obtained specimens from Georgia ; we have no doubt but further investigation will give it a more extensive geographical range.

GENERAL REMARKS

We have arranged this species under the sub-genus of Mr. Waterhouse, proposed in the Zoological Society of London, Feb. 17th, 1837, (see their transactions.) It is thus characterized ; "Sub-genus *Calomys*, (from Καλος beautiful and *mus*.) Fur, moderate, soft ; tarsus almost entirely clothed beneath the hair. Front molar, with three indentations of enamel on the inner side, and two on the outer ; and the last molar with one on each side. The type *mus* (*calomys*,) *bimaculatus*. Two other species have been described, from South America ; *mus* (*calomys*) *elegans*, and *m. gracilives*

Plate XCVI.

The Cougar

FELIS CONCOLOR.—Linn.

The Cougar.—Panther.

PLATE XCVI.—Male:—PLATE XCVII.—Female and young.

F. immaculata fulva; auriculis nigricantibus, cauda elongatâ, apice nigra neque floccosâ.

CHARACTERS.

Uniformly tawny-yellow; ears, blackish behind; tail, elongated, apex black, without a tuft.

SYNONYMES.

Felis Concolor, Linn. Syst. Nat., ed. Gmel., 1. p. 79.
    "   "   Schreb Saugth., p. 394.
    "   "   Buffon, Hist. Nat., t. 9.
    "   "   Gonazouara, D'Azara Anim. du Paraguay.
    "   "   Desmarest in Nouv. Dict., p. 90, 2.
Puma, Leo Americanus, Hernandez.
F. Concolor, Cuv. Regne Animal, vol. 1, p. 161
Brown Tiger, Pennant's Syn. p. 179.
Black Tiger,   "   "   180.
F. Concolor, Harlan, Fauna Am., p. 94.
    "   "   Godman, vol. 1, p. 291.
    "   "   Dekay's Nat. Hist. N. Y., p. 47.

DESCRIPTION.

Body, long and slender; head, small; neck, long; ears, rounded; legs, short and stout; tail, long, slender and cylindrical, sometimes trailing; fur, soft and short.

COLOUR.

Body and legs, of a uniform fulvous or tawny colour; under surface, reddish-white; around the eyes, grayish-yellow; hairs within the ears, yellowish-white; exterior of the ears, blackish; lips, at the moustache, black; throat, whitish; tail of the male, longer than that of the female, brown at tip, not tufted.

VOL. III.—19.

We have seen several specimens differing from the above in various shades of colour. These accidental variations, however, are not sufficient to warrant us in regarding these individuals as distinct species.

The young are beautifully spotted and barred with blackish-brown, and their hair is soft and downy.

DIMENSIONS.

Male, shot by J. W. AUDUBON, at Castroville, Texas 28th January, 1846.

	Feet.	Inches
From point of nose to root of tail - -	5	1
Tail - - - - - - - -	3	1
Height of ear posteriorly - - - -		3
Length of canine teeth, from gums - - -		1¾

Female, killed 26th January, 1846.

Length of head and body - - -	4	11
" Tail - - - - - -	2	8
" Height of ear - - - - -		3
" of canine teeth - - - - -		1½

Weight, 149 lbs.

HABITS.

The Cougar is known all over the United States by the name of the panther or painter, and is another example of that ignorance or want of imagination, which was manifested by the "Colonists," who named nearly every quadruped, bird, and fish, which they found on our continent, after species belonging to the Old World, without regard to more than a most slight resemblance, and generally with a total disregard of propriety. This character of the "Colonists," is, we are sorry to say, kept up to a great extent by their descendants, to the present day, who in designating towns and villages throughout the land, have seized upon the names of Rome, Carthage, Palmyra, Cairo, Athens, Sparta, Troy, Babylon, Jericho, and many other ancient cities, as well as those of Boston, Portsmouth, Plymouth, Bristol, Paris, Manchester, Berlin, Geneva, Portland, &c., &c., from which probably some of the founders of our country towns may have emigrated. We sincerely hope this system of nomenclature will henceforth be discarded; and now let us go back to the Cougar, which is but little more like the true *panther* than an opossum is like the kangaroo! Before, however, entirely quitting this subject, we may mention that for a long time the Cougar was thought to be the lion; the supposition was that all the skins of the animal that were brought into the settlements by the Indians were skins of females; and the lioness, having something the same colour and but little mane, it occurred to the colonists that the skins they saw could belong to no other animal!

The Cougar is found sparsely distributed over the whole of North America up to about latitude 45°. In former times this animal was more abundant than at present, and one was even seen a few miles from the city of New-York within the recollection of Dr. Dekay, who speaks of the consternation occasioned by its appearance in Westchester County, when he was a boy.

The Cougar is generally found in the very wildest parts of the country, in deep wooded swamps, or among the mountain cliffs and chasms of the Alleghany range. In Florida he inhabits the miry swamps and the watery everglades; in Texas, he is sometimes found on the open prairies, and his tracks may be seen at almost every cattle-crossing place on the slug-gish bayous and creeks with their quick-sands and treacherous banks. At such places the Cougar sometimes finds an unfortunate calf, or perhaps a cow or bullock, that has become fast in the oozy, boggy earth, and from exhaustion has given up its strugglings, and been drowned or suffocat-ed in the mire.

This species at times attacks young cattle, and the male from which our drawing was made, was shot in the act of feeding upon a black heifer which he had seized, killed, and dragged into the edge of a thicket close adjoining the spot. The Cougar, is however, generally compelled to sub-sist on small animals, young deer, skunks, raccoons, &c., or birds, and will even eat carrion when hard pressed by hunger. His courage is not great, and unless very hungry, or when wounded and at bay, he seldom attacks man.

J. W. Audubon was informed, when in Texas, that the Cougar would remain in the vicinity of the carcase of a dead horse or cow, retiring after gorging himself, to a patch of tall grasses, or brambles, close by, so as to keep off intruders, and from which lair he could return when his appetite again called him to his dainty food. In other cases he returns, after catch-ing a pig or calf, or finding a dead animal large enough to satisfy his hun-gry stomach, to his accustomed haunts, frequently to the very place where he was whelped and suckled.

Dr. Dekay mentions, that he was told of a Cougar in Warren County, in the State of New-York, that resorted to a barn, from whence he was repeatedly dislodged, and finally killed. "He shewed no fight whatever, His mouth was found to be filled with the spines of the Canada porcupine, which was probably the cause of his diminished wariness and ferocity, and would in all probability have finally caused his death."

The panther, or " painter," as the Cougar is called, is a nocturnal animal more by choice than necessity, as it can see well during the day time. It steals upon its intended prey in the darkness of night, with a silent, cautious step, and with great patience makes its noiseless way through the tangled thickets of the deepest forest. When the benighted traveller, or the wearied hunter may be slumbering in his rudely and hastily constructed bivouac at the foot of a huge tree, amid the lonely forest, his fire nearly out, and all around most dismal, dreary, and obscure, he may perchance be roused to a state of terror by the stealthy tread of the prowling Cougar; or his frightened horse, by its snortings and struggles to get loose, will awaken him in time to see the glistening eyes of the dangerous beast glaring upon him like two burning coals. Lucky is he then, if his coolness does not desert him, if his trusty rifle does not miss, through his agitation, or snap for want of better flint; or well off is he, if he can frighten away the savage beast by hurling at him a blazing brand from his nearly extinguished camp-fire. For, be sure the animal has not approached him without the gnawing hunger—the desire for blood, engendered by long fasting and gaunt famine. Some very rare but not well authenticated instances have been recorded in our public prints, where the Cougar at such times has sprang upon the sleeper. At other times the horses are thrown into such a fright, that they break all fastenings and fly in every direction. The late Mr. ROBERT BEST of Cincinnati, wrote to Dr. GODMAN, that one of these animals had surprised a party of travellers, sprung upon the horses, and so lacerated with its claws and teeth their flanks and buttocks, that they with the greatest difficulty succeeded in driving the poor creatures before them next morning, to a public house some miles off. This party, however, had no fire, and were unarmed.

A planter on the Yazoo river, some years ago, related the following anecdote of the Cougar to us. As he was riding home alone one night, through the woods, along what is called a " bridle-path" (i. e. a horse-track), one of these animals sprang at him from a fallen log, but owing to his horse making a sudden plunge forward, only struck the rump of the gallant steed with one paw, and could not maintain his hold. The gentleman was for a moment unable to account for the furious start his horse had made, but presently turning his head saw the Cougar behind, and putting spurs to his horse, galloped away. On examining the horse, wounds were observed on his rump corresponding with the claws of the Cougar's paw, and from their distance apart, the foot must have been spread widely when he struck the animal.

Another respectable gentleman of the State of Mississippi gave us the following account. A friend of his, a cotton planter, one evening, while at tea, was startled by a tremendous outcry among his dogs, and ran out

The Cougar

Lith Printed & Col^d by J.T. Bowen, Phil^a

Bowen & Young

to quiet them, thinking some person, perhaps a neighbour, had called to see him. The dogs could not be driven back, but rushed into the house ; he seized his horsewhip, which hung inside the hall door, and whipped them all out, as he thought, except one, which ran under the table. He then took a candle and looking down, to his surprise and alarm discovered the supposed refractory dog to be a Cougar. He retreated instanter, the females and children of his family fled frightened half out of their senses. The Cougar sprang at him, he parried the blow with the candlestick, but the animal flew at him again, leaping forward perpendicularly, striking at his face with the fore-feet, and at his body with the hind-feet. These attacks he repelled by dealing the Cougar straight-forward blows on its belly with his fist, lightly turning aside and evading its claws, as he best could. The Cougar had nearly overpowered him, when luckily he backed toward the fire-place, and as the animal sprang again at him, dodged him, and the panther almost fell into the fire ; at which he was so terrified that he endeavoured to escape, and darting out of the door was immediately attacked again by the dogs, and with their help and a club was killed.

Two raftsmen on the Yazoo river, one night encamped on the bank, under a small tent they carried with them, just large enough to cover two. They had a merry supper, and having made a large fire, retired, "turned in" and were soon fast asleep. The night waned, and by degrees a drizzling rain succeeded by a heavy shower pattering on the leaves and on their canvas roof, which sheltered them from its fury, half awakened one of them, when on a sudden the savage growl of a Cougar was heard, and in an instant the animal pounced upon the tent and overthrew it. Our raftsmen did not feel the full force of the blow, as the slight poles of the tent gave way, and the impetus of the spring carried the Panther over them ; they started up and scuffled out of the tent without further notice "to quit," and by the dim light of their fire, which the rain had nearly extinguished, saw the animal facing them and ready for another leap ; they hastily seized two of the burning sticks, and whirling them around their heads with loud whoops, scared away the midnight prowler. After this adventure they did not, however, try to sleep under their tent any more that night !

We have given these relations of others to show that at long intervals, and under peculiar circumstances, when perhaps pinched with hunger, or in defence of its young, the Cougar sometimes attacks men. These instances, however, are very rare, and the relations of an affrighted traveller must be received with some caution, making a due allowance for a natural disposition in man to indulge in the marvellous.

Our own experience in regard to the habits of this species is somewhat limited, but we are obliged to state that in the only three instances in which we observed it in its native forests, an impression was left on our minds that it was the most cowardly of any species of its size belonging to this genus. In our boyhood, whilst residing in the northern part of New-York, forty-eight years ago, on our way to school through a wood, a Cougar crossed the path not ten yards in front of us. We had never before seen this species, and it was, even at that early period, exceedingly rare in that vicinity. When the Cougar observed us he commenced a hurried retreat; a small terrier that accompanied us gave chase to the animal, which, after running about a hundred yards, mounted an oak and rested on one of its limbs about twenty feet from the ground. We approached and raised a loud whoop, when he sprang to the earth and soon made his escape. He was, a few days afterwards, hunted by the neighbours and shot. Another was treed at night, by a party on a raccoon hunt; supposing it to be a raccoon, one of the men climbed the tree, when the Cougar leaped to the ground, overturning one of the young hunters that happened to be in his way, and made his escape. A third was chased by cur-dogs in a valley in the vicinity of the Catskill mountains, and after half an hour's chase ascended a beech-tree. He placed himself in a crotch, and was fired at with duck-shot about a dozen times, when he was finally killed, and fell heavily to the ground. A Mr. RANDOLPH, of Virginia, related to us an amusing anecdote of a rencontre which he and a Kentuckian had in a valley of one of the Virginia mountains with a Cougar. This occurrence took place about thirty years ago. They had no guns, but meeting him near the road, they gave chase with their horses, and after a run of a few hundred yards he ascended a tree. RANDOLPH climbed the tree, and the Cougar sprang down, avoiding the Kentuckian, who stood ready to attack him with his club. The latter again followed, on his horse, when he treed him a second time. RANDOLPH again climbed after him, but found the animal was coming down, and disposed to fight his way to the ground. He stunned him with a blow, when the Cougar let go his hold, fell to the earth, and was killed by his comrade, who was waiting with his club below.

From all the conversations we have had with hunters who were in the habit of killing the Cougar, we have been brought to the conviction that a man of moderate courage, with a good rifle and a steady arm, accompanied by three or four active dogs, a mixture of either the fox-hound or grey-hound, might hunt the Cougar with great safety to himself, and with a tolerable prospect of success.

This animal, which has excited so much terror in the minds of the ignorant and timid, has been nearly exterminated in all our Atlantic States, and we do not recollect a single well authenticated instance where any hunter's life fell a sacrifice in a Cougar hunt.

Among the mountains of the head-waters of the Juniatta river, as we were informed, the Cougar is so abundant, that one man has killed for some years, from two to five, and one very hard winter, he killed seven. In this part of the country the Cougar is hunted with half-bred hounds, the full-blooded dogs lacking courage to attack so large and fierce looking an animal when they overtake it. The hunt is conducted much in the manner of a chase after the common wild-cat. The Cougar is "treed" after running about fifteen or twenty minutes, and generally shot, but sometimes it shews fight before it takes to a tree, and the hunters consider it great sport: we heard of an instance of one of these fights, in which the Cougar got hold of a dog, and was killing it, when the hunter in his anxiety to save his dog, rushed upon the Cougar, seized him by the tail and broke his back with a single blow of an axe.

According to the relations of old hunters, the Cougar has three or four young at a litter. We have heard of an instance of one being found, a very old female, in whose den there were five young, about as large as cats, we believe, however, that the usual number of young, is two.

The dens of this species are generally near the mouth of some cave in the rocks, where the animal's lair is just far enough inside to be out of the rain; and not in this respect like the dens of the bear, which are sometimes ten or twelve yards from the opening of a large crack or fissure in the rocks. In the Southern States, where there are no caves or rocks, the lair of the Cougar is generally in a very dense thicket, or in a cane-brake. It is a rude sort of bed of sticks, weeds, leaves, and grasses or mosses, and where the canes arch over it; as they are evergreen, their long pointed leaves turn the rain at all seasons of the year. We have never observed any bones or fragments of animals they had fed upon, at the lairs of the Cougar, and suppose they always feed on what they catch near the spot where they capture the prey.

The tales related of the cry of the Cougar in the forest in imitation of the call of a lost traveller, or the cry of a child, must be received with much caution, and may in many of their exaggerations be set down as vulgar errors. In a state of captivity, we have never heard the male uttering any other note than a low growl; the female, however, we have frequently heard uttering a kind of mewing like that of a cat, but a more prolonged and louder note, that could be heard at the distance of about two hundred yards. All the males, however, of the cat kind, at the season when the sexes seek each other, emit remarkable and startling cries, as is evidenced by the common cat, in what is denominated caterwauling. We have observed the same habit in the leopard, the ocelot, and in our two species of lynx. It is not impossible, therefore, that the male Cougar, may at the rutting season have some peculiar and startling notes. The cries, however, to which persons have from time to time directed our attention, as belonging to the Cougar, we were well convinced were uttered by other animals. In one instance, we ascertained them to proceed from a red fox which was killed in the hunt, got up for the purpose of killing the Cougar. In other cases the screams of the great horned, the barred, or the screech owl are mistaken for the cries of this animal.

The female Cougar is a most affectionate mother, and will not leave her young cubs, unless occasionally to procure food to support her own strength; she therefore often becomes very lean and poor. The female we have figured, was in this condition; we procured one of her cubs and figured it, presenting its beautiful spots, seldom before noticed. The other made its escape.

The whelps are suckled by the dam until about half grown, and then hunt with the old ones (which generally go in pairs) until the mother is with young again, or the young ones find mates for themselves, and begin to breed.

The period of gestation of the Cougar is ninety-seven days, as has been ascertained at the Zoological Society of London, (Proceedings, 1832, p. 62.) In the Northern and Middle States, the young are produced in the spring. In the Southern States, however, where the animal is supplied with an abundance of food, and not much incommoded by the cold, the the young have in some instances been discovered in autumn. J. W. AUDUBON found, in Texas, young Cougars nearly half grown in February.

GEOGRAPHICAL DISTRIBUTION.

This species has a wide geographical range. It was formerly found in all the Northern and Eastern States, and we have seen a specimen procured in Upper Canada. The climates of Lower Canada, New Foundland, and Labrador, appear to be too cold for its permanent residence. In all the Atlantic States it was formerly found, and a few still exist in the less cultivated portions. It is occasionally shot in the extensive swamps, along the river courses of Carolina, Georgia, Mississippi, and Louisiana; it is found sparingly on the whole range of the Alleghanies, running through a considerable portion of the United States. It has crossed the Rocky Mountains, and exists on the Pacific, in Oregon and California; it is quite abundant in Florida and Texas; is found within the tropics in Mexico, and Yucatan, and has penetrated through Panama into Guyana and South America, where it is sometimes called the Puma.

GENERAL REMARKS.

The variations of size, to which this species is subject, have created much confusion among our books of Natural History, and added a considerable number of supposed new species. After having examined very carefully very many specimens, both in a prepared state, and alive in menageries, procured in most parts of North and South America, we have arrived at the conclusion that the Cougar of North America and the Puma of our Southern Continent are one and the same species, and cannot even be regarded as varieties.

GENUS BASSARIS.—Lichtenstein.

DENTAL FORMULA.

$$Incisive \; \frac{6}{6}; \; Canine \; \frac{1-1}{1-1}; \; Molar, \; \frac{6-6}{6-6} = 40.$$

Body, long and rather slender; head, round; snout, attenuated like that of a fox; eyes, rather large: eyelids, oblong, lateral; ears, conspicuous, of moderate size, their points rounded.

There are five toes on each foot; tail, nearly the length of the body.

Hairs on the body, short and dense, much longer on the tail.

The specific name is derived from the Greek, Βασσαρις, (bassaris), a little fox.

This is the only species in the genus.

BASSARIS ASTUTA.—Licht.

RING-TAILED BASSARIS.

PLATE XCVIII.—Male.

B. Supra gilvus nigro-variegatus, auriculis, macula supra oculari et ventre flavido-albis; cauda, annulis octo albis nigrisque alternantibus, picta.

CHARACTERS.

Dull yellow, mixed with black, above; a spot above the eye, ears, and under surface, yellowish-white; tail, eight times ringed with black and white.

SYNONYMES.

Cacamitztli, Hernandez.

Tepe-Maxtlaton, Hernandez.

Bassaris Astuta, Lichtenstein, Darstellung neuer, oder wenig bekannter Säugethiere, Tafel 43, Berlin, 1827–1834.

DESCRIPTION.

The first impression made by this animal on the observer is, that he has met with a little fox; its erect ears, sharp nose, and cunning look, are

On Stone by W.ᵐ E. Hitchcock

Ring-Tailed Bassaris

Drawn from Nature by J. W. Audubon

Lith. Printed & Col.ᵈ by J. T. Bowen Phil.

all fox-like. It however, by its long and moveable muzzle approaches the civets, (*viverra,*) the genets, (*gennetta,*) and the coatis (*ictides.*)

The head is small; skull, not much flattened; nose, long; muzzle, pointed, naked; moustaches, numerous, long and rigid; ears, long, erect, ovate, clothed with short hair on the outer surface; sparingly within; neck and body, long; legs, longer than those of the martens, but shorter than those of the fox; nails, sharp and much hooked; toes, covered with hairs concealing them; palms, naked; tail, with long coarse hairs, containing scarcely any under fur; the inner hair on the back, is of moderate fineness, interspersed with rather coarser and longer hairs. The longer hairs on the back are about an inch in length, those on the tail, two inches, and the under-fur, on the back, half an inch.

COLOUR.

The hair on the back is grayish, for three-fourths of its length from the roots, then pale yellowish-white, then yellowish-brown, deepening into black at the tips; the under-fur is first plumbeous, then yellowish-white; this disposition of colours gives it a brindled brownish-black appearance on the head and upper surface. Moustaches, black; point of nose, dark brown. There is a light grayish spot above the eye; ears, chin, throat, neck and belly, yellowish-white. The tail is regularly and conspicuously ringed with bars of white and black, alternately; the upper white one very indistinct; the next black-obscure and increasing in more conspicuous bands of white and black to the end, which is broadly tipped with black; on the upper surface of tail, the black colours predominate, and on the under surface, the white.

DIMENSIONS.

	Feet.	Inches.
From point of nose to root of tail, - -	1	6
Tail, (vertebræ), - - - - -	1	2
" to end of hair, - - - -	1	4
From point of nose to head, between the ears, -		3¼
Height of ear, posteriorly, - - - - -		1¾
Breadth of ear at base, - - - - -		1
From shoulder to end of toes, - - - -		6
Length of longest moustache, - - - -		3¼

HABITS.

The greater portion of Texas is prairie-land, and it is chiefly along the water courses, that trees are found growing together in numbers sufficient to constitute a "wood." On certain level and clayey portions of the prairie, however, the land is swampy, and is covered with several kinds of oaks and a few other trees. The well-known musquit tree or bush is found generally distributed in the western parts of the State. It resembles the acacia in leaf, and has a small white pea-shaped blossom; at a distance it looks something like an old peach tree. Its wood is similar to coarse mahogany in appearance, and burns well, in fact beautifully, as the coals keep in for a long time; and the wood gives out little or no smoke. The musquit bottoms are furnished with these trees, they are small, about the size of the alder, and grow much in the same way; the musquit has sharp thorns. The musquit *grass*, (*Holcus lanatus*), resembles what is called, *guinea* grass, it is broader, shorter, softer, and more curly.

The general features of the State of Texas, as it will be seen by the foregoing, do not indicate a country where many tree-climbing animals could be found, and the present beautiful species, which Professor Lichtenstein most appropriately named Bassaris *astuta*, is by no means common. It is a lively, playful, and nimble creature, leaps about on the trees, and has very much the same actions as the squirrel, which it resembles in agility and grace, always having a hole in the tree upon which it resides, and betaking itself to that secure retreat at once if alarmed.

The Bassaris Astuta is shy and retired in its habits, and in the daytime often stays in its hole in some tree, so that we were only able to procure about half a-dozen of these animals during our stay in Texas; among which, to our regret, there was not a single female.

The food of this species is chiefly small animals, birds, and insects; they also eat nuts, as we were told, descending from their hiding place and travelling to the pecan and other trees, for the purpose of feeding on the nuts which, if true, is singular, as they are decidedly carnivorous in their dentition.

They are much attached to the tree on which they live, which is generally a post-oak, a live-oak, or other large tree, and they seldom quit the immediate vicinity of their hole, unless when driven out by thrusting a stick at them, when they ascend the trunk of the tree, and jump about among the higher branches so long as the pole is held close to their nest; as soon as this is withdrawn, they descend and at once re-enter their dwelling-place and hide themselves. These animals have a singular habit of eating or gnawing off the bark around the mouth of their holes, and where the bark does not appear freshly peeled off at their hole, you may be certain the animal is not at home, or has deserted the place. Their holes are generally the result of natural decay, and are situated on knobs, or at the ends of branches broken short off close to the main trunk.

They generally select a hole of this kind on the lower side of a leaning tree, probably for better protection from the rain ; their holes vary in depth, but are seldom more than about a foot or eighteen inches to the bottom ; they are usually furnished with moss or grass, for bedding. Sometimes pecan shells are found in these holes, which no doubt affords presumptive evidence that the Bassaris feeds upon this nut.

When scolding or barking at an intruder, the ring-tailed Raccoon, (as this animal is called by the Texans), holds the tail over its back, bending it squirrel fashion ; this animal, however, does not stand upon his hind feet like a squirrel, and cannot jump or leap so far. We have not heard of their springing from one branch to another beyond the distance of about ten feet, and when frightened at the presence of a man, they will sometimes run along a branch even toward him, in order to get within jumping distance of another, evincing more timidity than a squirrel exhibits in springing among the boughs, although they run up the bark with ease, holding on with their claws.

Sometimes the Ring-tailed Bassaris may be seen squatted on the top of a branch, basking in the sun, and half rolled up, appearing almost asleep. On the slightest manifestation of danger, however, he darts into his hole, (which is always within a foot or two of his basking place), and he is seen no more. We have the impression that only one of these singular animals is to be found on a tree at a time—they, therefore, are not very social in their habits, and, as the live-oak and other trees are generally very much scattered, and many of them have no holes suitable for residences for the Bassaris, it is very difficult to procure one. At the foot of many of the trees whereon they dwell, the cactus, brush-wood, and chapperal generally are so thick and tangled, that a man would be pretty well scratched should he attempt to penetrate the thorny, prickly thicket which surrounds the dwelling-place of this solitary and singular animal.

Notwithstanding the shyness and retired habits of this species, it is easily tamed, and when it has been confined in a cage a sufficient length of time, is frequently let loose in the houses of the Mexicans, where it answers the purpose of a playful pet, and catches mice and rats. We have seen one that was thus domesticated, running about the streets of a little Mexican village, and we were informed that one was kept as a great pet in a Camanche camp, visited by the Indian who hunted for us during our explorations of the western part of Texas. As far as we could ascertain, the northern limit of the range of this species is somewhere in the neighbourhood of the southern branches of Red river. As you travel south they are more abundant, and probably are found throughout all Mexico; we were informed by our friend, the celebrated Col. Hays, the Ranger, that he saw them more abundant in the mountainous region near the head-waters of the San Saba river than at any other place.

The Bassaris produces three or four young at a birth, as has been ascertained from the animal kept in confinement.

GEOGRAPHICAL DISTRIBUTION.

This animal exists in Mexico, and is common in the immediate vicinity of the capital of that name; our specimens were obtained in Texas, which appears to be its northern limit.

GENERAL REMARKS.

This species is called by the Mexicans caco-mixtle. It is mentioned no less than four times by Hernandez under the names of Cacamiztli and Tepe-Maxtlaton. The first specimens were sent to Berlin in 1826, by Mr. Deppe, and the earliest scientific description was given by Lichtenstein, who named it as above.

Plate XCIX

Drawn from Nature by J.J. Audubon, F.R.S.F.L.S.

On Stone by W.E. Hitchcock.

Lith. Printed & Col.d by J.T. Bowen, Phila.

Prairie Dog.— Prairie Marmot Squirrel

SPERMOPHILUS LUDOVICIANUS.—Ord.

PRAIRIE MARMOT-SQUIRREL.—WISHTONWISH.—PRAIRIE DOG.

PLATE XCIX.—1. MALE. 2. FEMALE. 3. YOUNG.

S. super cervinus pilis nigris interspersis ; subtus sordide albus, ungue pollicari conico majusculu, caudâ brevi apicem versus fusco torquatâ.

CHARACTERS.

Back, reddish brown, mixed with grey and black ; belly, soiled white ; tail, short, banded with brown near the tip ; thumb-nail, rather large, and conical.

SYNONYMES

PRAIRIE DOG, Lewis and Clark's Exp., 1st vol., p. 67.
WISHTONWISH, Pike's Expedition, &c., p. 156.
ARCTOMYS LUDOVICIANUS, Ord, in Guthrie, Geog., 2d, 302, 1815.
ARCTOMYS MISSOURIENSIS, Warden, Descr. des Etats Unis, vol. 5., p. 567.
ARCTOMYS LUDOVICIANUS, Say, Long's Exped., 1st vol., p. 451.
ARCT. LUDOVICIANUS, Harlan, p. 160.
    "   "   Godman, vol. 2, p. 114.

DESCRIPTION.

This animal in its external form has more the appearance of a marmot, than of a spermophile. It is short, thick, and clumsy, and is not possessed of the light, squirrel-like shape, which characterizes the *spermophili.* In its small cheek-pouches, however, being three-fourths of an inch in depth, and in the structure of its teeth, it approaches nearer the *spermophili,* and we have accordingly arranged it under that genus.

The head is broad and depressed ; nose short and blunt, hairy to the nostrils. Incisors, large, protruding beyond the lips ; eyes, large ; ears, placed far backwards, short, and oblong, being a mere flap nearly covered by the short fur ; neck, short and thick ; legs, short and stout. This species is pendactylous ; the rudimental thumb on the fore-feet protected by a sharp, conical nail ; nails, of medium size, scarcely channelled beneath, nearly straight, and sharp, extending beyond the hair ; tail, short and bushy; hair on the body, rather coarse; under-fur, of moderate fineness. The female has ten mammæ arranged along the sides of the belly.

COLOUR.

The hair on the back is, from the roots, for one-third of its length, bluish-black, then soiled-white—then light-brown; some of the hairs having yellowish-white, and others black, tips. The hairs on the under-surface, are at the roots bluish, and for nearly their whole length yellowish-white, giving the sides of face, cheeks, chin, and throat, legs, belly, and under-surface of tail a yellowish-white colour. Teeth, white; moustaches and eyes, black; nails, brown. The tail partakes of the colour of the back for three-fourths of its length, but is tipped with black, extending one inch from the end.

DIMENSIONS.

	MALE.	FEMALE.
Nose to root of tail, - - -	13 inches	12 3/8 inches.
" to end of tail, - - -	16 3/4 do	15 3/4 do
Tail, vertebræ, - - - -	2 5/8 do	2 1/4 do
" to end of hair, - - -	3 1/8 do	
Nose to anterior canthus, - -	1 1/4 do	1 1/8 do
Height of ear, - - - - -	7/16 do	7/16 do
Width between eyes, - - -	1 1/2 do	1 5/16 do
Length of fore-hand, - - -	1 5/16 do	1 1/4 do
" of heel and hind-foot - -	2 1/8 do	2 do
Depth of pouch, - - - - -	3/4 do	
Diameter of ditto, - - - - -	5/8 do	

Feet slightly webbed at base.

HABITS.

The general impression of those persons who have never seen the "Prairie Dog" called by the French Canadians "*petit chien*," would be far from correct in respect to this little animal, should they incline to consider it as a small "dog." It was probably only owing to the sort of yelp, chip, chip, chip, uttered by these marmots, that they were called Prairie *Dogs*, for they do not resemble the genus *Canis* much more than does a common gray squirrel!

This noisy *spermophile*, or marmot, is found in numbers, sometimes hundreds of families together, living in burrows on the prairies; and their galleries are so extensive as to render riding among them quite unsafe in many places. Their habitations are generally called "dog-towns," or villages, by the Indians and trappers, and are described as being intersected by streets (pathways) for their accommodation, and a degree of neatness and cleanliness is preserved. These villages, or communities, are, however, sometimes infested with rattle-snakes and other reptiles, which feed upon the marmots. The burrowing owl, (*Surnia cunicularia.*) is also found among them, and probably devours a great number of the defenceless animals.

The first of these villages observed by our party, when we were ascending the Missouri river in 1843, was near the "Great bend" of that stream. The mounds were very low, the holes mostly open, and but few of the animals to be seen.

Our friend EDWARD HARRIS, Mr. BELL and MICHAUX, shot at them, but we could not procure any, and were obliged to proceed, being somewhat anxious to pitch our camp for the night, before dark. Near Fort George, (a little farther up the river,) we again found a village of these marmots, and saw great numbers of them. They do not *bark*, but utter a chip, chip, chip, loud and shrill enough, and at each cry jerk their tail, not erecting it, however, to a perpendicular.

Their holes are not straight down, but incline downwards, at an angle of about forty degrees for a little distance and then diverge sideways or upwards. We shot at two of these marmots which were not standing across their holes apparently, but in front of them, the first one we never saw after the shot; the second we found dying at the entrance of the burrow, but at our approach it worked itself backward—we drew our ramrod and put the screw in its mouth, it bit sharply at this, but notwithstanding our screwing, it kept working backward, and was soon out of sight and beyond the reach of our ramrod.

Mr. BELL saw two enter the same hole, and Mr. HARRIS observed three. Occasionally these marmots stood quite erect, and watched our movements, and then leaped into the air, all the time keeping an eye on us. We found that by lying down within twenty or thirty steps of their holes, and remaining silent, the animals re-appeared in fifteen or twenty minutes. Now and then one of them, after coming out of its hole, issued a long and somewhat whistling note, perhaps a call, or invitation to his neighbours, as several came out in a few moments. The cries of this species are probably uttered for their amusement, or as a means of recognition, and not, especially, at the appearance of danger. They are, as we think, more in the habit of feeding by night than in the day time ; their droppings are scattered plentifully in the neighbourhood of their villages.

A few days after this visit to the Prairie Dogs, one of our hunters, who had been out a great part of the night, brought in three of them, but they had been killed with very coarse shot, and were so badly cut and torn by the charge, that they were of little use to us. We ascertained that these marmots are abundant in this part of the country, their villages being found in almost every direction.

From the number of teats in the female, the species is no doubt very prolific.

On our return down the river, we killed two Prairie Dogs on the 23d of August, their notes resembled the noise made by the Arkansas flycatcher precisely.

We have received an interesting letter from Col. ABERT of the Topographical bureau at Washington City, giving us an account of the quadrupeds and birds observed by Lieut. ABERT, on an exploratory journey in the south-west, in New Mexico, &c.   Lieut. ABERT observed the Prairie Dogs in that region of country, in the middle of winter; he says " our Prairie Dog (a marmot) does not hibernate, but is out all winter, as lively and as pert as on any summer day."

This is not in accordance with the accounts of authors, who have it that this animal does hibernate. We find it stated that it "closes accurately the mouth of the burrow, and constructs at the bottom of it a neat globular cell of fine dry grass, having an aperture at top sufficiently large to admit a finger, and so compactly put together that it might almost be rolled along the ground uninjured." We feel greatly obliged to Lieut. ABERT, for the information he gives us, which either explodes a long received error, or acquaints us with a fact of some importance in natural history—that changes of climate will produce so great an effect as to abrogate a provision of nature, bestowed upon some animals, to enable them to exist during the rigorous winters of the north ; so that, by migrating to a warmer region, species that would, in high latitudes be compelled to sleep out half their lives, could enjoy the air and light, and luxuriate in the sense of " being alive" all the circling year !   We have not been able to gather any information in relation to this subject since receiving the above-mentioned letter, but in our article on *Arctomys monax*, (vol. i., p. 20) some curious facts were related in respect to the effect of artificial heat, applied from time to time to that animal, when in a torpid state, which produced each time a temporary animation ; thus shewing that a certain absence of caloric causes hibernation immediately, while its presence arouses the powers of life in a few minutes.   The special construction of hibernating animals is not (as far as we have ascertained) yet explained by the researches of comparative anatomy.

Lewis and Clark give a very good description of the Prairie Dog, at page 67, vol. 1. They poured five barrels of water into one of their holes without filling it, but dislodged and caught the owner. They further say that after digging down another of the holes for six feet, they found on running a pole into it that they had not yet dug half-way to the bottom ; they discovered two frogs in the hole, and near it killed a dark rattlesnake, which had swallowed one of the Prairie Dogs.

Our friend Dr., now Sir John Richardson, (in the Fauna Boreali Americana,) has well elucidated the notices of this and other species described in Lewis and Clark's " Expedition," but, appears not to be certain whether this animal has cheek-pouches or not, and is puzzled apparently by the following : " the jaw is furnished with a *pouch* to contain his food, but not so large as that of the common squirrel." The Dr. in a note says—" It is not easy to divine what the " common squirrel is which has ample cheek-pouches." We presume that this passage can be made plain by inserting the word *ground* so that " common *ground*-squirrel" be the reading. The " common ground-squirrel" was doubtless well known to Lewis and Clark, and has ample cheek-pouches (see our account of *Tamias Lysterii*, vol. 1, p. 65.) This explanation would not be volunteered by us but for our respect for the knowledge and accuracy of Lewis and Clark, both of whom we had the pleasure of personally knowing many years ago.

For an amusing account of a large village of these marmots, we extract the following from Kendall's Narrative of the Texan Santa Fé Expedition, vol. 1, p. 189. " We had proceeded but a short distance, after reaching this beautiful prairie, before we came upon the outskirts of the commonwealth, a few scattering dogs were seen scampering in, their short, sharp yelps giving a general alarm to the whole community. The first brief cry of danger from the outskirts was soon taken up in the centre of the city, and now nothing was to be heard or seen in any direction but a barking, dashing, and scampering of the mercurial and excitable denizens of the place, each to his burrow.

Far as the eye could reach the city extended, and all over it the scene was the same. We rode leisurely along until we had reached the more thickly settled portion of the place. Here we halted, and after taking the bridles from our horses to allow them to graze, we prepared for a regular attack upon the inhabitants. The burrows were not more than ten or fifteen yards apart, with well trodden paths leading in different directions, and I even fancied I could discover something like regularity in the laying out of the streets.

We sat down upon a bank under the shade of a musquit, and leisurely surveyed the scene before us. Our approach had driven every one to his home in our immediate vicinity, but at the distance of some hundred yards the small mound of earth in front of each burrow was occupied by a Dog, sitting erect on his hinder legs, and coolly looking about for the cause of the recent commotion. Every now and then some citizen, more adventurous than his neighbour, would leave his lodgings on a flying visit to a friend, apparently exchange a few words, and then scamper back as fast as his legs would carry him.

By-and-by, as we kept perfectly still, some of our near neighbours were seen cautiously poking their heads from out their holes, and looking craftily, and, at the same time, inquisitively about them. Gradually a citizen would emerge from the entrance of his domicil, come out upon his observatory, perk his head cunningly, and then commence yelping somewhat after the manner of a young puppy—a quick jerk of the tail accompanying each yelp. It is this short bark alone that has given them the name of Dogs, as they bear no more resemblance to that animal, either in appearance, action, or manner of living, than they do to the hyena.

We were armed, one with a double-barrelled shot-gun, and another with one of Colt's repeating-rifles of small bore, while I had my short heavy rifle, throwing a large ball, and acknowledged by all to be the best weapon in the command. It would drive a ball completely through a buffalo at the distance of a hundred and fifty-yards, and there was no jumping off or running away by a deer when struck in the right place ; to use a common expression, " he would never know what had hurt him." Hit one of the Dogs where we would, with a small ball, he would almost invariably turn a peculiar somerset, and get into his hole, but by a ball, from my rifle, the entire head of the animal would be knocked off, and after this, there was no escape. With the shot-gun again, we could do nothing but waste ammunition. I fired it at one Dog not ten steps off, having in a good charge of buckshot, and thought I must cut him into fragments. I wounded him severely, but with perhaps three or four shot through him, he was still able to wriggle and tumble into his hole.

For three hours we remained in this commonwealth, watching the movements of the inhabitants and occasionally picking off one of the more unwary. No less than nine were got by the party ; and one circumstance I would mention as singular in the extreme, and shewing the social relationship which exists among these animals, as well as the kind regard they have for one another. One of them had perched himself upon the pile of earth in front of his hole, sitting up and exposing a fair mark, while a companion's head was seen poking out of the entrance, too timid, perhaps, to trust himself farther. A well-directed ball from my rifle carried away the entire top of the former's head, and knocked him some two or three feet from his post perfectly dead. While reloading, the other boldly came out, seized his companion by one of his legs, and before we could reach the hole had drawn him completely out of sight. There was a touch of feeling in the little incident, a something human, which raised the animals in my estimation, and ever after I did not attempt to kill one of them, except when driven by extreme hunger."

Mr. KENDALL says, further on, of these animals:—" They are a wild, frolicsome, madcap set of fellows when undisturbed, uneasy and ever on the move, and appear to take especial delight in chattering away the time, and visiting from hole to hole to gossip and talk over each other's affairs— at least, so their actions would indicate. When they find a good location for a village, and there is no water in the immediate vicinity, old hunters say, they dig a well to supply the wants of the community. On several occasions I crept close to their villages, without being observed, to watch their movements. Directly in the centre of one of them I particularly noticed a very large Dog, sitting in front of the door or entrance to his burrow, and by his own actions and those of his neighbours, it really seemed as though he was the president, mayor, or chief—at all events, he was the " big dog" of the place. For at least an hour I secretly watched the operations in this community. During that time the large Dog I have mentioned received at least a dozen visits from his fellow-dogs, which would stop and chat with him a few moments, and then run off to their domicils. All this while he never left his post for a moment, and I thought I could discover a gravity in his deportment not discernible in those by which he was surrounded. Far is it from me to say, that the visits he received were upon business, or had anything to do with the local government of the village; but it certainly appeared so. If any animal has a system of laws regulating the body politic, it is certainly the Prairie Dog."

This marmot tumbles, or rolls over, when he enters his hole, " with an eccentric bound and half-somerset, his hind-feet knocking together as he pitches headlong into the darkness below; and before the spectator has recovered from the half-laugh caused by the drollery of the movement, he will see the Dog slowly thrust his head from his burrow, and with a pert and impudent expression of countenance, peer cunningly about, as if to ascertain the effect his recent antic had caused."

Mr. KENDALL thinks that the burrowing owl, which he mentions as " a singular species of owl, invariably found residing in and about the dog towns," is on the best of terms with these marmots, and says, " as he is frequently seen entering and emerging from the same hole, this singular bird may be looked upon as a member of the same family, or at least, as a retainer whose services are in some way necessary to the comfort and well-being of the animal whose hospitality he shares." This idea is doubtless incorrect; and we would almost hazard the assertion that these owls prey upon the young, or even the adults, of these marmots; they also, probably, devour the bodies of those which die in their holes, and thus may stand toward the animals in the light of sexton and undertaker! Mr. KENDALL is entirely correct in what he says about the rattle-snakes, which dwell in the same lodges with the Dogs. "The snakes I look upon as loafers, not easily shaken off by the regular inhabitants, and they make use of the dwellings of the Dogs as more comfortable quarters than they can find elsewhere. We killed one a short distance from a burrow, which had made a meal of a half-grown Dog; and although I do not think they can master the larger animals, the latter are still compelled to let them pass in and out without molestation—a nuisance, like many in more elevated society, that cannot be got rid of."

Mr. KENDALL and his companions found the meat of this species "exceedingly sweet, tender, and juicy—resembling that of the squirrel, only that it was much fatter."

None of these animals were seen by J. W. AUDUBON in his journey through that part of Texas lying between Galveston and San Antonio, and he only heard of one village, to the northward and westward of Torrey's Lodge; they do not approach the coast apparently, being found only on the prairies beyond, or to the westward of the wooded portions of that State. A collector of animals and birds, who has passed the last three years in various parts of Mexico, and who showed us his whole collection, had none of these marmots, and we suppose their range does not extend as far south as the middle portions of that country.

GEOGRAPHICAL DISTRIBUTION.

This species is found on the banks of the Missouri and its tributaries. It also exists near the Platte river in great abundance. It was seen by J. W. AUDUBON in limited numbers in Sonora and on the sandy hills adjoining the Tulare Valley, and in other parts of California. We do not know whether it is an inhabitant of Oregon or not.

Plate C

Drawn from Nature by J.W.Audubon.

On Stone by W.E.Hitchcock

Lith Printed & Col by J.T.Bowen, Philad⁵

Missouri Mouse.

MUS MISSOURIENSIS.—Aud. and Bach.

Missouri Mouse.

PLATE C.—Females.

M. capite amplo, cruribus robustis, auriculis sub albidis, cauda curta, corpore supra dilute fusca, infra alba.

CHARACTERS.

Head, broad ; legs, stout ; ears, whitish ; tail, short, light fawn colour above, white beneath.

SYNONYME.

Mus Missouriensis, Aud. and Bach., Quads. North America, vol. 2, plates, pl. 100.

DESCRIPTION.

At first sight we might be tempted to regard this animal, as one of the endless varieties of the white-footed mouse. It is, however, a very different species, and when examined in detail, it will be discovered that the colour is the only point of resemblance. The body is stouter, shorter, and has a more clumsy appearance. The nose is less pointed ; ears, much shorter and more rounded ; and the tail, not one-third of the length.

Head, short and blunt ; nose, pointed ; eyes, large ; ears, short, broad at base and round, sparsely clothed with short hairs on both surfaces ; moustaches, numerous, long, bending forwards and upwards ; legs, stout ; four toes on the fore-feet, with the rudiment of a thumb, protected by a conspicuous nail ; nails, rather long, slightly bent, but not hooked. The hind-feet are pendactylous ; the palms are naked ; the other portions of the feet and toes, covered with short hairs, which do not, however, conceal the nails. The tail is short, round, stout at base, gradually diminishing to a point ; it is densely covered with very short hair ; the fur on both surfaces is short, soft and fine.

COLOUR.

Teeth, yellowish ; whiskers, nearly all white, a few black hairs interspersed. The fur on the back is plumbeous at the roots to near the points,

the hairs on the sides are broadly tipped with yellowish-fawn, and
on the back, are first fawn, and then slightly tipped with black ; on
the under surface, the hairs are at the roots plumbeous, broadly tipped
with white. The ears are nearly white, having a slight tinge of buff on
the outer and inner surfaces, edged with pure white ; on the sides of the
cheeks, and an irregular and indistinct line along the sides, the colours
are brighter than those on the flanks, and may be described as light yellowish-brown. The feet, on both surfaces, belly, and under surface of tail,
white ; from this admixture, this species is on the back, light fawn, with
an indistinct line on the back, and upper surface of tail, of a shade darker colour.

DIMENSIONS.

		Inches.
From point of nose to root of tail, - - - - -		$4\frac{1}{2}$
" " tail, - - - - - - -		$1\frac{1}{8}$
Height of ear, posteriorly, - - - - - -		$\frac{3}{8}$

HABITS.

We close our second volume with this new species of mouse, of which
we have given three figures. This pretty little animal was discovered for
us by Mr DENIG, during our sojourn at, and in the neighbourhood of Fort
Union in 1843. It was in full summer pelage, having been killed on the
14th of July. At that time being in quest of antelopes and large animals,
we did not give it that close attention, which we should have done. A
glance at our plate, or an examination of our description, will suffice to
convince any one of its being entirely new. This species is much larger,
and has a thicker and shorter tail than *mus leucopus.*

Expecting to get more of them we did not make any notes of the
habits of those killed at that time, and which had doubtless been observed
by the hunters, who procured them. The next day after they were
brought in, we left the fort on an expedition to the Yellow-Stone river,
from which we did not return for some time.

As a short description of our mode of travelling, &c., the first day's
journal is here given. " July 15, Saturday, we were all up pretty early,
making preparations for our trip to the Yellow-Stone river. After breakfast all the party who were going, announced themselves as ready, and
with a wagon, a cart, and two extra men from the fort, we crossed the
Missouri, and at 7 o'clock, were fairly under way ; HARRIS, BELL, CUL-
BERTSON, and ourself in the wagon, SQUIRES, PROVOST, and OWEN on horseback, while the cart brought a skiff, to be launched on the Yellow-Stone,

when we should arrive at that river. We travelled rather slowly until we had crossed a point and headed the ponds on the prairie at the foot of the hills opposite the fort. We saw one sharp-tailed grouse, but although Mr. HARRIS searched for it diligently, it could not be started. Soon after this we got one of the wheels of our wagon fast in a crack or crevice in the ground, and wrenched it so badly that we were obliged to get out and walk, while the men set to work to repair the wheels which were all in a rickety condition; after the needful fixing-up had been done, the wagon overtook us, and we proceeded on. Saw some antelopes on the prairie, and many more on the tops of the hills bounding our view to the westward. We stopped to water the horses at a "saline," where we observed that buffaloes, antelopes, and other animals had been to drink, and had been lying down on the margin. The water was too hot for us to drink. After sitting for nearly an hour to allow the horses to get cool enough to take a bait, for it was very warm, we again proceeded on until we came to the bed of a stream, which during spring overflows its banks, but now exhibits only pools of water here and there. In one of these pools we soaked our dry wagon wheels, by way of tightening the "tires," and here we refreshed ourselves and quenched our thirst. SQUIRES, PROVOST, and OWEN, started on before us to reconnoitre, and we followed at a pretty good pace, as the prairie was hereabouts firm and tolerably smooth. Shot a red-winged black-bird. Heard the notes of NUTTALL's short-billed marsh-wren,—supposed by some of our party to be those of a new bird. Saw nothing else; reached our camping-place at about 6 o'clock. Unloaded the wagon and cart, hobbled the horses, and turned them out to grass. Two or three of the men went off to a point above our camp, in search of something for supper. We took the red-winged black-bird, and a fishing-line, and went to the bank of the famed Yellow-Stone river, (near the margin of which our tent was pitched,) and in this stream of the far west, running from the bases of the Rocky Mountains, we threw our line, and exercised our piscatory skill so successfully as to catch some cat fish. These fish we found would not bite at pieces of their own kind, with which we tried them; after expending our bird bait, we therefore gave up fishing. One of our men took a bath, while two others, having launched the skiff rowed across the river to seek for deer or other game on the opposite shore. Toward dark the hunting parties all returned to camp without success; and we found the cat-fish the principal portion of our supper, having no fresh meat at all.

Our supper over, all parties shortly disposed themselves to sleep as they best could. About 10 o'clock, we were all disturbed by a violent thunder storm, accompanied by torrents of rain and vivid flashes of lightning;

the wind arose and blew a gale; all of us were a-foot in a few moments; and amid some confusion, our guns, loaded with ball, and our ammunition, were placed under the best covering we could provide, our beds huddled together under the tent along with them, and some of us crawled in on top of all, while others sought shelter under the shelving bank of the river. This storm benefitted us, however, by driving before the gale the mosquitoes, to keep off which we had in vain made a large fire, before we laid ourselves down for the night."

As there is little grain of any kind grown in this part of the country, the Missouri Mouse no doubt exists on the seeds and roots of wild plants entirely, of which it is able to lay up a store for the winter in holes in the ground. It may, however, possibly resort to the patches of corn planted by the squaws of some of the Indian tribes, at the time that grain is ripe. We brought with us from this country, when we returned home, some ears of a very small corn, (maize,) which ripens early, and bears its fruit near the ground. Having planted it on our place, we found that it was advanced enough to be eaten at table as a vegetable, several weeks before the ordinary kinds of corn known about New-York. We, therefore, distributed some of the seed among our farming neighbours, and likewise sent some to England to Lord DERBY and other friends, but this was unfortunately lost. We incline to believe that this corn would ripen well in the climate of England or Scotland. Unluckily, ours has become mixed by having been planted too near common corn, and is now depreciated or re duced to nearly the same thing as the latter.

GEOGRAPHICAL DISTRIBUTION.

This species was discovered in the State of Missouri.

GENERAL REMARKS.

The Missouri Mouse bears some resemblance to the common and very widely distributed White-footed Mouse. Its comparatively heavy and clumsy form—its large head and short tail have induced us to regard it as a distinct species. In the mice, shrews, and bats, we have no doubt several interesting species will yet be detected in our country.

INDEX.

TABLE OF CONTENTS.

TABLE OF GENERA DESCRIBED IN THIS VOLUME.